P9-ASA-587

A FIRST COURSE IN

Rational Continuum Mechanics

Volume 1

GENERAL CONCEPTS

Pure and Applied Mathematics

A Series of Monographs and Textbooks

Editors **Samuel Eilenberg and Hyman Bass**

Columbia University, New York

RECENT TITLES

E. R. Kolchin. Differential Algebra and Algebraic Groups

Gerald J. Janusz. Algebraic Number Fields

A. S. B. Holland. Introduction to the Theory of Entire Functions

Wayne Roberts and Dale Varberg. Convex Functions

A. M. Ostrowski. Solution of Equations in Euclidean and Banach Spaces, Third Edition of Solution of Equations and Systems of Equations

H. M. Edwards. Riemann's Zeta Function

Samuel Eilenberg. Automata, Languages, and Machines: Volumes A and B

Morris Hirsch and Stephen Smale. Differential Equations, Dynamical Systems, and Linear Algebra

Wilhelm Magnus. Noneuclidean Tesselations and Their Groups

François Treves. Basic Linear Partial Differential Equations

William M. Boothby. An Introduction to Differentiable Manifolds and Riemannian Geometry

Brayton Gray. Homotopy Theory: An Introduction to Algebraic Topology

Robert A. Adams. Sobolev Spaces

John J. Benedetto. Spectral Synthesis

D. V. Widder. The Heat Equation

Irving Ezra Segal. Mathematical Cosmology and Extragalactic Astronomy

J. Dieudonné. Treatise on Analysis: Volume II, enlarged and corrected printing; Volume IV; Volume V. *In preparation*

Werner Greub, Stephen Halperin, and Ray Vanstone. Connections, Curvature, and Cohomology: Volume III, Cohomology of Principal Bundles and Homogeneous Spaces

I. Martin Isaacs. Character Theory of Finite Groups

James R. Brown. Ergodic Theory and Topological Dynamics

C. Truesdell. A First Course in Rational Continuum Mechanics: Volume 1, General Concepts

K. D. Stroyan and W. A. J. Luxemburg. Introduction to the Theory of Infinitesimals

B. M. Puttaswamaiah and John D. Dixon. Modular Representations of Finite Groups

In preparation

Melvyn Berger. Nonlinearity and Functional Analysis: Lectures on Nonlinear Problems in Mathematical Analysis

George Gratzer. Lattice Theory

A FIRST COURSE IN

Rational Continuum Mechanics

C. TRUESDELL

VOLUME 1

General Concepts

ACADEMIC PRESS New York San Francisco London 1977
A Subsidiary of Harcourt Brace Jovanovich, Publishers

ACADEMIC PRESS, INC.
111 Fifth Avenue, New York, New York 10003

United Kingdom Edition published by
ACADEMIC PRESS, INC. (LONDON) LTD.
24/28 Oval Road, London NW1

Library of Congress Cataloging in Publication Data

Truesdell, Clifford Ambrose, Date
A first course in rational continuum mechanics.

(Pure and applied mathematics, a series of monographs and textbooks ;)
Earlier version published in French in 1973 has title: Introduction à la mécanique rationnelle des milieux continus.
CONTENTS: v. 1. General concepts. I. Continuum mechanics. I. Title. II. Series.
QA3.P8 [QA808.2] 531 75-40617
ISBN 0-12-701301-6

PRINTED IN THE UNITED STATES OF AMERICA

To the members of
The Society for Natural Philosophy,
founded at Baltimore in 1963.

. . . Deduction [and] Induction . . . render the indefinite definite; Deduction explicates; Induction evaluates: that is all. Over the chasm that yawns between the ultimate goal of science and such ideas of Man's environment as . . . he managed to communicate to some fellow, we are building a cantilever bridge of induction, held together by scientific struts and ties. Yet every plank of its advance is first laid by Retroduction alone, that is to say, by the spontaneous conjectures of instinctive reason . . . .

C. S. PEIRCE
Scientific Metaphysics (1908), ¶475

[While] the creative power of pure thought is at work, the outside world asserts itself again; through the real phenomena it forces new questions upon us; it opens up new fields of mathematical science; and while we try to gain these new fields of science for the realm of pure thought, we often find the answers to old unsolved problems and so at the same time best further the old theories. . . .

Besides, it is wrong to think that rigor in proof is the enemy of simplicity. Numerous examples establish the opposite, that the rigorous method is also the simpler and the easier to grasp. The pursuit of rigor compels us to discover simpler arguments; also, often it clears the path to methods susceptible of more development than were the old, less rigorous ones. . . .

While I insist upon rigor in proofs as a requirement for a perfect solution of a problem, I should like, on the other hand, to oppose the opinion that only the concepts of analysis, or even those of arithmetic alone, are susceptible of a fully rigorous treatment. This opinion, occasionally advocated by eminent men, I consider entirely mistaken. Such a one-sided interpretation of the requirement of rigor would soon lead us to ignore all concepts that derive from geometry, mechanics, and physics, to shut off the flow of new material from the outside world, and finally, indeed, as a last consequence to reject the concepts of the continuum and of the irrational number. What an important, vital nerve would be cut, were we to root out geometry and mathematical physics! On the contrary, I think that wherever mathematical ideas come up, whether from the theory of knowledge or in geometry, or from the theories of natural science, the task is set for mathematics to investigate the principles underlying these ideas and establish them upon a simple and complete system of axioms in such a way that in exactness and in application to proof the new ideas shall be no whit inferior to the old arithmetical concepts.

To new concepts correspond, necessarily, new symbols. These we choose in such a way that they remind us of the phenomena which gave rise to the formation of the new concepts. . . .

If we do not succeed in solving a mathematical problem, it is often because we have failed to recognize the more general standpoint from which the problem before us appears only as a single link in a chain of related problems. . . . This way to find general methods is certainly the most practicable and the surest, for he who seeks for methods without having a definite problem in mind mainly seeks in vain.

A role still more important than generalization's in dealing with mathematical problems is played, I believe, by specialization. Perhaps in most cases where we seek in vain for the answer to a question the cause of failure lies in our having not yet or not completely solved problems simpler and easier than the one in hand. Everything depends then on finding these easier problems and effecting the solution of them by use of tools as perfect as possible and of concepts susceptible to generalization. This rule is one of the most important levers for overcoming mathematical difficulties. . . .

[The] conviction that every mathematical problem can be solved is a powerful incentive to us as we work. We hear within us the perpetual call: *There is the problem. Seek its solution. You can find it by pure thinking, for in mathematics there is no ignorabimus!*

HILBERT
Mathematical Problems
Archiv für Mathematik und Physik (3) **1**, 44–63, 213–237 (1901).

Preface

The mechanics of finite systems of points and rigid bodies was given a fairly definitive form by LAGRANGE's exposition in his *Méchanique Analitique*, 1788. While that book covers only certain aspects of the rational mechanics created by LAGRANGE's great predecessors, it presents them systematically and as a branch of mathematics: "Ceux qui aiment l'Analyse, verront avec plaisir la Méchanique en devenir une nouvelle branche, ..." The physics and the applications are omitted. He who will apply and interpret the theory, or dwell upon the intricacies and mysteries of its place among the relations between mind and external nature, is expected to learn it first. While the knowledge he thus acquires does not of itself put applications into his hands, it gives him the tools to fashion them efficiently, or at least to classify, describe, and teach the applications already known. By consistently leaving applications to the appliers, LAGRANGE set them on common ground with the theorists who sought to pursue the mathematics further: Both had been trained in the same workshop and spoke the same jargon. Even today this comradeship of infancy lingers on, provided discrete systems and rigid bodies exhaust the universe of mechanical discourse.

In 1788 the mechanics of deformable bodies, which is inherently not only subtler, more beautiful, and grander but also far closer to nature than is the rather arid special case called "analytical mechanics", had been explored only in terms of isolated examples, brilliant but untypical. Unfortunately most of these fitted into LAGRANGE's scheme; those that did not, he passed over in silence. Further brilliant examples, feigned mainly upon the framework of NEWTON's and EULER's concepts and not easily subsumed under LAGRANGE's, were created in the next century but were studied mainly for their own sakes, separately, and did not lead to a general doctrine, despite the deep and original syntheses of stress and strain forged by CAUCHY.

A hundred years after CAUCHY died began a renascence of "classical" mechanics as a whole, taking the deformable continuum as the typical body and describing it in terms of an equally specific concept of material, which had been left nebulous and physical or metaphysical before then. This new general doctrine is now fit to be learned and used by mathematicians, experimentists, and engineers and to join the old analytical mechanics as an element of common education. Physicists should be able to understand it if they wished to. Like geometry, it is a part of mathematics.

In writing a textbook of continuum mechanics at this time I imitate the example of LAGRANGE in several ways. My book offers merely a selection from the wondrous harvest of the last few decades; leaving much else unmentioned, it bases that selection on criteria of naturalness, ease, and subsumption to a general method and conceptual frame. Thus it is a short book, designed for readers who know already that applications to further cases are numberless and possibilities for further mathematical study infinite. As LAGRANGE wrote, "On ne trouvera point de Figures dans cet Ouvrage. Les méthodes que j'y expose, ne demandent ni constructions, ni raisonnemens géométriques où méchaniques, mais seulement des opérations algébriques, assujetties à une marche regulière & uniforme." This claim is as true—or as false—of the present book as of LAGRANGE's. Of course, many proofs are easier to grasp if a figure is drawn, and both teacher and student should illumine and enrich the "regular and uniform course" by sketches. Finally—and here, perhaps, lies the greatest difference between this book and others with similar titles—it follows LAGRANGE's example in presuming that the reader commands the elementary mathematics of his own day[1], making no attempt to offer a shadowy substitute for

[1] The reader is expected to know the elements of measure theory. For almost everything else needed in "pure" mathematics, more than sufficient background is given in the book by H. K. NICKERSON, D. C. SPENCER, & N. E. STEENROD, *Advanced Calculus*, Princeton, Van Nostrand, 1959, reprinted in 1968 by Affiliated East-West Press Pvt. Ltd., New Delhi. The same may be said of the recent textbook by R. M. BOWEN & C.-C. WANG, designed especially for students of continuum mechanics: *Introduction to Vectors and Tensors*, 2 volumes, New York and London, Plenum Press, 1976. While in the text below certain more specialized works are cited in reference to some particular theorems, most of those may be found also in the two undergraduate texts just cited.

decent modern training in algebra and calculus or to appease the notorious reluctance of old men to learn anything new. The student may well find this book easier than his teacher does.

In three respects, however, I depart from LAGRANGE's model. First, I leave important if small pieces of the arguments, and some illustrations of them, as exercises for the reader, since my experience in teaching the new mechanics as it sprouted and grew has assured me that he who does not for himself re-create and digest the mathematics step by step will never master this doctrine. Second, while LAGRANGE's presentation bestowed upon the subject a gloss of closure and completeness which by the passage of time has been abundantly proved specious, in this book I try to present the science of "classical" mechanics even to the beginner as what it is: a magnificent array of ordered concepts and proved theorems, some of them old, even very old, and some on the frontiers of research into great unsolved problems and not yet distilled experience of nature as human eyes see it and human hands feel it. Third, the frequent attributions of major ideas and results to others will make it clear that I claim little of the substance for my own. The citations of other works, however, are intended not as acknowledgments of sources but as aids to the student. Those at the ends of the chapters direct him to places where further matters closely related to the text are developed; those in the footnotes, to specific details passed over in the text such as counterexamples, direct generalizations, proofs of theorems cited from other parts of mathematics, and tangent domains of modern mechanics.

Finally, I wish to thank those who have helped me to understand mechanics and to complete and purify this book. Thus above all I thank WALTER NOLL, and after him J. L. ERICKSEN, R. A. TOUPIN, B. D. COLEMAN, M. E. GURTIN, C.-C. WANG, W. O. WILLIAMS, L. SOLOMON, T. TOKUOKA, W.-L. YIN, R. C. BATRA, and D. EUVRARD. I am indebted to Mr. BATRA also for a full set of solutions to the exercises.

"Il Palazzetto"
Baltimore
May 1, 1972

Addendum. Parts 1 through 4 of this work, expertly translated into French by D. EUVRARD from my text of 1972, were published in December, 1973, by Masson et Cie in a single volume with the title *Introduction à la Mécanique Rationnelle des Milieux Continus.* Parts 1 through 5 appeared in 1975 in Russian, Первоначальный Курс Рациональной Механики Сплошных Сред, Moscow, Мир, translated from my text of

1973 by R. V. GOLDSHTEIN & V. M. ENTOV under the guidance of P. A. ZHILIN & A. I. LUR'E. Since that time I have been able to add some material and also to work through the text again and make numerous improvements, partly in response to criticisms and suggestions offered by readers of the French book.

A question has been raised regarding the knowledge of mechanics the student is expected to have already. A good treatise on the theory of functions of a real variable does not strictly require of its readers any previous acquaintance with the subject, even in the most elementary aspects of infinitesimal calculus, yet a student armed with no more than a naked, virgin mind is unlikely to survive the first few pages. In the same way, although this book does not call upon any previous knowledge of continuum mechanics, or even of schoolboy mechanics, it is designed for students not altogether innocent of hydrodynamics and elasticity. Much as a crude and awkward first affair may furnish knowledge that, however elementary, is indispensable to him who aspires toward Venus' ultimate refinements, a bad course—something nowadays cheaply found—will serve well enough here, too.

Some comments on the preliminary editions in French and Russian suggest need for reminder that this is a mathematical textbook, not a treatise or a history. In attaching names to a proposition I follow the commonest usage in the mathematical literature, proclaiming respect for those to whom I think we owe that proposition, be it in entirety, be it for discovery and proof of a pilot case, be it for clearest statement or most elegant proof; a second name never indicates rediscovery but always some major improvement, and of course it would not be feasible in any discipline so broadly cultivated as rational mechanics now is to list all the persons who have done something valuable, even if I knew of them all.

Volume 1 contains Part 1 only. Volume 2, containing Part 2 on fluid mechanics and Part 3 on elasticity, is presently being polished for the press. I plan to complete the textbook by a third volume, to concern fading memory, thermodynamics, statics, and thermostatics.

I thank Mr. BATRA for further suggestions and for checking the manuscript of this volume. I am deeply grateful to him and to Messrs. DAFERMOS, ERICKSEN, GURTIN, MUNCASTER, NOLL, and WILLIAMS for their generous gift of time and care in correcting the proofsheets so as to remove errors and obscurities even at the last moment. For such faults as, alas, surely remain I bear an uncommon charge, for seldom has an author had the benefit of such abundant and expert aid.

I owe a double debt of gratitude to the U.S. National Science Foundation for its continued and generous support: first, for the work of some of the great savants whose discoveries are incorporated here; second, for my own long effort to compose the essence of modern rational mechanics into an easy union with the magnificent tradition from which it sprang, so that beginners might learn both old and new together and in such a way as to see each illuminate and ennoble the other.

December 20, 1976 C.T.

Contents

Chapter II **Kinematics** 71

Chapter III **The Stress Tensor** 117

Chapter IV **Constitutive Relations** 157

Contents of Future Volumes

VOLUME 2 BODIES OF FLUID OR ELASTIC MATERIAL

PART 2 TOPICS IN FLUID MECHANICS

PART 3 TOPICS IN ELASTICITY

VOLUME 3 (TENTATIVE) FADING MEMORY, THERMOMECHANICS, STATICS, AND THERMOSTATICS

PART 4 FADING MEMORY

PART 5 THERMOMECHANICS

PART 6 STATICS AND THERMOSTATICS

PART 1

GENERAL CONCEPTS

In the following Chapters on Abstract Dynamics we confine ourselves mainly to the general principles, and the fundamental formulas and equations of the mathematics of this extensive subject; and neither seeking nor avoiding mathematical exercitations, we enter on special problems solely with a view to possible usefulness for physical science, whether in the way of the *material* of experimental investigation, or for illustrating physical principles, or for aiding in speculations of Natural Philosophy.

THOMSON & TAIT
Treatise on Natural Philosophy
(2nd ed., 1883), §453

Chapter I

Bodies, Forces, Motions, and Energies

I have restored to the concepts of *mass* and *force* their old rights. Beyond all doubt we need these *things*, for without them, there is no mechanics. Force is more than mass times acceleration, as may be seen from the basic equation itself, which always asserts that mass times acceleration equals the *sum* of the forces. Therefore, why not use the good old *words*? The concepts themselves are not unclear; it is just that the books described them often in a very metaphysical and dark way. And what matter, if the concepts are remarkably useful—perhaps a bit riddling?—if the concepts of mechanics are deeper than many find convenient, and cannot be disposed of with a few elegant words like convention and economy of thought, abstraction and idealisation?

HAMEL
On the foundations of mechanics,
Mathematische Annalen **66**, 350–397
(1909).

Space, time, and force are *a priori* forms; they can be derived only from contemplation and from general principles of research. Their common relation to each other in mechanics must be regarded as something inspired indeed by experience but in its generality fixed by convention.

HAMEL
Elementare Mechanik (1912), ¶5

1. Rational Mechanics

Rational Mechanics is the part of mathematics that provides and develops logical models for the enforced changes of position and shape which we see everyday things suffer. It describes also much of what is observed or inferred in the laboratories where professional scientists produce experiments. For example, it is always presumed as a part of the basis for design and control of scientific apparatus which is regarded by physicists as producing decisive experimental evidence that mechanics itself is only an "approximate" theory of nature.

The things mechanics represents by mathematical constructs include animals and plants, mountains and the atmosphere, oceans and the subterraneous riches, the whole orb which is the seat of our life and experience, heavenly objects both old and new, and the elements out of which these things seem to be composed: earth, water, air, and fire. As its name suggests, mechanics represents also the contrivances of man's artifice: fountains and engines and vehicles, bridges and fabrics, instruments of music and warfare, sewers and rockets. All these things mechanics models, but models crudely. Like any other branch of mathematics, it abstracts and evolves the common features of what it represents, setting aside most of the detail. As is necessary in any science which aims not merely to describe but also to predict, it seeks to select and correlate the simple out of the manifold and insuperable complexity of nature. Simplicity, while it does not ensure success in a branch of mechanics, is necessary there. A complicated theory in mechanics, although it may be socially or sociably useful at a particular time and place, does not enlighten and hence does not endure. Finally, since our experience grows with time and in proportion to our ingenuity, while the progress of mathematics enables us to manage easily and neatly mathematical ideas and operations of greater and greater scope, mechanics cannot be a closed science but must contain or at least be provided with means of improving or refining the models it presently possesses and also of constructing new ones.

Mechanics does not study natural things directly. Instead, it considers *bodies*, which are mathematical concepts designed to abstract some common features of many natural things. One such feature is the *mass* assigned to each body. Bodies are always found to occupy some *place*. The theory of places, which is called *geometry*, was created long ago and thus lies ready to hand for application in mechanics. The change of place undergone by a body in the course of *time* is called the *motion* of that body, and description of motion, or *kinematics*, is the second part of the foundation of mechanics. Third, motions of bodies are conceived as resulting from or at least being invariably accompanied by the action of *forces*. Fourth, the gain and loss of *heat* give rise to the concepts of the *energy*, *temperature*, and *calory* of a body. Thus mechanics is a *mathematical model*, or, better, an infinite class of models, for certain aspects of nature.

Kinematics, then, being presumed, mechanics rests upon four substructures: theories of bodies, forces, energies, and calories, in connection with places, times, and temperatures. These substructures provide the concepts mechanics is to connect. Relations among them are of two kinds: the general ones, common to all the bodies entering a given branch of mechanics, and particular ones, which distinguish one class of such bodies from another. The former kind constitute two theories: *statics*, which compares putative equilibria, and *dynamics*, which refers to motions of all sorts. Relations of the latter kind, which define particular bodies, are called *constitutive*.

In this book I cannot develop all of mechanics from explicit axioms.[1] So as to reach the level at which we may formulate and study constitutive relations, we shall pass lightly over the substructure of mechanics in general. However, although I do not here attempt to construct the theories of bodies, forces, energies, and calories with strict and detailed mathematics, we shall state clearly their defining assumptions and demonstrate their main properties.

While the presentation is lacunary and informal, it is abstract. The reader who is content to take bodies, the event world, frames of reference, motions, forces, and energies for granted may skip this chapter and pass to the next one, which begins the formal treatment of continuum mechanics along traditional lines. The traditional approach to mechanics is in no way incorrect, but it fails to satisfy modern standards of criticism and explicitness.

2. Bodies in General

The bodies with which mechanics deals are of many kinds: mass-points, which occupy but a single point at any one time; rigid bodies, which never deform; strings and rods and jets, which are 1-dimensional; membranes and shells, which sweep out mere surfaces; space-filling fluids and solids; and

[1] The sixth of the problems HILBERT set for the twentieth century to solve was the formulation of an axiomatic structure for physics, and especially for mechanics. Apart from a noteworthy attempt of HAMEL in 1909, this problem was given scarcely any attention until it was taken up by NOLL in 1957. The content of Chapter I of this book derives essentially from the work of NOLL and his collaborators.

many more. Bodies $\mathscr{A}$, $\mathscr{B}$, $\mathscr{C}$, ..., $\mathscr{X}$ of any one of these kinds, or several of them, conform to a particular mathematical structure called a *Boolean lattice* or *complemented distributive lattice*. The student who is familiar with this structure may pass directly to the next section. Here we shall simply list in order of their immediacy the properties common to all bodies in any theory of mechanics and prove some theorems concerning them.

The set Ω of all bodies is called the *universe*. At the very beginning in any branch of mechanics, a universe is specified, though in older work the reader was expected to infer the particular universe from the context. The sign $=$ means identity, "the same as". If the body $\mathscr{B}$ is a *part* of the body $\mathscr{C}$, we write $\mathscr{B} \prec \mathscr{C}$. The relation $\prec$ gives Ω the structure of a *partially ordered set*, defined by familiar axioms:

Axiom B1. $\mathscr{B} \prec \mathscr{B}$.

Axiom B2. $(\mathscr{X} \prec \mathscr{B})\ \&\ (\mathscr{B} \prec \mathscr{X}) \Rightarrow \mathscr{X} = \mathscr{B}$.

Axiom B3. $(\mathscr{B} \prec \mathscr{C})\ \&\ (\mathscr{C} \prec \mathscr{D}) \Rightarrow \mathscr{B} \prec \mathscr{D}$.

That is, $\mathscr{B}$ is a part of itself; $\mathscr{B}$ is not a part of any other of its parts; and if $\mathscr{B}$ is a part of $\mathscr{C}$, while $\mathscr{C}$ is a part of $\mathscr{D}$, then $\mathscr{B}$ is a part of $\mathscr{D}$. Another wording of Axiom B2 is, a body is the greatest of its own parts.

To visualize the relations among bodies, it may help to consider the case when Ω is the collection of all open sets in the Euclidean plane and to take $\prec$ as being the sign of inclusion, $\subset$, so that suggestive diagrams are easy to draw. This illustration is only one of many. Others, including the universes commonly presumed in mechanics, will be presented in the next section.

When the bodies $\mathscr{B}$ and $\mathscr{C}$ are given, neither need be a part of the other, but often they are both parts of a third one, $\mathscr{D}$. Such a $\mathscr{D}$ is called an *envelope* of $\mathscr{B}$ and $\mathscr{C}$. If, further, there is a body $\mathscr{A}$ that is an envelope of $\mathscr{B}$ and $\mathscr{C}$ and is itself a part of every envelope of $\mathscr{B}$ and $\mathscr{C}$, then $\mathscr{A}$ is called the *join* of $\mathscr{B}$ and $\mathscr{C}$. This relation among bodies is denoted as follows:

$$\mathscr{A} = \mathscr{B} \curlyvee \mathscr{C}. \tag{I.2-1}$$

Formally, this equation means that

$$(\mathscr{B} \prec \mathscr{A}, \mathscr{C} \prec \mathscr{A})\ \&\ (\mathscr{B} \prec \mathscr{D}, \mathscr{C} \prec \mathscr{D}) \quad \Rightarrow \quad \mathscr{A} \prec \mathscr{D}. \tag{I.2-2}$$

Thus the join of $\mathscr{B}$ and $\mathscr{C}$, if it exists, may be regarded as the "least" envelope of $\mathscr{B}$ and $\mathscr{C}$, since it is a part of every envelope of $\mathscr{B}$ and $\mathscr{C}$. Likewise, if

$$(\mathscr{A} \prec \mathscr{B}, \mathscr{A} \prec \mathscr{C}) \,\&\, (\mathscr{D} \prec \mathscr{B}, \mathscr{D} \prec \mathscr{C}) \quad \Rightarrow \quad \mathscr{D} \prec \mathscr{A}, \tag{I.2-3}$$

we write

$$\mathscr{A} = \mathscr{B} \wedge \mathscr{C} \tag{I.2-4}$$

and call $\mathscr{A}$ the *meet* of $\mathscr{B}$ and $\mathscr{C}$. If it exists, it is the "greatest" common part of $\mathscr{B}$ and $\mathscr{C}$, since every other common part of $\mathscr{B}$ and $\mathscr{C}$ is a part of it. Two bodies $\mathscr{B}$ and $\mathscr{C}$ may fail to have a meet or a join, or both, but, if they do have them, then plainly

$$\mathscr{B} \vee \mathscr{C} = \mathscr{C} \vee \mathscr{B}, \qquad \mathscr{B} \wedge \mathscr{C} = \mathscr{C} \wedge \mathscr{B}, \qquad \mathscr{B} \wedge \mathscr{C} \prec \mathscr{B} \prec \mathscr{B} \vee \mathscr{C}. \tag{I.2-5}$$

Also

$$\mathscr{B} \prec \mathscr{C} \quad \Leftrightarrow \quad \mathscr{B} \wedge \mathscr{C} = \mathscr{B} \quad \Leftrightarrow \quad \mathscr{B} \vee \mathscr{C} = \mathscr{C}, \tag{I.2-6}$$

whence

$$\mathscr{B} \wedge \mathscr{B} = \mathscr{B} \vee \mathscr{B} = \mathscr{B}, \tag{I.2-7}$$

and here, trivially, both the meet and the join exist.

Exercise I.2.1. Prove the following identities, where in each case it is assumed that the indicated meets and joins exist:

$$\mathscr{B} \prec \mathscr{C} \quad \Rightarrow \quad (\mathscr{B} \wedge \mathscr{A} \prec \mathscr{C} \wedge \mathscr{A}) \,\&\, (\mathscr{B} \vee \mathscr{A} \prec \mathscr{C} \vee \mathscr{A}). \tag{I.2-8}$$

If $\mathscr{B}_q$ are bodies from any collection, membership in which is indexed by subscripts q taken from any given set $\mathscr{k}$, meets and joins of the collection are defined in the same way and denoted by $\bigwedge_{q \in \mathscr{k}} \mathscr{B}_q$ and $\bigvee_{q \in \mathscr{k}} \mathscr{B}_q$. The former, for example, if it exists, is a body which is a part of each $\mathscr{B}_q$ and contains every other such body. For three bodies it is sometimes clearer to use the longer special notation $\mathscr{B} \wedge \mathscr{C} \wedge \mathscr{D}$, but then we must recall that the order of considering $\mathscr{B}$, $\mathscr{C}$, and $\mathscr{D}$ makes no difference, as is clear from the general notation and is illustrated in $(5)_{1,2}$.

To see that partial ordering does not ensure the existence of meets and joins, it suffices to consider the example of a universe Ω consisting in all non-empty half-open intervals $]a, b]$ and $[c, d[$ of real numbers, with $\prec$ defined as being inclusion in the sense of set theory. If $\mathscr{B} = [0, 2[$, $\mathscr{C} =]1, 4]$, and $\mathscr{D} = [3, 5[$, then $\mathscr{B}$ and $\mathscr{C}$ are common parts of infinitely many half-open intervals, yet they have no join, since if a certain half-open interval contains all the points of $\mathscr{B}$ and $\mathscr{C}$, we can find a shorter one that does so. The same may be said of the pair $\mathscr{C}$, $\mathscr{D}$. Nevertheless, $\mathscr{B} \curlyvee \mathscr{C} \curlyvee \mathscr{D} = [0, 5[$, which is a member of Ω. Note that $\mathscr{B} \curlyvee \mathscr{D} = \mathscr{B} \curlyvee \mathscr{C} \curlyvee \mathscr{D} \neq \mathscr{B} \cup \mathscr{D}$.

Exercise I.2.2. If $\mathscr{B} \curlywedge \mathscr{C}$ and $\mathscr{C} \curlywedge \mathscr{D}$ exist, and if either $(\mathscr{B} \curlywedge \mathscr{C}) \curlywedge \mathscr{D}$ or $\mathscr{B} \curlywedge (\mathscr{C} \curlywedge \mathscr{D})$ exists, then both do, and so does $\mathscr{B} \curlywedge \mathscr{C} \curlywedge \mathscr{D}$; also

$$(\mathscr{B} \curlywedge \mathscr{C}) \curlywedge \mathscr{D} = \mathscr{B} \curlywedge (\mathscr{C} \curlywedge \mathscr{D}) = \mathscr{B} \curlywedge \mathscr{C} \curlywedge \mathscr{D}. \tag{I.2-9}$$

A body $\mathscr{O} \in \Omega$ is called the *null body* if and only if it is a part of every body in Ω:

$$\mathscr{O} \prec \mathscr{B} \qquad \forall \mathscr{B} \in \Omega. \tag{I.2-10}$$

Ω need not contain such an element, but if it does, Axiom B2 shows that element to be unique. A body is called the *universal body* and denoted by ∞ if and only if every body in Ω is a part of it:

$$\mathscr{B} \prec \infty \qquad \forall \mathscr{B} \in \Omega. \tag{I.2-10A}$$

Such a body, if it exists, is obviously unique.

If the set Ω does not contain the null body or the universal body, we can formally adjoin either or both of these bodies so as to form the set $\overline{\Omega} \equiv \Omega \cup \{\mathscr{O}, \infty\}$, which is called the *closed universe* corresponding to Ω. By the following definitions we extend the partial order $\prec$ in Ω so as to form a partial order in $\overline{\Omega}$:

$$\begin{aligned} &\mathscr{O} \prec \mathscr{B} \quad \forall \mathscr{B} \in \overline{\Omega}, \qquad \mathscr{B} \prec \mathscr{O} \quad \Rightarrow \quad \mathscr{B} = \mathscr{O}, \\ &\mathscr{B} \prec \infty \quad \forall \mathscr{B} \in \overline{\Omega}, \qquad \infty \prec \mathscr{B} \quad \Rightarrow \quad \mathscr{B} = \infty. \end{aligned} \tag{I.2-11}$$

It is easy to verify that with the definitions (I.2-11) $\overline{\Omega}$ becomes a partially ordered set with partial order $\prec$, and that $\mathscr{O}$ and ∞ are the null body and the universal body of $\overline{\Omega}$.

In $\overline{\Omega}$ clearly

$$\mathscr{B} \curlywedge \mathscr{O} = \mathscr{O}, \qquad \mathscr{B} \curlyvee \mathscr{O} = \mathscr{B}, \qquad \mathscr{B} \curlywedge \infty = \mathscr{B}, \qquad \mathscr{B} \curlyvee \infty = \infty. \tag{I.2-12}$$

Any two bodies $\mathscr{B}$ and $\mathscr{C}$ in $\overline{\Omega}$ have at least one common part, namely $\mathscr{O}$. If they have no other common part, they are called *separate*. Thus $\mathscr{B}$ and $\mathscr{C}$ are separate if and only if

$$\mathscr{B} \curlywedge \mathscr{C} = \mathscr{O}. \tag{I.2-13}$$

Exercise I.2.3. Prove that

$$(\mathscr{B} \curlywedge \mathscr{C} = \mathscr{O}) \;\&\; (\mathscr{D} \prec \mathscr{C}) \quad \Rightarrow \quad \mathscr{B} \curlywedge \mathscr{D} = \mathscr{O}. \tag{I.2-14}$$

By setting $\mathscr{D} = \mathscr{B}$ and using (7), show that the only part of a body separate from that body is $\mathscr{O}$.

Next we need a concept of environment of a given body $\mathscr{B}$, so as to provide which we lay down a further axiom:

Axiom B4. *With each body $\mathscr{B}$ in $\overline{\Omega}$ is associated a unique body $\mathscr{B}^e$, which is called the* exterior *of $\mathscr{B}$, such that*

$$\mathscr{B} \curlywedge \mathscr{B}^e = \mathscr{O}, \qquad \mathscr{B} \curlyvee \mathscr{B}^e = \infty. \tag{I.2-15}$$

Thus $\mathscr{B}^e$ is separate from $\mathscr{B}$, and the only body that contains both $\mathscr{B}$ and $\mathscr{B}^e$ is ∞.

The example given just before Exercise I.2.2 shows that Axiom B4 cannot follow from Axioms B1, B2, and B3, since the points exterior to [0, 1[do not constitute a half-open interval.

Exercise I.2.4. Prove that

$$\mathscr{B} \prec \mathscr{B}^e \qquad \Rightarrow \qquad \mathscr{B} = \mathscr{O}. \tag{I.2-16}$$

By putting $\mathscr{B} = \mathscr{O}$ in $(12)_{3,4}$ and comparing the result with (15), we see that

$$\mathscr{O}^e = \infty, \qquad \infty^e = \mathscr{O}. \tag{I.2-17}$$

Likewise

$$(\mathscr{B}^e)^e = \mathscr{B}. \tag{I.2-18}$$

Also, putting $\mathscr{B} = \mathscr{C}^e$ in (14) shows that

$$\mathscr{D} \prec \mathscr{C} \qquad \Rightarrow \qquad \mathscr{D} \curlywedge \mathscr{C}^e = \mathscr{O}. \tag{I.2-19}$$

We now postulate that the converse of this proposition holds:

Axiom B5. *The only bodies separate from $\mathscr{C}^e$ are the parts of $\mathscr{C}$.*

While it has been proved[1] that Axiom B5 does not follow from Axioms B1–B4, no simple example to illustrate this fact seems to be known.

Formally, we may combine (19) with Axiom B5 as follows:

$$\mathscr{B} \prec \mathscr{C} \qquad \Leftrightarrow \qquad \mathscr{B} \curlywedge \mathscr{C}^e = \mathscr{O}. \tag{I.2-20}$$

By (18), then,

$$\mathscr{B} \prec \mathscr{C} \qquad \Leftrightarrow \qquad \mathscr{C}^e \curlywedge (\mathscr{B}^e)^e = \mathscr{O}. \tag{I.2-21}$$

[1] R. P. DILWORTH, "Lattices with unique complements," *Transactions of the American Mathematical Society* **57**, 123–154 (1945).

If we now replace $\mathscr{B}$ by $\mathscr{C}^e$ and $\mathscr{C}$ by $\mathscr{B}^e$ in (20) and compare the result with (21), we see that

$$\mathscr{B} \prec \mathscr{C} \quad \Leftrightarrow \quad \mathscr{C}^e \prec \mathscr{B}^e. \tag{I.2-22}$$

Hence it follows that if $\bigvee_{q \in k} \mathscr{B}_q$ exists, so does $\bigwedge_{q \in k} (\mathscr{B}_q)^e$, and

$$\bigwedge_{q \in k} (\mathscr{B}_q)^e = \left(\bigvee_{q \in k} \mathscr{B}_q \right)^e, \tag{I.2-23}$$

while if $\bigwedge_{q \in k} \mathscr{B}_q$ exists, so does $\bigvee_{q \in k} (\mathscr{B}_q)^e$, and

$$\bigvee_{q \in k} (\mathscr{B}_q)^e = \left(\bigwedge_{q \in k} \mathscr{B}_q \right)^e. \tag{I.2-24}$$

Drawing a diagram will make evident the statement and proof of the following theorem, where all meets and joins indicated are assumed to exist:

Theorem. *If*

$$\mathscr{A}_1 \wedge \mathscr{B} \prec \mathscr{C}, \qquad \mathscr{A}_2 \wedge \mathscr{B} \prec \mathscr{C}, \qquad \mathscr{D} \prec \mathscr{B}, \qquad \mathscr{D} \prec \mathscr{A}_1 \vee \mathscr{A}_2, \tag{I.2-25}$$

then

$$\mathscr{D} \prec \mathscr{C}. \tag{I.2-26}$$

Proof. By $(25)_{1,2}$ and (20),

$$(\mathscr{A}_q \wedge \mathscr{B}) \wedge \mathscr{C}^e = \mathscr{O}, \qquad q = 1, 2. \tag{I.2-27}$$

Let $\mathscr{E}$ be a common part of $\mathscr{D}$ and $\mathscr{C}^e$:

$$\mathscr{E} \prec \mathscr{D}, \qquad \mathscr{E} \prec \mathscr{C}^e. \tag{I.2-28}$$

Then $\mathscr{E} \prec \mathscr{B}$ by $(25)_3$. Let $\mathscr{F}_q$ be a common part of $\mathscr{E}$ and $\mathscr{A}_q$:

$$\mathscr{F}_q \prec \mathscr{E}, \qquad \mathscr{F}_q \prec \mathscr{A}_q, \qquad q = 1, 2. \tag{I.2-29}$$

Then $\mathscr{F}_q \prec \mathscr{B}$, and hence

$$\mathscr{F}_q \prec \mathscr{A}_q \wedge \mathscr{B}, \qquad q = 1, 2. \tag{I.2-30}$$

Using this conclusion and (27) in (14), we find that

$$\mathscr{F}_q \wedge \mathscr{C}^e = \mathscr{O}. \tag{I.2-31}$$

But $\mathscr{F}_q \prec \mathscr{E} \prec \mathscr{C}^e$, so by the result of Exercise I.2.3 we see that $\mathscr{F}_q = \mathscr{O}$. In view of the hypothesis (29), then,

$$\mathscr{E} \wedge \mathscr{A}_q = \mathscr{O}, \qquad q = 1, 2. \tag{I.2-32}$$

By (20), then, $\mathscr{A}_q \prec \mathscr{E}^e$, and hence

$$\mathscr{A}_1 \curlyvee \mathscr{A}_2 \prec \mathscr{E}^e. \tag{I.2-33}$$

But by $(25)_4$ and $(28)_1$

$$\mathscr{E} \prec \mathscr{A}_1 \curlyvee \mathscr{A}_2. \tag{I.2-34}$$

Hence $\mathscr{E} \prec \mathscr{E}^e$, so by (16) $\mathscr{E} = \mathscr{O}$. The hypothesis (28) has thus led to the conclusion that $\mathscr{D} \curlywedge \mathscr{C}^e = \mathscr{O}$, and by (20) we obtain (26). △

The theorem enables us to prove the distributive laws of Boolean algebra. First, if $\mathscr{A}_1 \curlywedge \mathscr{B}$, $\mathscr{A}_2 \curlywedge \mathscr{B}$, and $\mathscr{A}_1 \curlyvee \mathscr{A}_2$ exist, then

$$(\mathscr{A}_1 \curlyvee \mathscr{A}_2) \curlywedge \mathscr{B} = (\mathscr{A}_1 \curlywedge \mathscr{B}) \curlyvee (\mathscr{A}_2 \curlywedge \mathscr{B}), \tag{I.2-35}$$

provided either side exist. Indeed, the theorem tells us that any body which is a part of both $\mathscr{B}$ and $\mathscr{A}_1 \curlyvee \mathscr{A}_2$ is also a part of any body of which $\mathscr{A}_1 \curlywedge \mathscr{B}$ and $\mathscr{A}_2 \curlywedge \mathscr{B}$ are parts. Thus the body on the right-hand side of (35), if it exists, contains every common part of $\mathscr{A}_1 \curlyvee \mathscr{A}_2$ and $\mathscr{B}$. Since it is trivially a common part of $\mathscr{A}_1 \curlyvee \mathscr{A}_2$ and $\mathscr{B}$, by the definition of "meet" it is $(\mathscr{A}_1 \curlyvee \mathscr{A}_2) \curlywedge \mathscr{B}$. Similar reasoning applies if the body on the left-hand side is assumed to exist.

If we replace the bodies occurring in (35) by their exteriors and use (23) and (24), we obtain the second distributive law: If $\mathscr{A}_1 \curlyvee \mathscr{B}$, $\mathscr{A}_2 \curlyvee \mathscr{B}$, and $\mathscr{A}_1 \curlywedge \mathscr{A}_2$ exist, then

$$(\mathscr{A}_1 \curlywedge \mathscr{A}_2) \curlyvee \mathscr{B} = (\mathscr{A}_1 \curlyvee \mathscr{B}) \curlywedge (\mathscr{A}_2 \curlyvee \mathscr{B}), \tag{I.2-36}$$

provided either side exist.

If in (35) we take $\mathscr{A}_1$ and $\mathscr{A}_2$ as being $\mathscr{A}$ and $\mathscr{A}^e$, we see that if $\mathscr{A} \curlywedge \mathscr{B}$ and $\mathscr{A}^e \curlywedge \mathscr{B}$ exist, then

$$\mathscr{B} = (\mathscr{A} \curlywedge \mathscr{B}) \curlyvee (\mathscr{A}^e \curlywedge \mathscr{B}). \tag{I.2-37}$$

Finally, we have the ***basic decomposition theorem***, which enables us to express any body $\mathscr{B}$ as the join of any one of its parts $\mathscr{A}$ with a certain uniquely determined separate body $\mathscr{C}$:

$$(\mathscr{B} = \mathscr{A} \curlyvee \mathscr{C}) \;\&\; (\mathscr{A} \curlywedge \mathscr{C} = \mathscr{O}) \Leftrightarrow (\mathscr{A} \prec \mathscr{B}) \;\&\; (\mathscr{C} = \mathscr{B} \curlywedge \mathscr{A}^e). \tag{I.2-38}$$

To prove this implication, we assume first that the decomposition exists. Then $\mathscr{A} \prec \mathscr{B}$, and by (20) and (18) $\mathscr{C} \prec \mathscr{A}^e$; equivalently, by (6), $\mathscr{C} \curlywedge \mathscr{A}^e = \mathscr{C}$. By $(38)_1$ and (35), then,

$$\begin{aligned} \mathscr{B} \curlywedge \mathscr{A}^e &= (\mathscr{A} \curlyvee \mathscr{C}) \curlywedge \mathscr{A}^e, \\ &= (\mathscr{A} \curlywedge \mathscr{A}^e) \curlyvee (\mathscr{C} \curlywedge \mathscr{A}^e), \\ &= \mathscr{O} \curlyvee \mathscr{C} = \mathscr{C}, \end{aligned} \tag{I.2-39}$$

so that the implication forward in (38) is proved. Now suppose, conversely, that $\mathscr{A} \prec \mathscr{B}$ and $\mathscr{C} = \mathscr{B} \wedge \mathscr{A}^e$. Then $\mathscr{A} \wedge \mathscr{B} = \mathscr{A}$, so that (37) yields $\mathscr{B} = \mathscr{A} \vee \mathscr{C}$. Since $\mathscr{C} \prec \mathscr{A}^e$, it follows that $\mathscr{A} \wedge \mathscr{C} = \mathscr{O}$. $\triangle$

The final axiom for bodies asserts the existence of the meet:

Axiom B6. *For any two bodies $\mathscr{B}$ and $\mathscr{C}$, the meet $\mathscr{B} \wedge \mathscr{C}$ exists.*

In the next section we shall see by example that Axiom B6 is not a consequence of Axioms B1–B5. By assuming it, we may omit the qualifications hitherto expressed regarding the existence of meets and joins, for (23) shows that $(\mathscr{A}^e \wedge \mathscr{B}^e)^e = \mathscr{A} \vee \mathscr{B}$.

Axioms B1–B6 may be described as specifying that the universe $\overline{\Omega}$ shall be a *Boolean lattice*.[1]

3. Examples of Universes

We now construct systems satisfying Axioms B1–B6 and hence providing possible universes for mechanics. In all these, Ω is a class of sets, the symbol $\prec$ is taken as being $\subset$, the sign of inclusion, but only in the first one are $\wedge$ and $\vee$ the same as $\cap$ and $\cup$, the symbols of intersection and union.

Example 1. Let $\overline{\Omega}$ consist in all subsets of an arbitrary set $\mathscr{Y}$, and let $\prec$ be inclusion in the sense of set theory, denoted by the symbol $\subset$. Then $\mathscr{A} \wedge \mathscr{B} = \mathscr{A} \cap \mathscr{B}$, $\mathscr{A} \vee \mathscr{B} = \mathscr{A} \cup \mathscr{B}$, and $\mathscr{A}^e$ is the complement of $\mathscr{A}$ in $\mathscr{Y}$.

When, in this example, $\mathscr{Y}$ is a finite set with elements, say, X_1, X_2, ..., X_n, we obtain the universe of classical analytical dynamics. The bodies $\{X_k\}$ are called "particles" or "mass-points", but Ω contains other bodies as well: not only $\mathscr{O}$ but also $\{X_k\} \cup \{X_p\}$, $\{X_k\} \cup \{X_p\} \cup \{X_q\}$, *etc*. Of course $\infty = \bigcup_{k=1}^{n} \{X_k\}$.

Example 2. Let $\overline{\Omega}$ consist in all closures of open sets in a topological space, and let $\prec$ be inclusion in the sense of set theory. The exterior $\mathscr{B}^e$ of $\mathscr{B}$ is the closure of the complement of $\mathscr{B}$. The meet of $\mathscr{B}$ and $\mathscr{C}$ is not, in

[1] The material presented above is self-contained. For further information on Boolean lattices the student may consult R. SIKORSKI's *Boolean Algebra*, 2nd ed., New York, Academic Press, 1964.

general, their intersection, but rather the closure of the intersection of their interiors $\mathring{\mathscr{B}}$ and $\mathring{\mathscr{C}}$:

$$\mathscr{B} \curlywedge \mathscr{C} = \overline{\mathring{\mathscr{B}} \cap \mathring{\mathscr{C}}}. \tag{I.3-1}$$

Exercise I.3.1. Prove (I.3-1) as a consequence of the general definition of meet, given in §I.2.

For any collection of bodies $\mathscr{B}_k$, the join is given by

$$\bigvee_k \mathscr{B}_k = \overline{\bigcup_k \mathring{\mathscr{B}}_k}. \tag{I.3-2}$$

If the collection is finite, the join is simply the union, but for infinite collections of bodies such is not always the case.

The common universes of continuum mechanics are subcollections taken from particular universes of this kind. The dimensions 1, 2, and 3 correspond to rods, shells, and ordinary bodies. Only the last will be considered in this book.

In this universe the join of an infinite collection of bodies need not be their union. Indeed, let the topological space be the real line, and consider the particular bodies $\mathscr{B}_k \equiv [1/k, 1]$, $k = 1, 2, 3, \ldots$. Then $\bigcup_{k=1}^{\infty} \mathscr{B}_k =]0, 1]$ which is not a body, but by (2) we see that $\bigvee_{k=1}^{\infty} \mathscr{B}_k = [0, 1]$.

Example 3. Let Ω consist in all sets in a Euclidean space that are closures of regions regular in the sense of KELLOGG,[1] and let $\prec$ be taken as inclusion in the sense of set theory. The meet, if it exists, is defined as in Example 2. Axioms B1–B5 are satisfied by $\overline{\Omega}$, but Axiom B6 is not.

To see this last, we need only remark that the intersection of two sets with piecewise smooth boundaries need not itself have a piecewise smooth boundary. Suppose, for example, the elements of Ω are sets in the plane; let $\prec$ be taken as $\subset$; and let $\mathscr{B}_1$ be the closed square $-1 \leqq y \leqq 0$, $0 \leqq x \leqq 1$, while $\mathscr{B}_2$ is the closure of the set of points such that $0 < x < 1$, $-1 < y < x^2 \sin x^{-1}$. While $\mathscr{B}_1$ and $\mathscr{B}_2$ have infinitely many common parts, they have no meet.

Often in continuum mechanics it would be desirable to use Ω as defined in Example 3, since it allows a more natural class of elements than does

[1] Regular surfaces and regions in a Euclidean point space are defined by GURTIN, §5 of "The linear theory of elasticity," FLÜGGE's *Handbuch der Physik* **VIa/2,** ed. C. TRUESDELL, Berlin, Heidelberg and New York, Springer-Verlag, 1973.

Example 2 when specialized to a Euclidean space. As the definition suggests, "bodies" of this kind can be identified with regions to which the divergence theorem applies for all smooth vector fields.[1] Unfortunately, a specific and rigorous theory for this "universe" has not yet been developed. That the difficulty is a real one, may be seen from the example just above, which shows that the intersection of two regions, in each of which the divergence theorem holds, is itself not necessarily such a region.

4. Mass

The bodies of interest in mechanics have mass; as we may say, they are *massy*. The massy bodies form a non-empty subclass Ω_M of the closed universe $\overline{\Omega}$, the properties of which were set forth in §2. The mass of $\mathscr{B}$ is the value $M(\mathscr{B})$ of a non-negative *mass function* M defined over Ω_M:

Axiom M1. $0 \leqq M(\mathscr{B}) \leqq \infty \quad \forall \mathscr{B} \in \Omega_M .$

Further, we lay down

Axiom M2.

$$\mathscr{B} \in \Omega_M \quad \Rightarrow \quad \mathscr{B}^e \in \Omega_M ,$$
$$\mathscr{B}_1 \;\&\; \mathscr{B}_2 \in \Omega_M \quad \Rightarrow \quad \mathscr{B}_1 \curlyvee \mathscr{B}_2 \in \Omega_M .$$

That is, the exteriors[2] and joins of massy bodies also are massy bodies. In particular, $\mathscr{O}$ and ∞ are massy. Because $(\mathscr{B}_1^e \curlyvee \mathscr{B}_2^e)^e = \mathscr{B}_1 \curlywedge \mathscr{B}_2$, it follows from Axiom M2 that

$$\mathscr{B}_1 \;\&\; \mathscr{B}_2 \in \Omega_M \quad \Rightarrow \quad \mathscr{B}_1 \curlywedge \mathscr{B}_2 \in \Omega_M . \tag{I.4-1}$$

Thus the meet of two massy bodies is massy. Moreover, we assume *mass is additive*:

Axiom M3. *If $\mathscr{B}_1$ and $\mathscr{B}_2$ are separate massy bodies, then*

$$M(\mathscr{B}_1 \curlyvee \mathscr{B}_2) = M(\mathscr{B}_1) + M(\mathscr{B}_2).$$

[1] *Cf.* §§9–10 of Chapter IV of O. D. Kellogg, *Foundations of Potential Theory*, Berlin, Springer, 1929, or other works in which a rigorous proof of the divergence theorem (Green's transformation) is given. See also §II.1.

[2] If the requirement that $\mathscr{B}^e$ should be massy seems artificial, the student should recall that the possibility that $M(\mathscr{B}^e) = 0$ is not excluded.

Hence

$$M(\mathcal{O}) = 0, \qquad M(\infty) = M(\mathscr{B}) + M(\mathscr{B}^e) \quad \forall \mathscr{B} \in \Omega_M. \qquad \text{(I.4-2)}$$

The mass assigned to the infinite body ∞ need not be ∞. If $M(\infty) = \infty$, then $M(\mathscr{B}) < \infty \Rightarrow M(\mathscr{B}^e) = \infty$. A body of mass 0 is called *massless*. Thus $\mathcal{O}$, the null body, is massless, but of course there may be other massless bodies. That is, $M(\mathscr{B}) = 0 \not\Rightarrow \mathscr{B} = \mathcal{O}$. Also, by (I.2-38),

$$\mathscr{A} \prec \mathscr{B} \quad \Rightarrow \quad M(\mathscr{B}) = M(\mathscr{A}) + M(\mathscr{B} \curlywedge \mathscr{A}^e) \geqq M(\mathscr{A}). \qquad \text{(I.4-3)}$$

Exercise I.4.1. Prove that

$$M(\mathscr{B} \curlyvee \mathscr{C}) \leqq M(\mathscr{B}) + M(\mathscr{C}). \qquad \text{(I.4-4)}$$

While these properties reflect the obvious requirements of the idea of mass, they do not suffice to define it effectively. As is well known, if we are to obtain the convenient mathematical structure known as *measure theory*, further assumptions must be laid down. While we shall assume that Ω is rich enough in bodies to contain a massy part Ω_M, we shall not attempt to *construct* a measure based on Ω_M, for at present there seems to be no entirely satisfactory way of doing so in general.

While the notions of mass and electric charge, along with volume and area, were distilled to form the basis of measure theory, that theory in its present state accounts satisfactorily only for the latter two, not for the former. Indeed, the mass function is a measure, but measure theory does not suffice for constructing a mass function. This is so because measure theory refers to *sets*, while, as we have seen in §I.3, the notions of meet $\curlywedge$ and join $\curlyvee$ of bodies generally are not the same as intersection $\cap$ and union $\cup$ in the algebra of sets, even in the case when bodies are indeed sets. A good mathematical theory of mass would be purely algebraic, assuming of bodies no more than the axioms B1–B6 (and preferably not the last).[1] The defect here is more one of clarity and elegance than application, since, as we shall see more clearly in Chapter II, the concepts of shape and motion enable us to use in continuum mechanics the common theory of Borel measure.

Henceforth in dealing with bodies we shall assume that they are closures of open sets in some topological space, and that the mass M defined over

[1] A purely algebraic theory was indeed developed by C. CARATHÉODORY in his last book, *Mass und Integral und ihre Algebraisierung*, Basel, Birkhäuser, 1956, translated as *Algebraic Theory of Measure and Integration*, Bronx, New York, Chelsea, 1963. While the σώματα over which CARATHÉODORY defines a measure formalize a concept of "body", he uses again and again the axiom that an enumerable collection of bodies has a join, which for applications in continuum mechanics is not always true.

Ω_M can be extended so as to be a Borel measure defined over all the Borel sets[1] of that space. There will be no confusion if we denote also this extended measure by M, even though most of the Borel sets are not bodies. The assumption is more confining than it may appear at first glance, for if M is a measure on the Borel sets, it is additive on disjoint unions of them. Our basic Axiom M3 requires only that it be additive on the joins of separate bodies.

Once a non-negative mass function M is given, clearly KM is also a non-negative mass function, where K is any positive constant. To any one particular body $\mathscr{B}$ that is not massless we may assign any positive mass we please, and the ratios of the masses of bodies are unaffected by this choice. In physics the assignment of a particular mass to some one body in the universe is called "fixing the unit of mass".

Henceforth we shall consider only Ω_M, not any greater universe $\overline{\Omega}$, and we shall use the symbol $\overline{\Omega}$ to denote Ω_M, thus excluding tacitly from our discourse any bodies that are not massy. Our assumptions enable us to write

$$M(\mathscr{A}) = \int_{\mathscr{A}} dM \qquad \text{if} \quad \mathscr{A} \in \overline{\Omega}, \tag{I.4-5}$$

and the integral

$$\int_{\mathscr{A}} f\, dM \tag{I.4-6}$$

of any continuous function f can be defined in the way shown in books on the theory of measure and integration.

By assigning masses directly to the bodies of the universe we express the *principle of conservation of mass.*

This principle is nowadays considered appropriate to mathematical models for phenomena in which chemical or nuclear reactions may be neglected and the speeds associated with bodies are small in comparison with the speed of light. In theories of chemical reactions the principle still holds, but only for sufficiently large bodies, among the parts of which mass generally is exchanged.

5. Force

A *system of forces* on a universe Ω is an assignment of vectors in some finite-dimensional inner-product space $\mathscr{V}_{\mathbf{f}}$ to all pairs of separate bodies of Ω. Vectors are denoted by bold-faced letters.

[1] The collection of Borel sets in a topological space is the smallest σ-algebra that contains all of the open sets. Thus all open sets, all closed sets, and all unions and intersections of enumerable collections of open sets or closed sets are Borel sets. That M is a Borel measure, means that the measure of every compact set is finite, a fact which is important for some of the arguments in Chapter III.

Let $(\Omega \times \Omega)_0$ be the collection of such pairs. The first axiom of forces is

Axiom F1. $\mathbf{f}: (\Omega \times \Omega)_0 \to \mathscr{V}_{\mathbf{f}}$.

The vector $\mathbf{f}(\mathscr{B}, \mathscr{C})$ is called the *force exerted on* $\mathscr{B}$ *by* $\mathscr{C}$. Since we are here considering Ω rather than $\overline{\Omega}$, no force need be assigned to pairs one member of which is ∞ or $\mathcal{O}$. Moreover, the force exerted by two separate bodies on a third body separate from both is the sum of the forces exerted by each, and the force exerted by a body on the join of two separate parts of a separate body is the sum of the forces exerted on each. That is, the function $\mathbf{f}$ is additive in each of its variables:

Axiom F2. $\mathbf{f}(\mathscr{C}_1 \vee \mathscr{C}_2, \mathscr{B}) = \mathbf{f}(\mathscr{C}_1, \mathscr{B}) + \mathbf{f}(\mathscr{C}_2, \mathscr{B})$.

Axiom F3. $\mathbf{f}(\mathscr{B}, \mathscr{C}_1 \vee \mathscr{C}_2) = \mathbf{f}(\mathscr{B}, \mathscr{C}_1) + \mathbf{f}(\mathscr{B}, \mathscr{C}_2)$.

Both of these axioms refer to *pairwise separate* bodies $\mathscr{C}_1$, $\mathscr{C}_2$, and $\mathscr{B}$.

If $\mathbf{f} \cdot \mathbf{g}$ denotes an inner product in $\mathscr{V}_{\mathbf{f}}$, then $K\mathbf{f} \cdot \mathbf{g}$ is also an inner product if $K > 0$. Choice of a particular K is called "fixing the unit of force".

It is easy to extend $\mathbf{f}$ from $(\Omega \times \Omega)_0$ to $(\overline{\Omega} \times \overline{\Omega})_0$, since Axioms F2 and F3 allow no other value but $\mathbf{0}$ for the force exerted by or on the null body. Thus we must set

$$\mathbf{f}(\mathscr{B}, \mathcal{O}) \equiv \mathbf{f}(\mathcal{O}, \mathscr{B}) \equiv \mathbf{0} \qquad \forall \mathscr{B} \in \overline{\Omega}. \tag{I.5-1}$$

The choices $\mathscr{B} = \infty$ or $\mathscr{B} = \mathcal{O}$ are not excluded here.

Since Axioms F2 and F3 are statements of additivity, we may draw at once the following trivial but nevertheless important conclusion: *If* $\mathbf{f}_1$ *and* $\mathbf{f}_2$ *are systems of forces on* $(\Omega \times \Omega)_0$, *then* $A\mathbf{f}_1 + B\mathbf{f}_2$ *is a system of forces, where* A *and* B *are any real numbers.*

Since every body in $\overline{\Omega}$ is separate from its exterior $\mathscr{B}^e$, Axiom F1 enables us to form $\mathbf{f}(\mathscr{B}, \mathscr{B}^e)$, the force exerted on $\mathscr{B}$ by its exterior. We call this particular force the *resultant force* on $\mathscr{B}$. Resultant forces are subject to a fundamental identity:

$$\mathbf{f}(\mathscr{B}, \mathscr{C}) + \mathbf{f}(\mathscr{C}, \mathscr{B}) = \mathbf{f}(\mathscr{B}, \mathscr{B}^e) + \mathbf{f}(\mathscr{C}, \mathscr{C}^e) - \mathbf{f}(\mathscr{B} \vee \mathscr{C}, (\mathscr{B} \vee \mathscr{C})^e), \tag{I.5-2}$$

for all pairs of separate bodies $\mathscr{B}$ and $\mathscr{C}$ in $\overline{\Omega}$. To prove the identity, suppose first that $\mathscr{C} = \mathscr{B}^e$. Then mere statement of (2) requires extension of $\mathbf{f}$ to $\overline{\Omega}$ and hence leads to (1), whence (2) follows trivially. If $\mathscr{C} \neq \mathscr{B}^e$, extension of $\mathbf{f}$ to $\overline{\Omega}$ is not needed, and the following argument holds in Ω as well as in $\overline{\Omega}$, provided only $\mathscr{B}$ and $\mathscr{B} \vee \mathscr{C}$ have exteriors. Since

$\mathscr{B} \wedge \mathscr{C} = \mathscr{O}$ by hypothesis, from (I.2-20), (I.2-37), and (I.2-23) we see that $\mathscr{B}^e$ may be decomposed into separate parts as follows:

$$\mathscr{B}^e = \mathscr{C} \vee (\mathscr{B} \vee \mathscr{C})^e \qquad \forall \mathscr{C} \prec \mathscr{B}^e. \tag{I.5-3}$$

By Axiom F3

$$\begin{aligned} \mathbf{f}(\mathscr{B}, \mathscr{B}^e) &= \mathbf{f}(\mathscr{B}, \mathscr{C}) + \mathbf{f}(\mathscr{B}, (\mathscr{B} \vee \mathscr{C})^e), \\ \mathbf{f}(\mathscr{C}, \mathscr{C}^e) &= \mathbf{f}(\mathscr{C}, \mathscr{B}) + \mathbf{f}(\mathscr{C}, (\mathscr{B} \vee \mathscr{C})^e), \end{aligned} \tag{I.5-4}$$

while by Axiom F2

$$\mathbf{f}(\mathscr{B} \vee \mathscr{C}, (\mathscr{B} \vee \mathscr{C})^e) = \mathbf{f}(\mathscr{B}, (\mathscr{B} \vee \mathscr{C})^e) + \mathbf{f}(\mathscr{C}, (\mathscr{B} \vee \mathscr{C})^e). \tag{I.5-5}$$

Adding $(4)_1$ to $(4)_2$ and subtracting (5) from the sum yields (2). △

If the force exerted by $\mathscr{C}$ on $\mathscr{B}$ is of magnitude equal and of sign opposite to that exerted by $\mathscr{B}$ on $\mathscr{C}$, that is,

$$\mathbf{f}(\mathscr{B}, \mathscr{C}) = -\mathbf{f}(\mathscr{C}, \mathscr{B}) \qquad \forall (\mathscr{B}, \mathscr{C}) \in (\Omega \times \Omega)_0, \tag{I.5-6}$$

the system of forces $\mathbf{f}$ is said to be *pairwise equilibrated*. From (2) we may read off the following

Theorem (NOLL, GURTIN & WILLIAMS). *A system of forces is pairwise equilibrated if and only if the resultant force* $\mathbf{f}(\mathscr{B}, \mathscr{B}^e)$, *regarded as a function of* $\mathscr{B}$, *is additive on the separate bodies of* $\overline{\Omega}$.

A system of forces such that the resultant force on every body vanishes:

$$\mathbf{f}(\mathscr{B}, \mathscr{B}^e) = \mathbf{0} \qquad \forall \mathscr{B} \in \overline{\Omega}, \tag{I.5-7}$$

is said to be *balanced*. Since the function whose value is $\mathbf{0}$ is additive, NOLL's theorem has the following

Corollary (NOLL). *Every balanced system of forces is pairwise equilibrated.*

As is clear from NOLL's theorem, the converse of his corollary does not hold. Indeed, there are many systems of forces that are pairwise equilibrated but not balanced. One important case is presented as an example later on in this section; another is furnished by the contact forces in continuum mechanics, as will be explained in §III.1.

In the past, special cases of (6) were often inferred from a vague "axiom" called the law of "action and reaction", which was regarded as stating the content of NEWTON's Third Law of Motion: "To an action there is always a contrary and equal reaction; or, the actions of two bodies mutually upon one another are always equal and directed toward contrary parts." If, indeed, what NEWTON meant by "action" is what we here call "force", which is by no means clear from his own words or the contexts in which he applied them, then the above argument shows that axiom to be equivalent, as far as pairs of separate bodies are concerned, to the additivity of resultant forces on separate bodies. This fact is independent of whatever relations there may be between forces and motions.

Axiom F2 states, among other things, that the forces exerted by the exterior $\mathscr{B}^e$ of a body $\mathscr{B}$ on the separate parts of that body are additive:

$$(\mathscr{P}_1 \prec \mathscr{B})\ \&\ (\mathscr{P}_2 \prec \mathscr{B})\ \&\ (\mathscr{P}_1 \curlywedge \mathscr{P}_2 = \mathscr{O}) \quad \Rightarrow \quad \mathbf{f}(\mathscr{P}_1 \curlyvee \mathscr{P}_2, \mathscr{B}^e) = \mathbf{f}(\mathscr{P}_1, \mathscr{B}^e) + \mathbf{f}(\mathscr{P}_2, \mathscr{B}^e). \tag{I.5-8}$$

This fact suggests that for every particular body $\mathscr{B}$ the forces exerted by $\mathscr{B}^e$ on a certain set of parts of $\mathscr{B}$ might define a vector-valued measure over $\mathscr{B}$, a measure which we could denote formally thus:

$$\mathbf{f}(\mathscr{A}, \mathscr{B}^e) = \int_{\mathscr{A}} d\mathbf{f}_{\mathscr{B}^e} \qquad \text{if} \quad \mathscr{A} \prec \mathscr{B}. \tag{I.5-9}$$

It would be desirable to construct an abstract theory of integration with respect to systems of forces, as defined only by the above axioms and some further ones of a technical nature, but since no such general theory is presently available, we shall simply assume that our systems of forces are of this kind:

Axiom F4. *For each $\mathscr{B}$ in $\overline{\Omega}$, the function $\mathbf{f}(\cdot, \mathscr{B}^e)$ is a vector-valued measure over $\mathscr{B}$.*

Theorem. *If $\mathscr{A}$ and $\mathscr{B}$ are separate, then $\mathbf{f}(\cdot, \mathscr{B})$ is a measure over $\mathscr{A}$.*

Proof. By (I.2-18), every body $\mathscr{B}$ is the exterior of another one, namely, $\mathscr{B}^e$. By Axiom F4, $\mathbf{f}(\cdot, \mathscr{B})$ is a measure over $\mathscr{B}^e$. If $\mathscr{A}$ and $\mathscr{B}$ are separate, $\mathscr{A} \prec \mathscr{B}^e$ by Axiom B5, so Axiom F4 yields the theorem at once. △

In fact the theorem merely rephrases Axiom F4.

In this book we shall make the assumption, already mentioned at the end of §I.4, that $\overline{\Omega}$ is a collection of sets which may be extended in the common way so as to include all Borel sets. Thus the mass M is Borel measure or an extension of it such as Lebesgue measure. We shall suppose also that $\overline{\Omega} = \Omega_M$, excluding any bodies that are not massy. Integration with respect to systems of forces on $(\overline{\Omega} \times \overline{\Omega})_0$ can then be constructed by the usual theory of vector-valued measures[1] in such a way as to satisfy Axiom F4.

We may then introduce the Stieltjes integral of a continuous real function ϕ over $\mathscr{B}$ with respect to the measure $\mathbf{f}(\cdot, \mathscr{B}^e)$; we denote this integral by

$$\int \phi \, d\mathbf{f}_{\mathscr{B}^e} \tag{I.5-10}$$

and call it "the integral of ϕ with respect to $\mathbf{f}_{\mathscr{B}^e}$".

For example, if $\mathscr{B}$ is a discrete set consisting in the elements $X_1, X_2, \ldots, X_n$, then

$$\int_{\mathscr{B}} \phi \, d\mathbf{f}_{\mathscr{B}^e} = \sum_{k=1}^{n} \phi(X_k)\mathbf{f}(\{X_k\}, \mathscr{B}^e). \tag{I.5-11}$$

This is the kind of system of forces used in analytical dynamics.

If

$$\mathbf{w}: \mathscr{B} \to \mathscr{V}_{\mathbf{f}}, \tag{I.5-12}$$

so that $\mathbf{w}(X) \in \mathscr{V}_{\mathbf{f}}$, and if $\mathbf{a} \in \mathscr{V}_{\mathbf{f}}$, then $\mathbf{a} \cdot \mathbf{w}$ is a scalar field over $\mathscr{B}$, and $\int_{\mathscr{B}} (\mathbf{a} \cdot \mathbf{w}) \, d\mathbf{f}_{\mathscr{B}^e}$, if it exists, is a linear function of $\mathbf{a}$. Consequently there is a linear transformation on $\mathscr{V}_{\mathbf{f}}$ whose value is $\int_{\mathscr{B}} (\mathbf{a} \cdot \mathbf{w}) \, d\mathbf{f}_{\mathscr{B}^e}$. Denoting the transpose of this transformation by $\int_{\mathscr{B}} \mathbf{w} \otimes d\mathbf{f}_{\mathscr{B}^e}$, we have

$$\left[\int_{\mathscr{B}} \mathbf{w} \otimes d\mathbf{f}_{\mathscr{B}^e}\right]^{\mathrm{T}} \mathbf{a} = \int_{\mathscr{B}} (\mathbf{a} \cdot \mathbf{w}) \, d\mathbf{f}_{\mathscr{B}^e}. \tag{I.5-13}$$

The trace of this linear transformation will be written as follows:

$$\int_{\mathscr{B}} \mathbf{w} \cdot d\mathbf{f}_{\mathscr{B}^e} \equiv \operatorname{tr}\left[\int_{\mathscr{B}} \mathbf{w} \otimes d\mathbf{f}_{\mathscr{B}^e}\right]. \tag{I.5-14}$$

For example, if $\mathscr{B}$ is a discrete set of n elements X_k, then

$$\int_{\mathscr{B}} \mathbf{w} \cdot d\mathbf{f}_{\mathscr{B}^e} = \sum_{k=1}^{n} \mathbf{w}(X_k) \cdot \mathbf{f}(\{X_k\}, \mathscr{B}^e). \tag{I.5-15}$$

We have given as (11) and (15) two formulae valid when $\mathscr{B}$ is a discrete set. In analytical dynamics the universe is supposed to be a "system" of such discrete, mass-bearing elements, called *mass-points*, along with a single further body X_0,

[1] By choice of a particular basis a vector-valued measure may be regarded as an ordered finite set of scalar-valued measures, so the properties of vector-valued measures become rather obvious.

separate from all the rest and called "the environment of the system". To simplify the notation, henceforth in dealing with mass-points we shall write X_k for $\{X_k\}$. Then

$$\infty = \bigvee_{k=0}^{n} X_k. \tag{I.5-16}$$

It is the usage of analytical dynamics to apply the word "body" only to sub-collections of $\{X_1, X_2, \dots, X_n\}$, excluding $\mathcal{O}$, X_0, and ∞. That is, the bodies treated are those defined as follows by a subset $\imath_{\mathscr{B}}$ of $\{1, 2, \dots, n\}$:

$$\mathscr{B} = \bigvee_{k \in \imath_{\mathscr{B}}} X_k. \tag{I.5-17}$$

The forces $\mathbf{f}(X_k, X_q)$ are called *mutual*; the forces $\mathbf{f}(\mathscr{B}, X_0)$ are called *extrinsic.* The traditional notations, more or less, are as follows:

$$\begin{aligned} M_k &\equiv M(X_k), \\ \mathbf{f}_{kq} &\equiv \mathbf{f}(X_k, X_q), \\ \mathbf{f}_k^{\mathrm{e}} &\equiv \mathbf{f}(X_k, X_0); \end{aligned} \tag{I.5-18}$$

here k and q run from 1 to n, and, so as to set aside trivial exceptions, we assume that $M_k > 0$. Supposing assigned the quantities $\mathbf{f}_{kq}$ and $\mathbf{f}_k^{\mathrm{e}}$, we define the entire system of forces by the requirement that Axioms F1, F2, and F3 be satisfied. The resultant force $\mathbf{f}_k$ acting on X_k is given thus:

$$\mathbf{f}_k \equiv \mathbf{f}(X_k, X_k^{\mathrm{e}}) = \mathbf{f}_k^{\mathrm{e}} + \sum_{q=1}^{n}{}' \mathbf{f}_{kq}; \tag{I.5-19}$$

the symbol $\sum'$ indicates a sum omitting the term for which $q = k$. If by $i_{\mathscr{B}}^{-1}$ we denote the complement of $\imath_{\mathscr{B}}$ in $\{1, 2, \dots, n\}$, the resultant force on $\mathscr{B}$ is obtained as follows:

$$\mathbf{f}(\mathscr{B}, \mathscr{B}^{\mathrm{e}}) = \sum_{k \in \imath_{\mathscr{B}}} \left(\mathbf{f}_k^{\mathrm{e}} + \sum_{q \in \imath_{\mathscr{B}}^{-1}} \mathbf{f}_{kq} \right). \tag{I.5-20}$$

The double sum is the *resultant mutal force* on $\mathscr{B}$; the single sum is the *resultant extrinsic force* on $\mathscr{B}$. In particular, if $\mathscr{B} = \{1, 2, \dots, n\}$, then $i_{\mathscr{B}}^{-1}$ is empty, so for this particular $\mathscr{B}$,

$$\mathbf{f}(\mathscr{B}, \mathscr{B}^{\mathrm{e}}) = \sum_{k=1}^{n} \mathbf{f}_k^{\mathrm{e}}. \tag{I.5-21}$$

If the system of forces is balanced, then (6) holds, so that in particular

$$\mathbf{f}_{kq} = -\mathbf{f}_{qk}, \qquad q \neq k. \tag{I.5-22}$$

In this case (20) may be written as

$$\mathbf{f}(\mathscr{B}, \mathscr{B}^{\mathrm{e}}) = \sum_{k \in \imath_{\mathscr{B}}} \left(\mathbf{f}_k^{\mathrm{e}} + \sum_{q=1}^{n}{}' \mathbf{f}_{kq} \right), \tag{I.5-23}$$

since the terms by which the right-hand side differs from that of (20) cancel each other in pairs. By choosing $\mathscr{B}$ as X_k we conclude that

$$\mathbf{f}_k^e + \sum_{q=1}^{n}{}' \mathbf{f}_{kq} = \mathbf{0}. \tag{I.5-24}$$

If, conversely, (24) and (22) hold when $k = 1, 2, \ldots, n$, then (20) shows that $\mathbf{f}(\mathscr{B}, \mathscr{B}^e) = \mathbf{0}$. Thus *the conditions* (24) *and* (22) *are necessary and sufficient that the system of forces* $\mathbf{f}$ *on the universe of analytical dynamics be balanced*, provided we agree that also $\mathbf{f}(X_0, X_k) = -\mathbf{f}_k^e$, $k = 1, 2, \ldots, n$. Either by summing (24) on k or by inspection of (21) we conclude that

$$\sum_{k=1}^{n} \mathbf{f}_k^e = \mathbf{0}: \tag{I.5-25}$$

the total extrinsic force acting on the system is null. These simple theorems provide the standard basis for analytical dynamics.

It is easy to extend the foregoing to arbitrary pairs of bodies, which need not be distinct. If we introduce the *self-force* $\mathbf{f}_{kk}$ of X_k, the force exerted by X_k on itself, then we can define as follows the force exerted by the arbitrary body $\mathscr{C}$ on the arbitrary body $\mathscr{B}$:

$$\mathbf{f}(\mathscr{B}, \mathscr{C}) \equiv \sum_{\substack{h \in \imath_{\mathscr{B}} \\ q \in \imath_{\mathscr{C}}}} \mathbf{f}_{hq}, \tag{I.5-26}$$

$\imath_{\mathscr{B}}$ and $\imath_{\mathscr{C}}$ being the sets of integers that define $\mathscr{B}$ and $\mathscr{C}$ according to (17). When $\mathscr{B}$ and $\mathscr{C}$ are separate, this function $\mathbf{f}$ reduces to the $\mathbf{f}$ defined by the requirements F2 and F3 on the basis of (18). We may call $\mathbf{f}(\mathscr{B}, \mathscr{B})$ the *self-force* of $\mathscr{B}$. From (22) we see at once that *in a balanced system of forces,*

$$\mathbf{f}(\mathscr{B}, \mathscr{B}) = \mathbf{0} \quad \forall \mathscr{B} \qquad \Leftrightarrow \qquad \mathbf{f}_{kk} = \mathbf{0}, \qquad k = 0, 1, \ldots, n. \tag{I.5-27}$$

Exercise I.5.1. Prove that

$$\mathbf{f}(\mathscr{B}, \infty) = \mathbf{f}(\mathscr{B}, \mathscr{B}) + \mathbf{f}(\mathscr{B}, \mathscr{B}^e). \tag{I.5-28}$$

In analytical dynamics it is customary to assume both that $\mathbf{f}_{kk} = \mathbf{0}$ and that the system of forces is balanced. From (27) and (28) it then follows that $\mathbf{f}(\mathscr{B}, \infty) = \mathbf{f}(\infty, \mathscr{B}) = \mathbf{0}\ \forall \mathscr{B}$. We may express this fact as a statement that the *universal body of analytical dynamics is* passive: The body ∞ exerts null force upon its parts.

As their statements suggest, (28) and the theorem stated just after it are not limited to discrete systems. RIZZO has proposed, in effect, the following axioms as a natural extension of NOLL's:

Axiom FE1. $\mathbf{f}: \overline{\Omega} \times \overline{\Omega} \rightarrow \mathscr{V}_{\mathbf{f}}$.

Axiom FE2.

$$\mathbf{f}(\mathscr{C}_1 \curlyvee \mathscr{C}_2, \mathscr{B}) = \mathbf{f}(\mathscr{C}_1, \mathscr{B}) + \mathbf{f}(\mathscr{C}_2, \mathscr{B}) - \mathbf{f}(\mathscr{C}_1 \curlywedge \mathscr{C}_2, \mathscr{B}),$$
$$\mathbf{f}(\mathscr{O}, \mathscr{D}) = \mathbf{0}.$$

Axiom FE3.

$$\mathbf{f}(\mathscr{B}, \mathscr{C}_1 \curlyvee \mathscr{C}_2) = \mathbf{f}(\mathscr{B}, \mathscr{C}_1) + \mathbf{f}(\mathscr{B}, \mathscr{C}_2) - \mathbf{f}(\mathscr{B}, \mathscr{C}_1 \curlywedge \mathscr{C}_2),$$
$$\mathbf{f}(\mathscr{D}, \mathscr{O}) = \mathbf{0}.$$

If $\mathscr{B}$, $\mathscr{C}_1$, and $\mathscr{C}_2$ are separate, these axioms reduce to NOLL's, so any $\mathbf{f}$ that satisfies them is an extension from $(\overline{\Omega} \times \overline{\Omega})_0$ to $\overline{\Omega} \times \overline{\Omega}$ of an $\mathbf{f}$ that satisfies NOLL's axioms and (1). The formula (26) effects such an extension explicitly for a discrete universe. Axioms FE2 and FE3 are easy to motivate intuitively.

Exercise I.5.2. Prove that

$$\begin{aligned} \mathbf{f}(\mathscr{B}, \infty) &= \mathbf{f}(\mathscr{B}, \mathscr{C}) + \mathbf{f}(\mathscr{B}, \mathscr{C}^e), \\ \mathbf{f}(\infty, \mathscr{B}) &= \mathbf{f}(\mathscr{C}, \mathscr{B}) + \mathbf{f}(\mathscr{C}^e, \mathscr{B}). \end{aligned} \tag{I.5-29}$$

Hence show that (28) holds, that

$$\mathbf{f}(\infty, \mathscr{B}) = \mathbf{f}(\mathscr{B}, \mathscr{B}) + \mathbf{f}(\mathscr{B}^e, \mathscr{B}), \tag{I.5-30}$$

and that

$$\begin{aligned} \mathbf{f}(\mathscr{B}, \mathscr{B}) &= \mathbf{f}(\infty, \mathscr{B}) - \mathbf{f}(\mathscr{B}^e, \mathscr{B}), \\ \mathbf{f}(\mathscr{B}, \mathscr{B}) &= \mathbf{f}(\mathscr{B}, \infty) - \mathbf{f}(\mathscr{B}, \mathscr{B}^e). \end{aligned} \tag{I.5-31}$$

From (28), now proved in generality, we see that if $\mathbf{f}(\mathscr{B}, \infty) = \mathbf{0}$, then

$$\mathbf{f}(\mathscr{B}, \mathscr{B}^e) = -\mathbf{f}(\mathscr{B}, \mathscr{B}): \tag{I.5-32}$$

If the universal body is passive, the resultant force on each body is the negative of its self-force. Thus, in such a universe, *the system of forces is balanced if and only if the self-force of every body is* $\mathbf{0}$. More generally, if $\mathbf{f}(\mathscr{B}, \infty) \neq \mathbf{0}$ for some $\mathscr{B}$, (32) does not hold, and we cannot easily infer anything about $\mathbf{f}$ on $\overline{\Omega} \times \overline{\Omega}$ from the statements obtained above about its restriction to $(\overline{\Omega} \times \overline{\Omega})_0$. In particular, it is not obvious how to infer (6) for pairs of bodies that are not separate.

As a first step in this direction, we find on the basis of the extended axioms FE1–FE3 a counterpart for the theorem of NOLL and GURTIN & WILLIAMS concerning pairs of separate bodies.

Lemma. *A system of forces on* $\overline{\Omega} \times \overline{\Omega}$ *is pairwise equilibrated for* separate *bodies if and only if the self-force* $\mathbf{f}(\mathscr{B}, \mathscr{B})$, *regarded as a function of* $\mathscr{B}$, *is additive on all bodies of* $\overline{\Omega}$.

Proof. We apply Axiom FE2 when $\mathscr{B} = \mathscr{C}_1 \vee \mathscr{C}_2$ and $\mathscr{C}_1 \wedge \mathscr{C}_2 = \mathscr{O}$, then expand the result by use of Axiom FE3. Thus

$$\begin{aligned} \mathbf{f}(\mathscr{B}, \mathscr{B}) &= \mathbf{f}(\mathscr{C}_1, \mathscr{C}_1 \vee \mathscr{C}_2) + \mathbf{f}(\mathscr{C}_2, \mathscr{C}_1 \vee \mathscr{C}_2), \\ &= \mathbf{f}(\mathscr{C}_1, \mathscr{C}_1) + \mathbf{f}(\mathscr{C}_2, \mathscr{C}_2) + \mathbf{f}(\mathscr{C}_1, \mathscr{C}_2) + \mathbf{f}(\mathscr{C}_2, \mathscr{C}_1). \quad \text{(I.5-33)} \end{aligned}$$

The basic decomposition theorem following (I.2-38) assures us that for any $\mathscr{B}$ we may choose $\mathscr{C}_1$ as any of its parts. △

Since $\mathbf{0}$ is an additive function, the lemma has the following

Corollary (RIZZO). *If the self-force of every body is* $\mathbf{0}$, *the system of forces is pairwise equilibrated for separate bodies.*

Exercise I.5.3. Prove that

$$\begin{aligned} \mathbf{f}(\mathscr{B}, \mathscr{C}) + \mathbf{f}(\mathscr{C}, \mathscr{B}) = \mathbf{f}(\mathscr{B} \vee \mathscr{C}, \mathscr{B} \wedge \mathscr{C}) &+ \mathbf{f}(\mathscr{B} \wedge \mathscr{C}, \mathscr{B} \vee \mathscr{C}) \\ + \mathbf{f}(\mathscr{B} \vee \mathscr{C}, \mathscr{B} \vee \mathscr{C}) &+ \mathbf{f}(\mathscr{B} \wedge \mathscr{C}, \mathscr{B} \wedge \mathscr{C}) \qquad \text{(I.5-34)} \\ - \mathbf{f}(\mathscr{B}, \mathscr{B}) &- \mathbf{f}(\mathscr{C}, \mathscr{C}). \end{aligned}$$

Theorem (RIZZO). *In order that*

$$\mathbf{f}(\mathscr{B}, \mathscr{C}) + \mathbf{f}(\mathscr{C}, \mathscr{B}) = \mathbf{0} \qquad \forall (\mathscr{B}, \mathscr{C}) \in \overline{\Omega} \times \overline{\Omega}, \qquad \text{(I.5-35)}$$

it is necessary and sufficient that

$$\mathbf{f}(\mathscr{B}, \mathscr{B}) = \mathbf{0} \qquad \forall \mathscr{B} \in \overline{\Omega}. \qquad \text{(I.5-36)}$$

Proof. Necessity is obvious. To prove sufficiency, we note that the bodies $\mathscr{B} \wedge \mathscr{C}$ and $(\mathscr{B} \vee \mathscr{C}) \wedge (\mathscr{B} \wedge \mathscr{C})^e$ are separate, and that

$$\mathscr{B} \vee \mathscr{C} = (\mathscr{B} \wedge \mathscr{C}) \vee [(\mathscr{B} \vee \mathscr{C}) \wedge (\mathscr{B} \wedge \mathscr{C})^e]. \qquad \text{(I.5-37)}$$

By use of Axioms FE2 and FE3, with the aid of (36) we show that

$$\begin{aligned} \mathbf{f}(\mathscr{B} \vee \mathscr{C}, \mathscr{B} \wedge \mathscr{C}) &= \mathbf{f}((\mathscr{B} \vee \mathscr{C}) \wedge (\mathscr{B} \wedge \mathscr{C})^e, \mathscr{B} \wedge \mathscr{C}), \\ \mathbf{f}(\mathscr{B} \wedge \mathscr{C}, \mathscr{B} \vee \mathscr{C}) &= \mathbf{f}(\mathscr{B} \wedge \mathscr{C}, (\mathscr{B} \vee \mathscr{C}) \wedge (\mathscr{B} \wedge \mathscr{C})^e). \end{aligned} \qquad \text{(I.5-38)}$$

Substituting (38) into (34), shortened by use of (36), we obtain a result of the form

$$\mathbf{f}(\mathscr{B}, \mathscr{C}) + \mathbf{f}(\mathscr{C}, \mathscr{B}) = \mathbf{f}(\mathscr{D}, \mathscr{E}) + \mathbf{f}(\mathscr{E}, \mathscr{D}), \qquad \mathscr{D} \wedge \mathscr{E} = \mathscr{O}. \qquad \text{(I.5-39)}$$

The preceding corollary assures us that the right-hand side is null. △

NOLL's corollary, derived above, asserts that a balanced system of forces is pairwise equilibrated on separate bodies. We may ask if the same holds for all pairs of bodies. The answer is no. From (28) we see that in a balanced system of forces $\mathbf{f}(\mathscr{B}, \infty) = \mathbf{f}(\mathscr{B}, \mathscr{B})$. Only if the universal body is passive does it follow that $\mathbf{f}(\mathscr{B}, \mathscr{B}) = \mathbf{0}\ \forall \mathscr{B}$. Since this last condition is a necessary one in order for the forces to be pairwise equilibrated on $\overline{\Omega} \times \overline{\Omega}$, we conclude that *in order for a balanced system of forces to be pairwise equilibrated for all bodies, it is necessary and sufficient that the universal body be passive.*

Nothing said about forces in this section restricts the dimension of $\mathscr{V}_{\mathbf{f}}$. For most parts of classical mechanics forces are 3-dimensional vectors, and in this book we shall so consider them. Furthermore, we shall not need to consider systems of forces defined for pairs of bodies that are not separate.

6. The Event World. Framings

In common life we regard ourselves and other objects as occupying *places*, which are points in a 3-dimensional space, the properties of which are given once and for all and are not changed by our own presence or absence. Moreover, the changes we perceive in ourselves and in our environment we regard as occurring at specific *instants*, which are points in a 1-dimensional space altogether independent of the space of places. The totality of places and instants is a topological space which we call the *event world* $\mathscr{W}$. It is a space a typical point e of which we represent by an ordered pair $(\mathbf{x}, T)$, in which $\mathbf{x}$ is a *place* and T is an *instant*.

The event world is the blank canvas on which pictures of nature may be painted, the quarry for blocks from which statues of nature may be carved. This canvas, this quarry, must be chosen by the artist before he sets to work. It lays limitations upon his art, but it by no means determines the pictures or the statues he will fashion.

For the event world of classical mechanics we presume that the places and the instants used to represent events are themselves elements of Euclidean point spaces. We assume further that the space of instants, $\mathscr{T}$, has dimension 1; for the purposes of this book it is sufficient to regard the space of places, $\mathscr{E}$, as 3-dimensional, though only in a few contexts does the dimension of $\mathscr{E}$ make any difference in the nature of the concepts and apparatus we shall develop. Finally, we assume the structure of the event world $\mathscr{W}$ such that it may be mapped homeomorphically[1] onto the product $\mathscr{E} \times \mathscr{T}$:

$$\begin{aligned} \oint : \mathscr{W} &\to \mathscr{E} \times \mathscr{T}, \\ e &\mapsto (\mathbf{x}, T). \end{aligned} \tag{I.6-1}$$

A mapping $\oint$ of this kind we shall call a *framing*; the space $\mathscr{E} \times \mathscr{T}$ is the *frame of reference* onto which $\oint$ maps $\mathscr{W}$. We shall call $\mathbf{x}$ and T, respectively, the *place* and the *instant* of the event e in $\oint$.

For the purposes of classical mechanics, a frame of reference is interpreted by means of a rigid body and a clock.[2] As far as the theory is concerned, the descriptive terms "observer" and "frame of reference of an observer" are synonymous with "as assigned by a framing". So as to further the desired interpretation, we shall occasionally say that $\mathbf{x}$ is the place and T is the instant of the event e "as observed" in $\oint$ and other things of a like kind, but the verb "observe", like the noun "observer", does not enter the mathematical structure.

The space of instants $\mathscr{T}$ may be oriented. One particular orientation is selected and called positive and interpreted as pointing from "past" to "future".

In assuming that a framing exists, we impute a certain structure to $\mathscr{W}$. If there is one such mapping, there are also infinitely many others, but not all these are of interest. In §9, by deciding to consider as being framings only those homeomorphisms of $\mathscr{W}$ that share the sense of time and preserve the metrics in $\mathscr{E}$ and $\mathscr{T}$, we shall choose the part of the structure of $\mathscr{W}$ which we shall interpret, leaving unspecified such further structure as $\mathscr{W}$ may have. For the time being, however, we assume given one single framing $\oint$, and we define in terms of the frame of reference $\mathscr{E} \times \mathscr{T}$ various relations between bodies and the event world.

We may assign a co-ordinate system to $\mathscr{T}$. The co-ordinate t of an instant T is called the *time* of that instant. Only those assignments of times to instants that preserve the orientation of $\mathscr{T}$ are allowed. The oriented distance between instants whose times are t_1 and t_2 is $t_2 - t_1$. It is called the *time interval* between those instants, and if that interval is positive, the time t_2 is said to be *later* than t_1, whereas t_1 is *earlier* than t_2.

As seems to have been understood by the Ancients, only the ratios of time intervals, not the intervals themselves, can be checked against any experience of man. The most general transformation of $\mathscr{T}$ that preserves the ratios of time intervals and the sense of time is of the form

$$T \mapsto K(T - T_0) + L, \tag{I.6-2}$$

[1] A mapping $\oint$ of one topological space into another is homeomorphic if it is bijective and continuous, and if its inverse $\oint^{-1}$ is continuous.

[2] Rigid motion will be defined in §I.10. A rigid body is one susceptible of rigid motions only.

where T_0 and L are particular instants and K is a positive constant. Choice of a particular number K is called in physics "fixing the unit of time", while choice of particular instants L and T_0 is called "fixing the origin of time". Fixing the unit of time is equivalent to choosing one of those oriented co-ordinate systems on $\mathscr{T}$ that make the interval of time between some particular pair of distinct instants a given number.

Commonly, the unit and origin of time are regarded as fixed in advance, and $\mathscr{T}$ is identified with the real line $\mathscr{R}$ according to this choice of co-ordinates and metric. That is, the instants, which are 1-dimensional vectors, are confounded with times, which are the co-ordinates of instants. We shall follow this custom in the rest of this book.

We denote the translation space of $\mathscr{E}$ by $\mathscr{V}$. The term *vector* will refer always to an element $\mathbf{v}$ in $\mathscr{V}$, and $|\mathbf{v}|$ will denote the magnitude of $\mathbf{v}$. The given inner product of vectors $\mathbf{v}$ and $\mathbf{w}$ in $\mathscr{V}$ will be denoted by $\mathbf{v} \cdot \mathbf{w}$, and linear transformations of $\mathscr{V}$ into itself, which we shall call *tensors* over $\mathscr{V}$, will be denoted by bold-faced letters $\mathbf{T}$, $\mathbf{S}$, ... The Euclidean distance between the places $\mathbf{x}$ and $\mathbf{y}$ in $\mathscr{E}$ will be written as $|\mathbf{x} - \mathbf{y}|$, since $\mathbf{x} - \mathbf{y}$ is the vector in $\mathscr{V}$ that translates $\mathbf{y}$ into $\mathbf{x}$.

If $\mathbf{v} \cdot \mathbf{w}$ denotes an assigned inner product in $\mathscr{V}$, and if $K > 0$, then $K\mathbf{v} \cdot \mathbf{w}$ is also an inner product, and such a change from one inner product to another is in fact the most general one that preserves ratios of distances in $\mathscr{E}$. Only such ratios, not the distances themselves, can be checked against any experience of man, as the Ancients well knew. Choice of a particular positive constant K is called in physics "fixing the unit of length". Choosing the unit of length is equivalent to assigning a particular distance to a particular pair of distinct places in $\mathscr{E}$.

The most general transformation $\mathbf{Q}$ of $\mathscr{V}$ that preserves the inner product itself is a linear one:

$$\bar{\mathbf{v}} = \mathbf{Q}\mathbf{v}, \tag{I.6-3}$$

and a linear transformation preserves the inner product if and only if the tensor $\mathbf{Q}$ is *orthogonal*[1]:

$$\mathbf{Q}\mathbf{Q}^{\mathsf{T}} = \mathbf{1}, \tag{I.6-4}$$

$\mathbf{1}$ being the unit tensor.

[1] The student will know that in order for a transformation $\mathbf{Q}$ to preserve the inner product in a vector space, that is, $\mathbf{Q}(\mathbf{u}) \cdot \mathbf{Q}(\mathbf{v}) = \mathbf{u} \cdot \mathbf{v} \; \forall \, \mathbf{u}, \mathbf{v}$, it is necessary and sufficient that $\mathbf{Q}$ be a tensor (namely, a linear transformation) and that it satisfy (4). From (4) we see at once that $\det \mathbf{Q} = \pm 1$. If $\det \mathbf{Q} = +1$, $\mathbf{Q}$ is a *rotation*. Every orthogonal tensor on a space of odd dimension is either a rotation or the product of a rotation by the central inversion $-\mathbf{1}$; that is, there is one and only one rotation $\mathbf{R}$ such that either $\mathbf{Q} = \mathbf{R}$ or $\mathbf{Q} = -\mathbf{R}$, and the only possible proper numbers of $\mathbf{Q}$ are $+1$ and -1. If, as we always suppose, $\dim \mathscr{V} = 3$, then 1 is a proper number of every $\mathbf{R}$, and the corresponding proper space is one-dimensional unless $\mathbf{R} = \mathbf{1}$. This last statement is the content of a famous theorem of EULER: Every non-identical rotation about a point is in fact a rotation about a single line. The *axis* of $\mathbf{Q}$ is the proper line of the one and only $\mathbf{R}$ to which $\mathbf{Q}$ is proportional.

A *world-line* is a curve[1] in $\mathscr{W}$ whose image in $\mathscr{E} \times \mathscr{R}$ associates one place to each time, so we may represent a world-line as follows:

$$\lambda : \mathscr{I} \to \mathscr{E}, \tag{I.6-5}$$

$\mathscr{I}$ being an interval of $\mathscr{R}$. A collection of world-lines defined over $\mathscr{I}$ is a *world-tube*. The places on a world-tube at a fixed time t form a set $\mathscr{S}_t$, and for any two times t' and t'' in $\mathscr{I}$, every place in $\mathscr{S}_{t'}$ is connected with one or more places in $\mathscr{S}_{t''}$ by world-lines of the world-tube. Thus we may regard a world-tube τ as a mapping of times into the set of all subsets of $\mathscr{E}$, which is commonly denoted by $P(\mathscr{E})$:

$$\begin{aligned} \tau : \mathscr{I} &\to P(\mathscr{E}), \\ t &\mapsto \mathscr{S}_t . \end{aligned} \tag{I.6-6}$$

Intersections of world-lines represent collisions or the creation or destruction of bodies or elements of bodies. In specific mechanical theories such intersections are usually excluded altogether or allowed as exceptional cases subject to specified conditions.

Experiences are to be correlated with world-lines and world-tubes. We think of these as progressing "through" the event world $\mathscr{W}$ as time goes on.

Thus in this book we shall assume, informally, that there is a single, "absolute" or "unmoved" instantaneous space $\mathscr{E}$ of places, change of place within which constitutes experience. However, while some writers have claimed that the event world must be of this kind in classical physics, that is not so. As the reader of this book will be able to see for himself, every statement in classical mechanics retains meaning if the sections of the event world at different instants T are different Euclidean 3-dimensional spaces $\mathscr{E}_T$ rather than a single, assigned space $\mathscr{E}$. To visualize the distinction here, it is easiest to think of an event world of dimension 3, so that the instantaneous space is a Euclidean plane, and $\mathscr{W}$ is a stack of such planes, one and only one of which a given world-line crosses at each instant. A second but different event world of exactly the same kind may be constructed by rotating each of these planes through some angle about some one of its points, that point and that angle being functions of the particular plane—that is, of the particular instant. No statement made in classical mechanics depends upon the choice of these points and these rotations, and thus no result in classical mechanics can be used to establish or contradict any relation between the instantaneous spaces $\mathscr{E}_T$. Far from imposing an "absolute space" upon nature, classical mechanics requires no relation at all among the infinitely many instantaneous spaces $\mathscr{E}_T$.

[1] Of course, a curve is a piecewise differentiable one-parameter family of events, $e = f(s)$, and s varies over some real interval.

7. Motions

A mapping $\boldsymbol{\mu}$ of the universe Ω into the set $P(\mathscr{W})$ of all subsets of the event world $\mathscr{W}$,

$$\boldsymbol{\mu}: \Omega \to P(\mathscr{W}), \tag{I.7-1}$$

is called a *motion* if for each body $\mathscr{B}$ in Ω, $\boldsymbol{\mu}(\mathscr{B})$ is a world-tube. In this definition we continue to suppose given a particular framing $\mathfrak{f}$. Thus a motion may be represented alternatively as a mapping χ_Ω of $\Omega \times \mathscr{I}$ into $P(\mathscr{E})$:

$$\chi_\Omega: \Omega \times \mathscr{I} \to P(\mathscr{E}). \tag{I.7-2}$$

$\mathscr{I}$ is again some interval in $\mathscr{R}$, such as for example $]-\infty, t_0[$ for some t_0. The value $\chi_\Omega(\mathscr{B}, t)$ of χ_Ω, which is a set in $\mathscr{E}$, is called the *shape*[1] of $\mathscr{B}$ at the time t. When thinking of t as being the present time we shall call $\chi_\Omega(\mathscr{B}, t)$ the *present shape* of $\mathscr{B}$.

As we have said already, henceforth we shall consider only bodies that are sets in some topological space, or some extension thereof:

$$\mathscr{B} = \{X, Y, \ldots\}. \tag{I.7-3}$$

The points X of $\mathscr{B}$ were called "particles" until recently, but in order to avoid any possible confusion with physics we shall call them *body-points*. The motion of a body composed of body-points is engendered by the motions of those points. Using the symbol χ for this more detailed motion, we write

$$\chi: \mathscr{B} \times \mathscr{I} \to \mathscr{E}, \tag{I.7-4}$$

and, explicitly,

$$\mathbf{x} = \chi(X, t), \qquad X \in \mathscr{B}, \quad t \in \mathscr{I}. \tag{I.7-5}$$

In words, $\mathbf{x}$ is the place in $\mathscr{E}$ that the body-point X *occupies* at the time t in the motion χ. Moreover, the shape of $\mathscr{B}$ at the time t is the set of places its body-points occupy then:

$$\chi_\Omega(\mathscr{B}, t) = \{\chi(X, t) : X \in \mathscr{B}\}. \tag{I.7-6}$$

Each body-point X is thus associated with a world-line, and the world-lines of all the points of $\mathscr{B}$ constitute the world-tube of $\mathscr{B}$.

[1] In the literature usually both a map from the set of bodies into $\mathscr{E}$ and the value of such a map for a given body are called "configurations".

It is customary in mechanics, with specifically stated exceptions, to consider only such motions χ as are differentiable at least twice, and often as many times as desired, with respect to t, for each X. Denoting the derivatives of χ with respect to t when X is held fixed by $\dot{\chi}$, $\ddot{\chi}$ ..., $\overset{(n)}{\chi}$, so that in particular $\dot{\chi} = \overset{(1)}{\chi}$ and $\ddot{\chi} = \overset{(2)}{\chi}$, we call the values of these derivatives the *velocity* $\mathbf{v}$, *acceleration* $\mathbf{a}$, ..., the nth *velocity* ${}_n\mathbf{v}$ of the body-point at the time t:

$$\begin{aligned} \mathbf{v} &= \dot{\chi}(X, t), \\ \mathbf{a} &= \ddot{\chi}(X, t), \quad \ldots, \\ {}_n\mathbf{v} &= \overset{(n)}{\chi}(X, t). \end{aligned} \tag{I.7-7}$$

Thus ${}_1\mathbf{v} = \mathbf{v}$, and ${}_2\mathbf{v} = \mathbf{a}$. It is easy to show that, for any given χ, the velocities of a given body-point are vectors:

$${}_n\mathbf{v} \in \mathscr{V}, \qquad n = 1, 2, 3 \ldots, \tag{I.7-8}$$

and therefore for each time t the function $\overset{(n)}{\chi}(\cdot, t)$ is a vector field defined over $\mathscr{B}$.

The metric in the Euclidean point space $\mathscr{E}$ is determined by the inner product in the translation space $\mathscr{V}$. The magnitudes of the vectors ${}_n\mathbf{v}$ are not thereby determined, for they are fixed only by the choice of metric in the space of instants. We describe this fact by saying that "the units of ${}_n\mathbf{v}$ are those of (length) $\div$ (time)^n."

While the restriction $\chi(X, \cdot)$ of the mapping χ to a particular body-point X has been assumed smooth, nothing in the way of smoothness has been imputed to the restriction $\chi(\cdot, t)$ to a fixed time. For mechanics in its most general form, $\chi(\cdot, t)$ need not even be a one-to-one mapping of body-points onto places in $\mathscr{E}$. Indeed, in the example furnished by analytical dynamics, the motion χ carries the several mass-points into a discrete set of places $\mathbf{x}_i$ at each time t, but the restricted mapping $\chi(\cdot, t)$ is not always one-to-one, for at a collision the world-lines of two or more mass-points intersect, and it is possible even that two world-lines coalesce for an interval of time and then split asunder again. In continuum mechanics, however, usually χ is assumed smooth in both arguments; in particular, the mapping $\chi(\cdot, t)$: $\mathscr{B} \rightarrow \chi_\Omega(\mathscr{B}, t)$ is bijective. This statement, which asserts that two distinct body-points never come to occupy the same place at the same time, is sometimes called the *Axiom of Impenetrability*.

Of course it is possible to relax the Axiom of Impenetrability at singular points, curves, or surfaces so as to represent shock waves, slip sheets, tears, welds, and fractures, but in this book we do not consider those.

In a particular branch of mechanics a particular universe Ω is laid down once and for all. There is then no danger of confusion if we write χ for χ_Ω in (2) while retaining also the sense (4).

8. Linear Momentum. Rotational Momentum. Kinetic Energy. Working. Torque

We continue to suppose given a particular framing ϕ, in terms of which a motion χ of a body $\mathscr{B}$ is defined and is described by (I.7-2). The vector fields defined over $\mathscr{B}$ at the time t by means of the motion χ of $\mathscr{B}$ give rise to certain additive set functions defined by integration with respect to mass over $\mathscr{B}$. The most important of these are, first, the *linear momentum* of $\mathscr{B}$:

$$\mathbf{m}(\mathscr{B}; \chi(\cdot, t)) \equiv \int_{\mathscr{B}} \dot{\chi}(\cdot, t)\, dM; \tag{I.8-1}$$

second, the *rotational momentum* of $\mathscr{B}$ with respect to the place $\mathbf{x}_0$:

$$\mathbf{M}(\mathscr{B}; \chi(\cdot, t))_{\mathbf{x}_0} \equiv \int_{\mathscr{B}} (\chi(\cdot, t) - \mathbf{x}_0) \wedge \dot{\chi}(\cdot, t)\, dM = -\mathbf{M}(\mathscr{B}; \chi(\cdot, t))_{\mathbf{x}_0}^{\mathrm{T}}; \tag{I.8-2}$$

and, third, the *kinetic energy* of $\mathscr{B}$:

$$K(\mathscr{B}; \chi(\cdot, t)) \equiv \frac{1}{2} \int_{\mathscr{B}} |\dot{\chi}(\cdot, t)|^2\, dM. \tag{I.8-3}$$

From the definitions of $\mathbf{m}$, $\mathbf{M}_{\mathbf{x}_0}$, and K we see that for a given motion χ of a given body $\mathscr{B}$ the values of these functions at a given time t are vectors, skew tensors, and scalars, respectively.

To lighten the notation we shall henceforth usually leave $\mathscr{B}$, χ, and t unwritten in formulae involving $\mathbf{m}$, $\mathbf{M}_{\mathbf{x}_0}$, and K. We shall always remember that these important functions of t are associated to $\mathscr{B}$ by a motion χ.

It is obvious from (2) that

$$\mathbf{M}_{\mathbf{x}_0} = \mathbf{M}_{\mathbf{x}_1} + (\mathbf{x}_1 - \mathbf{x}_0) \wedge \mathbf{m}. \tag{I.8-4}$$

For a given motion χ of a given body $\mathscr{B}$, the quantities $\mathbf{m}$, $\mathbf{M}_{\mathbf{x}_0}$, and K are functions of time alone. Denoting the derivative with respect to time by a superimposed dot, we see that

$$\begin{aligned} \dot{\mathbf{m}} &= \int_{\mathscr{B}} \ddot{\chi}\, dM, \\ \dot{\mathbf{M}}_{\mathbf{x}_0} &= \int_{\mathscr{B}} (\chi - \mathbf{x}_0) \wedge \ddot{\chi}\, dM, \\ \dot{K} &= \int_{\mathscr{B}} \dot{\chi} \cdot \ddot{\chi}\, dM, \end{aligned} \tag{I.8-5}$$

on the assumption, which we shall make throughout this book except where the contrary is stated, that the acceleration field $\ddot{\chi}$ is continuous over $\mathscr{B}$. Furthermore, in $(5)_2$ the place $\mathbf{x}_0$ is taken as a stationary one in the framing ϕ.

Exercise 1.8.1. Let the place $\mathbf{x}_0$ be stationary, and let $\mathbf{x}_1(\cdot)$ be any place-valued differentiable function of time. Prove that

$$\begin{aligned} \dot{\mathbf{M}}_{\mathbf{x}_0} &= \dot{\mathbf{M}}_{\mathbf{x}_1} + (\mathbf{x}_1 - \mathbf{x}_0) \wedge \dot{\mathbf{m}} + \dot{\mathbf{x}}_1 \wedge \mathbf{m}, \\ &= \int_{\mathscr{B}} (\chi - \mathbf{x}_1) \wedge \ddot{\chi}\, dM + (\mathbf{x}_1 - \mathbf{x}_0) \wedge \dot{\mathbf{m}}. \end{aligned} \tag{I.8-6}$$

These definitions and relations are introduced here for later convenience. The basic principles of mechanics and energetics relate the rates of change $\dot{\mathbf{m}}$, $\dot{\mathbf{M}}_{\mathbf{x}_0}$, and $\dot{K}$ to the forces acting on $\mathscr{B}$, as we shall explain in §§I.12 and I.14.

In §I.5 we have defined a system of forces and an integration over $\mathscr{B}$ with respect to the forces exerted on the parts of $\mathscr{B}$ by its exterior, $\mathscr{B}^e$. The system of forces may be assigned differently at different instants. We will constantly remember this fact, though it is not reflected in the notation.

Forces are elements of $\mathscr{V}_{\mathbf{f}}$, while velocities are elements of $\mathscr{V}$. Henceforth we suppose that $\dim \mathscr{V}_{\mathbf{f}} = 3$. Thus the inner-product space $\mathscr{V}_{\mathbf{f}}$ is isomorphic to the inner-product space $\mathscr{V}$. By using a particular isomorphism we may form inner products of forces and other vectors such as velocities or accelerations. That there are infinitely many different isomorphisms of this kind, reflects the fact that units of force are not yet related to units of length and time. We shall consider any one isomorphism and by using the definition (I.5-14) introduce as follows the *working* W of the system of forces $\mathbf{f}_{\mathscr{B}^e}$ in the motion χ of $\mathscr{B}$ at the time t:

$$W(\mathscr{B}; \chi(\cdot, t); \mathbf{f}_{\mathscr{B}^e}) \equiv \int_{\mathscr{B}} \dot{\chi}(\cdot, t) \cdot d\mathbf{f}_{\mathscr{B}^e}. \tag{I.8-7}$$

The units of working are those of (force)(length) $\div$ (time). When there is no fear of confusion, we shall drop from the notation the arguments of W.

Forces are conceived as acting upon bodies, and when those bodies undergo motions and hence take shapes in $\mathscr{E}$, the forces are carried over into those shapes in some specified way. Since the shapes themselves depend upon the choice of framing, so also must any transference to those shapes of the forces acting on bodies. Consequently the definition (7) of the working W rests also upon a particular choice of framing. In §I.11 we shall impose as the basic axiom of mechanics the requirement that such dependence of

W be only apparent: that is, that the working, although it is defined by (7) in terms of a framing $\mathfrak{f}$, shall have the same value for all framings.

Since by means of the isomorphism selected we may in effect say that $\mathbf{f} \in \mathscr{V}$, we may define also the tensor product $\mathbf{v} \otimes \mathbf{f}$ and the exterior product $\mathbf{v} \wedge \mathbf{f}$, provided $\mathbf{v} \in \mathscr{V}$. In particular, the skew tensor $(\mathbf{x} - \mathbf{x}_0) \wedge \mathbf{f}$ is called the "moment at $\mathbf{x}$ of $\mathbf{f}$ with respect to $\mathbf{x}_0$". More generally, the moment $\mathbf{F}_{\mathbf{x}_0}$ of a system of forces $\mathbf{f}_{\mathscr{B}^e}$ on a part $\mathscr{A}$ of $\mathscr{B}$ in the motion of $\mathscr{B}$, with respect to $\mathbf{x}_0$, is defined thus:

$$\mathbf{F}(\mathscr{A}, \mathscr{B}^e; \chi(\cdot, t))_{\mathbf{x}_0} \equiv \int_{\mathscr{A}} [\chi(\cdot, t) - \mathbf{x}_0] \wedge d\mathbf{f}_{\mathscr{B}^e}. \tag{I.8-8}$$

Although the moment is only a special case of what is called a *torque*, in this book we shall regard the two terms as interchangeable and prefer to use the monosyllable. The particular torque $\mathbf{F}(\mathscr{B}, \mathscr{B}^e; \chi(\cdot, t))_{\mathbf{x}_0}$ is called the *resultant torque* of the system of forces on $\mathscr{B}$ with respect to $\mathbf{x}_0$ in the motion χ at the time t.

The moment of a system of forces acting on a body $\mathscr{B}$ is defined in terms of the shape of $\mathscr{B}$ in the motion χ, a particular frame of reference $\mathfrak{f}$ being presupposed. The moment is a skew tensor having the dimensions of (force) $\times$ (length). More generally, any skew tensor having these dimensions is called a *torque*, and a torque-valued function $\mathbf{F}(\mathscr{B}, \mathscr{C})$ of pairs of bodies is called a *system of torques* if it satisfies axioms obtained from Axioms F1–F4 by replacing $\mathbf{f}$ by $\mathbf{F}$ throughout. Torques that are not moments of forces are sometimes called *couples*. When all torques are moments of forces, as we shall assume in this book, the system of torques is called *simple*.

The importance of the torque will appear in §I.12. For the time being, we shall remark only that

$$\mathbf{F}(\mathscr{B}, \mathscr{B}^e)_{\mathbf{x}_0} = \mathbf{F}(\mathscr{B}; \mathscr{B}^e)_{\mathbf{x}_1} + (\mathbf{x}_1 - \mathbf{x}_0) \wedge \mathbf{f}(\mathscr{B}, \mathscr{B}^e). \tag{I.8-9}$$

That is, at the time t the resultant torque with respect to $\mathbf{x}_0$ differs from that with respect to $\mathbf{x}_1$ by the moment at $\mathbf{x}_1$, with respect to $\mathbf{x}_0$, of the resultant force on $\mathscr{B}$. Here we have dropped $\chi(\cdot, t)$ from the notation.

In §I.5 we have defined a balanced system of forces as one in which the resultant force on each body is $\mathbf{0}$. By (9) we see that *if the system of forces is balanced, the resultant torque it exerts on any body is the same with respect to all places.*

In view of what has just been said, the following definition makes sense: The torques arising from a balanced system of forces are said themselves to be *balanced* if the resultant torque of every body vanishes.

In a system of torques more generally, NOLL's corollary of §I.5 applies with merely verbal changes, enabling us to conclude that *in a balanced system of torques,* $\mathbf{F}(\mathscr{B}, \mathscr{C}) = -\mathbf{F}(\mathscr{C}, \mathscr{B})$.

As we shall see in §I.11, the basic axiom of mechanics will imply that a certain basic system of forces and system of torques be balanced and will in fact be equivalent to that statement.

In the universe of analytical dynamics, where each body $\mathscr{B}$ considered is the union of a subset of n mass-points X_k,

$$\begin{aligned} \mathbf{m}(\mathscr{B}; \chi(\cdot, t)) &= \sum_{k \in i_{\mathscr{B}}} M_k \dot{\mathbf{x}}_k, \\ \mathbf{M}(\mathscr{B}; \chi(\cdot, t))_{\mathbf{x}_0} &= \sum_{k \in i_{\mathscr{B}}} (\mathbf{x}_k - \mathbf{x}_0) \wedge M_k \dot{\mathbf{x}}_k, \\ K(\mathscr{B}; \chi(\cdot, t)) &= \tfrac{1}{2} \sum_{k \in i_{\mathscr{B}}} M_k |\dot{\mathbf{x}}_k|^2, \\ W(\mathscr{B}; \chi(\cdot, t); \mathbf{f}_{\mathscr{B}^e}) &= \sum_{k \in i_{\mathscr{B}}} \dot{\mathbf{x}}_k \cdot \left[\mathbf{f}_k^e + \sum_{q \in i_{\mathscr{B}}^{-1}} \mathbf{f}_{kq} \right]. \end{aligned} \tag{I.8-10}$$

Here $\mathbf{x}_k \equiv \chi(X_k, t)$, $\dot{\mathbf{x}}_k \equiv \dot{\chi}(X_k, t)$, and the other notations are those introduced in connection with analytical dynamics in §I.5. To obtain $(10)_4$, we have used (I.5-15) and (I.5-20). Two special cases of $(10)_4$ are of major interest. First, suppose $\mathscr{B}$ consists in X_k alone. Then

$$W(X_k; \chi(\cdot, t); \mathbf{f}_{X_k^e}) = \dot{\mathbf{x}}_k \cdot \mathbf{f}_k \tag{I.8-11}$$

By (I.5-24) we see that *if the system of forces is balanced,*

$$W(X_k; \chi(\cdot, t); \mathbf{f}_{X_k^e}) = 0. \tag{I.8-12}$$

Second, if $\mathscr{B} = \{X_1, X_2, \ldots, X_n\}$, then $(10)_4$ reduces to

$$W(\mathscr{B}; \chi(\cdot, t); \mathbf{f}_{\mathscr{B}^e}) = \sum_{k=1}^{n} \dot{\mathbf{x}}_k \cdot \mathbf{f}_k^e = -\sum_{k,q=1}^{n}{}' \dot{\mathbf{x}}_k \cdot \mathbf{f}_{kq}, \tag{I.8-13}$$

where the last expression holds if the system of forces is balanced. Thus, in general, the working of the system of forces on the dynamical system does not vanish.

The torque $\mathbf{F}(\mathscr{B}, \mathscr{C})_{\mathbf{x}_0}$ exerted by $\mathscr{C}$ on $\mathscr{B}$ with respect to $\mathbf{x}_0$ is defined by

$$\mathbf{F}(\mathscr{B}, \mathscr{C})_{\mathbf{x}_0} \equiv \sum_{k \in i_{\mathscr{B}}} (\mathbf{x}_k - \mathbf{x}_0) \wedge \sum_{q \in i_{\mathscr{C}}} \mathbf{f}_{kq}, \tag{I.8-14}$$

where $\mathscr{B}$ and $\mathscr{C}$ need not be separate. Likewise the torque exerted by the environment X_0 on the mass-point X_k is defined by

$$\mathbf{F}(X_k, X_0)_{\mathbf{x}_0} = (\mathbf{x}_k - \mathbf{x}_0) \wedge \mathbf{f}_k^e. \tag{I.8-15}$$

These definitions square with (8), and the resultant torque on $\mathscr{B}$ is given by

$$\mathbf{F}(\mathscr{B}, \mathscr{B}^e)_{\mathbf{x}_0} = \sum_{k \in i_\mathscr{B}} (\mathbf{x}_k - \mathbf{x}_0) \wedge \left(\mathbf{f}_k^e + \sum_{q \in i_\mathscr{B}^{-1}} \mathbf{f}_{kq}\right). \tag{I.8-16}$$

The *self-torque* of $\mathscr{B}$ is the torque it exerts on itself. By (14), this is

$$\mathbf{F}(\mathscr{B}, \mathscr{B})_{\mathbf{x}_0} = \sum_{k \in i_\mathscr{B}} (\mathbf{x}_k - \mathbf{x}_0) \wedge \sum_{q \in i_\mathscr{B}}' \mathbf{f}_{kq}. \tag{I.8-17}$$

In a balanced system of forces, (I.5-22) holds, and hence

$$\mathbf{F}(\mathscr{B}, \mathscr{B})_{\mathbf{x}_0} = \tfrac{1}{2} \sum_{k,\, q \in i_\mathscr{B}} (\mathbf{x}_k - \mathbf{x}_q) \wedge \mathbf{f}_{kq}. \tag{I.8-18}$$

If the force $\mathbf{f}_{kq}$ exerted by X_q on X_k is parallel to the vector $\mathbf{x}_k - \mathbf{x}_q$ that translates the place $\mathbf{x}_q$ occupied by X_q into the place $\mathbf{x}_k$ occupied by X_k, the mutual forces are called *central*. For central mutual forces each summand in (18) vanishes, and we have the

Theorem (POISSON). *For a balanced system of forces on the universe of analytical dynamics, the self-torque of every body vanishes if the mutual forces are central.*

When the self-torque of $\mathscr{B}$ vanishes, the resultant torque (16) may be written in the form

$$\mathbf{F}(\mathscr{B}, \mathscr{B}^e)_{\mathbf{x}_0} = \sum_{k \in i_\mathscr{B}} (\mathbf{x}_k - \mathbf{x}_0) \wedge \left(\mathbf{f}_k^e + \sum_{q=1}^{n}{}' \mathbf{f}_{kq}\right). \tag{I.8-19}$$

That is, the resultant torque on $\mathscr{B}$ is the sum of the moments of the resultant forces acting on the mass-points that make up $\mathscr{B}$. In a balanced system of forces, each of those resultant forces vanishes, so $\mathbf{F}(\mathscr{B}, \mathscr{B}^e) = \mathbf{0}$. That is, the system of torques is balanced.

Suppose, conversely, that the system of torques be balanced. Then by the analogue of (I.5-6),

$$\mathbf{F}(\mathscr{B}, \mathscr{C})_{\mathbf{x}_0} = -\mathbf{F}(\mathscr{C}, \mathscr{B})_{\mathbf{x}_0} \tag{I.8-20}$$

for all separate bodies $\mathscr{B}$ and $\mathscr{C}$. In particular, then,

$$\mathbf{F}(X_k, X_q)_{\mathbf{x}_0} = -\mathbf{F}(X_q, X_k)_{\mathbf{x}_0}. \tag{I.8-21}$$

That is,

$$(\mathbf{x}_k - \mathbf{x}_0) \wedge \mathbf{f}_{kq} = -(\mathbf{x}_q - \mathbf{x}_0) \wedge \mathbf{f}_{qk}. \tag{I.8-22}$$

If the forces are balanced, by (I.5-22) we obtain

$$(\mathbf{x}_k - \mathbf{x}_q) \wedge \mathbf{f}_{kq} = \mathbf{0}, \tag{I.8-23}$$

so that the mutual forces are central. In summary of the argument in this paragraph and the preceding one, we have the following

Theorem (NOLL). *If a system of forces on the universe of analytical dynamics is balanced, the corresponding system of torques is balanced if and only if the mutual forces are central.*

Exercise I.8.2. From (I.5-26) prove that in a balanced system of forces

$$\mathbf{F}(\mathscr{B}, \mathscr{B}^e)_{\mathbf{x}_0} + \mathbf{F}(\mathscr{B}, \mathscr{B})_{\mathbf{x}_0} = 0. \tag{I.8-24}$$

Hence the system of torques is balanced if and only if the self-torque of every body vanishes, and again NOLL's theorem follows.

Thus in analytical dynamics the balance of torques is *equivalent* to the hypothesis that the mutual forces are central, on the assumption that the system of forces is balanced. As should be plain from the arguments leading to NOLL's theorem, no such reduction of the balance of torques to the balance of forces can be expected in the more general and typical universes of mechanics. In continuum mechanics central forces, and indeed mutual forces of all kinds, play no special part, and the general approach of analytical dynamics is untypical and next to useless.

The position vector of a place $\mathbf{x}$ in $\mathscr{E}$ is the vector that translates some given "origin" $\mathbf{x}_0$ into $\mathbf{x}$. Thus the *position vector field* $\mathbf{p}$ corresponding to a motion χ is given by

$$\mathbf{p} \equiv \chi - \mathbf{x}_0 . \tag{I.8-25}$$

Commonly the origin $\mathbf{x}_0$ is a fixed place. In that case the time derivatives of $\mathbf{p}$ when X is held fixed are the same as the corresponding time derivatives of the motion itself:

$$\dot{\mathbf{p}} = \dot{\chi}, \qquad \ddot{\mathbf{p}} = \ddot{\chi}, \qquad etc. \tag{I.8-26}$$

The *center of mass* of a body $\mathscr{B}$ of positive mass $M(\mathscr{B})$ in a shape $\chi_\Omega(\mathscr{B}, t)$ is that place whose position vector $\bar{\mathbf{p}}$ is the mean, in the sense of the mass M, of the position vectors of all the body-points of $\mathscr{B}$:

$$\bar{\mathbf{p}}(\mathscr{B}) \equiv \frac{1}{M(\mathscr{B})} \int_{\mathscr{B}} \mathbf{p} \, dM. \tag{I.8-27}$$

Of course $\bar{\mathbf{p}}$ generally varies in time for a given body $\mathscr{B}$, but we do not indicate this fact in the notation. While $\bar{\mathbf{p}}$ depends upon the choice of the fixed place $\mathbf{x}_0$, its time derivative $\dot{\bar{\mathbf{p}}}$ does not, and by (1) we see that

$$\mathbf{m}(\mathscr{B}; \chi) = M(\mathscr{B}) \dot{\bar{\mathbf{p}}}(\mathscr{B}). \tag{I.8-28}$$

Comparison with $(10)_1$ yields the following

Theorem (KELVIN & TAIT). *The linear momentum of a body $\mathscr{B}$ is the same as that of a mass-point having the same mass as $\mathscr{B}$ and moving so as always to occupy the center of mass of $\mathscr{B}$.*

Exercise I.8.3. Let the place $\mathbf{x}_0$ with respect to which the position vector is calculated be fixed, and let $\mathbf{x}_c$ denote the center of mass of $\chi(\mathscr{B}, t)$. Prove that

$$\dot{\mathbf{M}}_{\mathbf{x}_0} = \dot{\mathbf{M}}_{\mathbf{x}_c} + \bar{\mathbf{p}} \wedge \dot{\mathbf{m}}, \tag{I.8-29}$$

and interpret the result.

9. Changes of Frame

Once a framing ϕ has been laid down as in §I.6, we may wish to consider another one ϕ^*:

$$\begin{aligned} \phi &: \mathscr{W} \to \mathscr{E} \times \mathscr{R}, \\ \phi^* &: \mathscr{W} \to \mathscr{E} \times \mathscr{R}. \end{aligned} \tag{I.9-1}$$

Since both these mappings are homeomorphisms, the composition $\phi^* \circ \phi^{-1}$ is a homeomorphism of the frame $\mathscr{E} \times \mathscr{R}$ into itself:

$$\phi^* \circ \phi^{-1}: \mathscr{E} \times \mathscr{R} \to \mathscr{E} \times \mathscr{R}. \tag{I.9-2}$$

We shall not call ϕ^* a framing unless it gives time the same sense as does ϕ. While this requirement is a natural one, it is not easy to express. For a given framing ϕ, let $\mathscr{P}_t$ be the set of events occurring at a certain time t. If ϕ^* is another framing, we wish it to map $\mathscr{P}_t$ onto a set of events which have a common time, s, relative to ϕ^*. That is, if $\phi(\mathscr{P}_t) = (\mathscr{E}, t)$, then $\phi^*(\mathscr{P}_t) = (\mathscr{E}, s)$ for some real s. It can then be shown[1] that the transformation $\phi^* \circ \phi^{-1}$ is of the form

$$\begin{aligned} (\mathbf{x}^*, t^*) &= \phi^* \circ \phi^{-1}(\mathbf{x}, t), \\ \mathbf{x}^* = \mathbf{f}(\mathbf{x}, t), &\qquad t^* = g(t). \end{aligned} \tag{I.9-3}$$

That is, the time t determines the time t^*, though it may require both the time t and the place $\mathbf{x}$ to determine the place $\mathbf{x}^*$.

Moreover, so as to suit the intended interpretation, we shall consider, once a single framing ϕ has been set down, only such other framings ϕ^* as yield mappings (2) that preserve the metrics in $\mathscr{E}$ and $\mathscr{R}$ separately—in

[1] W. NOLL, "Euclidean geometry and Minkowskian chronometry," *American Mathematical Monthly* **71**, 129–144 (1964). Reprinted in W. NOLL, *The Foundations of Mechanics and Thermodynamics*, New York, Heidelberg, and Berlin, Springer-Verlag, 1974.

the interpretation, mappings under which the distances between places and the lapses of time between instants are invariant. Such mappings, and only such, are called *changes of frame*.

We shall sometimes say "$\mathbf{x}^*$ and $\mathbf{x}$ are the places at which the same event is observed in $\mathfrak{f}^*$ and $\mathfrak{f}$, respectively." The student is expected to recognize this and like statements as pointing toward the interpretation of the mathematical structure in terms of experience. Similarly, the velocity of a body-point defined with respect to the framing $\mathfrak{f}$ will be called its velocity *in* $\mathfrak{f}$, and a similar usage will be followed for all other quantities defined in terms of framings: acceleration, momentum, *etc.*

According to a theorem of geometry,[1] a change of frame may be represented as a translation of $\mathscr{R}$ combined with a uniquely determined time-dependent orthogonal transformation of $\mathscr{V}$:

$$\begin{aligned} t^* &= t + a, \\ \mathbf{x}^* &= \mathbf{x}_0^*(t) + \mathbf{Q}(t)(\mathbf{x} - \mathbf{x}_0), \end{aligned} \tag{I.9-4}$$

where a is a particular time, $\mathbf{x}_0$ is a fixed place, $\mathbf{x}_0^*$ maps times into places, and $\mathbf{Q}$ maps times into orthogonal tensors over $\mathscr{V}$. Regarding $\mathbf{x}_0$ as the place and 0 as the time of a certain fixed event as observed in $\mathfrak{f}$, we interpret $\mathbf{x}_0^*(0)$ and a as the place and the time, respectively, of the same event as observed in $\mathfrak{f}^*$, while $\mathbf{Q}$ is a rotation of all the lines through $\mathbf{x}_0$ as observed in $\mathfrak{f}$ into lines through $\mathbf{x}_0^*$ as observed in $\mathfrak{f}^*$. The value $\mathbf{Q}(t)$ of $\mathbf{Q}$ is sometimes called the *relative orientation* of $\mathfrak{f}^*$ with respect to $\mathfrak{f}$ at time t. At a time t_0 when $\mathbf{Q}(t_0) = \mathbf{1}$ and $\mathbf{x}_0^*(t_0) = \mathbf{x}_0$, the two framings are said to *coincide*.

From (4) we see that one particular event, to which a place and time $(\mathbf{x}_0, t)$ are assigned by the framing $\mathfrak{f}$, may be assigned an arbitrary place $\mathbf{x}_0^*(t)$ and time t^* by some other framing $\mathfrak{f}^*$. The vector that translates the fixed place $\mathbf{x}_0$ into a general place $\mathbf{x}$ in $\mathfrak{f}$ is then rotated in $\mathscr{V}$ into the vector that translates $\mathbf{x}_0^*(t)$ into the corresponding place $\mathbf{x}^*$ in $\mathfrak{f}^*$, the rotation being the same for all places $\mathbf{x}$ at any one time.

If we like, we may picture a change of frame in terms of a motion (§I.7). If we suppose a body to be given such a shape that one of its points remains at the place $\mathbf{x}$ in $\mathfrak{f}$, then (4) is the motion of that point in $\mathfrak{f}^*$.

Exercise I.9.1. Prove that in (4) we may take as the constant place $\mathbf{x}_0$ in $\mathfrak{f}$ any one we please, or, if we prefer, we may substitute for it any place-valued function of time: $\mathbf{x}_0(\cdot)$. Hence the class of all changes of frame forms a group.

[1] W. NOLL, *op. cit.*

From $(4)_1$ we see that the definition of "world-line" in §I.6 is independent of the choice of framing.

We interpret the possibility of a change of frame as meaning that two observers who have chosen the same units of length and time may set their clocks differently and may be in arbitrary rigid motion with respect to one another, yet both are equally qualified to describe the phenomena represented by classical mechanics, whether they be right-handed or left-handed, so any statement made by one is equivalent to a certain statement made by the other. For example, if a function $f(\mathbf{x}^*, t^*)$ is given, substitution of (4) into it yields a certain function $g(\mathbf{x}, t)$ with the same value:

$$g(\mathbf{x}, t) = f(\mathbf{x}^*, t^*), \tag{I.9-5}$$

and any two functions so related are regarded as equivalent under the change of frame.

A change of frame induces a transformation of the translation space $\mathscr{V}$. Indeed, suppose that

$$\mathbf{v} \equiv \mathbf{x}_1 - \mathbf{x}_2 . \tag{I.9-6}$$

Then by (4)

$$\begin{aligned} \mathbf{v}^* \equiv \mathbf{x}_1^* - \mathbf{x}_2^* &= \mathbf{Q}(t)(\mathbf{x}_1 - \mathbf{x}_2) \\ &= \mathbf{Q}(t)\mathbf{v}. \end{aligned} \tag{I.9-7}$$

Likewise, a change of frame induces a transformation of the tensor space over $\mathscr{V}$. If $\mathbf{w} \in \mathscr{V}$ and $\mathbf{v} \in \mathscr{V}$, and if

$$\mathbf{w} = \mathbf{T}\mathbf{v}, \tag{I.9-8}$$

where $\mathbf{T}$ is a tensor, then by (7)

$$\mathbf{w}^* = \mathbf{Q}(t)\mathbf{w} = \mathbf{Q}(t)\mathbf{T}\mathbf{Q}(t)^{\mathrm{T}}\mathbf{v}^*; \tag{I.9-9}$$

that is, $\mathbf{w}^* = \mathbf{T}^*\mathbf{v}^*$, where

$$\mathbf{T}^* = \mathbf{Q}(t)\mathbf{T}\mathbf{Q}(t)^{\mathrm{T}}. \tag{I.9-10}$$

Rules of just the same form are induced for vector-valued and tensor-valued functions of time $\mathbf{v}(t)$ and $\mathbf{T}(t)$, respectively.

A change of frame (4) induces also a change in the motion (I.7-5) of a body-point X of a body $\mathscr{B}$. Namely, in $\mathfrak{f}^*$ the place $\mathbf{x}^*$ occupied by X at the time t^* is given by the relations

$$\begin{aligned} \mathbf{x}^* = \chi^*(X, t^*) &= \mathbf{x}_0^*(t) + \mathbf{Q}(t)(\chi(X, t) - \mathbf{x}_0), \\ t^* &= t + a. \end{aligned} \tag{I.9-11}$$

We shall regard χ^* as being the *same* motion as observed in $\oint^*$. We regard $\mathbf{x}_0$ and t as the place and time assigned by $\oint$ to some particular event and $\mathbf{x}_0^*(t)$ and t^* as the place and time assigned by $\oint^*$ to that same event. When we need to emphasize the role of a frame, we shall call χ *the motion of* $\mathscr{B}$ *in* $\oint$, and χ^* *the same motion of* $\mathscr{B}$ *in* $\oint^*$, as explained above. We shall refer to the transformation (11), which relates the motion in $\oint$ with that in $\oint^*$, by the same name as the transformation (4) of the frame $\mathscr{E} \times \mathscr{R}$, namely, a *change of frame*.

If a certain prescription defines vectors in terms of a framing, and if the prescription itself is independent of the choice of framing, it will deliver vectors $\mathbf{v}^*$ and $\mathbf{v}$, respectively, according as $\oint^*$ or $\oint$ is used, and generally these two vectors will not be the same. Consider, for example, the prescriptions $(\text{I}.7.7)_{1,2}$, which define the velocity and the acceleration in any framing:

$$\begin{aligned} \mathbf{v} &\equiv \dot{\chi}(X, t), & \mathbf{v}^* &\equiv \dot{\chi}^*(X, t^*), \\ \mathbf{a} &\equiv \ddot{\chi}(X, t), & \mathbf{a}^* &\equiv \ddot{\chi}^*(X, t^*). \end{aligned} \tag{I.9-12}$$

The dots in the second column indicate derivatives with respect to t^*, and the motion χ^* in $\oint^*$ is related to the motion χ in $\oint$ by (11). Hence if the change of frame is smooth enough for the time derivatives $\dot{\mathbf{x}}_0^*$ and $\dot{\mathbf{Q}}$ to exist,

$$\dot{\chi}^*(X, t^*) = \dot{\mathbf{x}}_0^*(t) + \mathbf{Q}(t)\dot{\chi}(X, t) + \dot{\mathbf{Q}}(t)(\chi(X, t) - \mathbf{x}_0). \tag{I.9-13}$$

Therefore the velocity $\dot{\chi}^*$ in $\oint^*$ is related to the velocity $\dot{\chi}$ in $\oint$ by the formula

$$\dot{\chi}^* - \mathbf{Q}\dot{\chi} = \dot{\mathbf{x}}_0^* + \mathbf{A}(\chi^* - \mathbf{x}_0^*), \tag{I.9-14}$$

in which

$$\mathbf{A} \equiv \dot{\mathbf{Q}}\mathbf{Q}^{\mathsf{T}} = -\mathbf{A}^{\mathsf{T}}. \tag{I.9-15}$$

The skew tensor $\mathbf{A}$ is the *spin*[1] of $\oint$ with respect to $\oint^*$. In (11) we regarded $\mathbf{x}_0^*(t)$ and t^* as the place and time assigned by $\oint^*$ to a certain reference event to which $\oint$ assigned the place $\mathbf{x}_0$ and the time t. Thus the value of the function $\dot{\mathbf{x}}_0^*$ is the rate of change of the place $\mathbf{x}_0^*(t)$ assigned by $\oint^*$ to that reference event. On the other hand by (13), the velocity in $\oint^*$ of the body-point that occupies the place $\mathbf{x}_0$ at the time t is $\dot{\mathbf{x}}_0^*(t) + \mathbf{Q}(t)\dot{\mathbf{x}}(\chi^{-1}(\mathbf{x}_0, t), t)$, which reduces to $\dot{\mathbf{x}}_0^*(t)$ if and only if the body-point is at rest in $\oint$.

[1] The old term "angular velocity" is gradually falling out of use, since not only is it an awkward polysyllable but also it suggests we should look for angles, which in general considerations we are better advised not to do.

Heretofore we have regarded the relative orientation $\mathbf{Q}$ as a known, differentiable function of t. Suppose instead we know the spin $\mathbf{A}$, which we assume to be a continuous function whose values are skew tensors. Considering the first-order linear differential equation

$$\dot{\mathbf{Y}} - \mathbf{A}\mathbf{Y} = \mathbf{0}, \tag{I.9-16}$$

we observe first that it has a unique solution $\mathbf{Y}$ such that $\mathbf{Y}(t_0)$ assumes an assigned value.

Exercise I.9.2. If $\mathbf{Z} \equiv \mathbf{Y}\mathbf{Y}^{\mathrm{T}}$ and $\mathbf{Y}$ satisfies (16), show that

$$\dot{\mathbf{Z}} = \mathbf{A}\mathbf{Z} - \mathbf{Z}\mathbf{A}. \tag{I.9-17}$$

By an appeal to the uniqueness theorem for ordinary differential equations, show that a solution $\mathbf{Y}$ of (16) which is orthogonal when $t = t_0$ is orthogonal for all t.

The argument completed in the foregoing exercise is summarized in the following

Theorem. *Let the spin* $\mathbf{A}$ *of* $\oint$ *with respect to* $\oint^*$ *be a continuous function of time, and let the relative orientation* $\mathbf{Q}(t)$ *be prescribed at some one time* t_0. *Then a unique change of frame is determined by assignment of the place* $\mathbf{x}_0^*(t)$ *occupied in* $\oint^*$ *at the time t by some one place* $\mathbf{x}_0$ *in* $\oint$.

Exercise I.9.3. Letting $\mathbf{A}^*$ denote the spin of $\oint^*$ with respect to $\oint$, prove that

$$\mathbf{A}^* = -\mathbf{Q}^{\mathrm{T}}\mathbf{A}\mathbf{Q}. \tag{I.9-18}$$

More generally, if $\mathbf{Q}_1$ and $\mathbf{Q}_2$ correspond to changes of frame from $\oint$ to $\oint_1$ and from $\oint_1$ to $\oint_2$, respectively, and if $\mathbf{Q}_3$ corresponds to the change from $\oint$ to $\oint_2$, then

$$\begin{aligned} \mathbf{Q}_3 &= \mathbf{Q}_2\mathbf{Q}_1, \\ \mathbf{A}_3 &= \mathbf{A}_2 + \mathbf{Q}_2\mathbf{A}_1\mathbf{Q}_2^{\mathrm{T}}. \end{aligned} \tag{I.9-19}$$

Hence if the framings $\oint_2$ and $\oint_1$ coincide at some instant, the spin of $\oint_2$ with respect to $\oint$ is the sum of its spin with respect to $\oint_1$ and the spin of $\oint_1$ with respect to $\oint$ at that instant.

The results (19) are commonly described as asserting that while rotations are multiplicative, spins are additive.

The axis of the $\mathbf{Q}(t)$ in the change of frame (4) is the *axis of rotation* of that change of frame at the time t. The *angle of rotation* of the change of frame is that root $\theta(t)$ of the equation $\cos\theta(t) = \frac{1}{2}(\operatorname{tr}\mathbf{R}(t) - 1)$ which lies between 0 and 2π; here, of course, $\mathbf{R}$ is the rotation such that $\mathbf{Q} = \mathbf{R}$ or $\mathbf{Q} = -\mathbf{R}$. Since $\mathbf{A}(t)$ is skew, its null space, likewise, is a single line except in the trivial case when $\mathbf{A}(t) = \mathbf{0}$. This line is called the *axis* *of spin.* The corresponding proper number of $\mathbf{A}(t)$ is 0. Since $\mathbf{A}\cdot\mathbf{A} = \dot{\mathbf{Q}}\cdot\dot{\mathbf{Q}}$, the magnitude of $\mathbf{A}$ is the same as the magnitude of $\dot{\mathbf{Q}}$. The value of $\sqrt{\frac{1}{2}\mathbf{A}\cdot\mathbf{A}}$, namely, the value of $|\mathbf{A}|/\sqrt{2}$, is called the *angular speed* ω at which $\mathfrak{f}$ is rotating with respect to $\mathfrak{f}^*$ at the time t.

Exercise I.9.4. Prove that if the axis of rotation is independent of t, then it is also the axis of spin. Hence if the angle of rotation is $\theta(t)$, then $\omega = |\dot{\theta}(t)|$.

With the aid of a convention of sign, we can define a vector $\boldsymbol{\omega}$ such that $\boldsymbol{\omega}\times\mathbf{b} = \mathbf{A}\mathbf{b}$ for all vectors $\mathbf{b}$. Here the sign $\times$ denotes the cross-product of 3-dimensional vector analysis. The vector $\boldsymbol{\omega}$ is called the *angular velocity* of the rotation $\mathbf{R}$. Of course $|\boldsymbol{\omega}| = \omega$.

Exercise I.9.5. (GALLETTO). Prove that if $\theta \neq 0$ and if $\mathbf{e}$ is a suitably selected unit vector in the axis of rotation,

$$\boldsymbol{\omega}\cdot\mathbf{e} = \dot{\theta}. \tag{I.9-20}$$

We turn now to the acceleration. If we differentiate (14) with respect to t, we obtain by $(12)_2$ and (15) the following relation between the acceleration $\ddot{\chi}^*$ in $\mathfrak{f}^*$ and the acceleration $\ddot{\chi}$ in $\mathfrak{f}$:

$$\begin{aligned}\ddot{\chi}^* - \mathbf{Q}\ddot{\chi} &= \dot{\mathbf{Q}}\dot{\chi} + \ddot{\mathbf{x}}_0^* + \dot{\mathbf{A}}(\chi^* - \mathbf{x}_0^*) + \mathbf{A}(\dot{\chi}^* - \dot{\mathbf{x}}_0^*),\\ &= \ddot{\mathbf{x}}_0^* + 2\mathbf{A}(\dot{\chi}^* - \dot{\mathbf{x}}_0^*) + (\dot{\mathbf{A}} - \mathbf{A}^2)(\chi^* - \mathbf{x}_0^*).\end{aligned} \tag{I.9-21}$$

The first term on the right-hand side is the acceleration of the place in $\mathfrak{f}^*$ assigned at the time t to the place $\mathbf{x}_0$ in $\mathfrak{f}$. The second term, named after CORIOLIS, is the acceleration in $\mathfrak{f}^*$ that corresponds, by (14), to the velocity of the body-point with respect to $\mathbf{x}_0^*$ in $\mathfrak{f}^*$ and to the spin of $\mathfrak{f}$ with respect to $\mathfrak{f}^*$. The third term has two parts, the first of which, named after EULER, corresponds to the rate of change of the angular velocity, while the second, called the *centripetal acceleration*, expresses the acceleration caused by the pure transport of the body-point with respect to $\mathfrak{f}^*$.

Exercise I.9.6. If $\mathbf{p} \equiv \mathbf{x} - \mathbf{x}_0$, prove that for any skew tensor $\mathbf{A}$

$$-\mathbf{A}^2\mathbf{p} = \nabla(-\tfrac{1}{2}\mathbf{p}\cdot\mathbf{A}^2\mathbf{p}), \tag{I.9-22}$$

∇ being the gradient operator, and interpret this result in the context of the centripetal acceleration. (Note that $\mathbf{A}^2$ is a symmetric tensor having the axis of spin as its nullspace and having $-\omega^2$ and 0 as its proper numbers.)

The linear momentum, rotational momentum, and kinetic energy of a body depend likewise upon the framing. The transformations of these quantities and their rates of change induced by a change of frame are easy to calculate by substitution of (14) and (21) into appropriate formulae of §I.8.

10. Rigid Motion

A motion of a body is called *rigid* if there is a framing $\oint^*$ such as to make its velocity field vanish. The framing $\oint^*$ is called a *rest framing* for that motion, and quantities referred to it are described as belonging to a *rest frame*. To calculate the velocity field of a rigid motion in a general $\oint$, we need only set $\dot{\chi}^* = \mathbf{0}$ in (I.9-14), generalized to allow $\mathbf{x}_0$ to depend on t as in Exercise I.9.1, and then by use of (I.9-15) obtain the following

Theorem (EULER). *A motion is rigid if and only if its velocity field in any, and hence every, framing $\oint$ is of the form*

$$\begin{aligned}\dot{\chi} &= \dot{\mathbf{x}}_0 - \mathbf{Q}^{\mathrm{T}}\dot{\mathbf{x}}_0^* - \mathbf{Q}^{\mathrm{T}}\mathbf{A}(\chi^* - \mathbf{x}_0^*),\\ &= \mathbf{c} + \mathbf{W}(\chi - \mathbf{x}_0);\end{aligned} \tag{I.10-1}$$

here $\mathbf{x}_0(t)$ is an arbitrary place in $\mathscr{E}$, $\mathbf{c}(t)$ is a vector, and $\mathbf{W}(t)$ is a skew tensor.

Of course $\mathbf{W} = \mathbf{A}^*$, the spin of a rigid framing $\oint^*$ with respect to $\oint$, related to $\mathbf{A}$ through (I.9-18). We use the special symbol $\mathbf{W}$ to remind the reader that we refer to a particular kind of motion of a body, or, if we like, a particular framing, while $\mathbf{A}$ is defined for any pair of framings, irrespective of whatever motion of a body may be taking place with respect to them. We call $\mathbf{W}$ the *spin* of the rigid motion.

The nullspace of $\mathbf{W}(t)$ is called the *axis* of the rigid motion in $\oint$ at the time t. Body-points lying upon a line through $\mathbf{x}_0$ and parallel to the axis of the motion are moving with the common velocity $\mathbf{c}(t)$.

Exercise I.10.1. For a given rigid motion at a given time, show that $\mathbf{W}$ is unique if and only if the shape of $\mathscr{B}$ is not part of a straight line.

In §I.9 we showed that the function $\mathbf{A}$ determines the function $\mathbf{Q}$ uniquely if $\mathbf{Q}(t_0)$ is prescribed. By (I.9-18) we may use the function $\mathbf{W}$ to determine $\mathbf{Q}$ in the same way, as we may see equally well by writing the differential equation (I.9-16) in the form

$$\dot{\mathbf{Y}} + \mathbf{Y}\mathbf{W} = \mathbf{0} \tag{I.10-2}$$

and seeking the solution $\mathbf{Y}$ that assumes the value $\mathbf{Q}(t_0)$ when $t = t_0$.

From $(1)_2$ we see that the vector $\mathbf{c}(t)$ is the velocity of the body-point currently occupying the place $\mathbf{x}_0$ in $\oint$; from $(1)_1$, that $\mathbf{c}$ is expressed as follows in terms of the functions $\mathbf{x}_0$ and $\mathbf{x}_0^*$ in (I.9-11):

$$\mathbf{c} = \dot{\mathbf{x}}_0 - \mathbf{Q}^{\mathrm{T}}\dot{\mathbf{x}}_0^*. \tag{I.10-3}$$

Thus, once $\mathbf{Q}$ has been determined and $\mathbf{x}_0$ assigned, $\mathbf{c}$ determines the function $\mathbf{x}_0^*$ to within an arbitrary constant place, on the presumption that the function $\mathbf{c}$ is continuous. If we choose as $\mathbf{x}_0(t)$ the place occupied in $\oint$ by a certain body-point X_0, then $\dot{\mathbf{x}}_0(t)$ is its velocity in $\oint$, and, since $\oint^*$ is a rest framing, $\dot{\chi}^*(X_0, t) = \mathbf{0}$, so in this way we recover the conclusion with which this paragraph began.

Exercise I.10.2. Directly from (1), without use of the general concepts and framework of the earlier sections but by choosing $\mathbf{x}_0(t)$ as $\chi(X_0, t)$ for some body-point X_0, show that if $\mathbf{p}_2$ and $\mathbf{p}_1$ are the position vectors with respect to $\mathbf{x}_0$ of the body-points X_2 and X_1 in a rigid motion at time t, then in fact $\mathbf{p}_2 \cdot \mathbf{p}_1$ is constant in time.

In summary of the foregoing argument we have the

Theorem (EULER). *Let the motion of single point X_0 of $\mathscr{B}$ be given as a differentiable function of time in $\oint$, and let $\mathbf{W}$ be a continuous function of time whose values are skew tensors. Then choice of the relative orientation $\mathbf{Q}(t_0)$ at some one time t_0 determines a unique rest frame and hence a unique rigid motion of $\mathscr{B}$ corresponding to the spin $\mathbf{W}$. If $\mathbf{W} = \mathbf{0}$, all points of $\mathscr{B}$ move with the same velocity as does X_0. Otherwise, the only points to share the velocity of X_0 at the time t are those lying on the single line through the place $\mathbf{x}_0(t)$ occupied by X_0 and parallel to the axis of $\mathbf{W}(t)$.*

In rough terms, EULER's theorem states that a rigid motion of $\mathscr{B}$ is composed instantaneously of a translation of $\mathscr{B}$ with the velocity of any one of its points and a rotation of $\mathscr{B}$ about a certain, generally time-dependent, axis through that point.

The rotational momentum of a body undergoing rigid motion has an especially simple form in terms of the *Euler tensor* $\mathbf{E}_{\mathbf{x}_0}$ with respect to $\mathbf{x}_0$, defined as follows:

$$\mathbf{E}_{\mathbf{x}_0} \equiv \int_{\mathscr{B}} \mathbf{p} \otimes \mathbf{p} \, dM, \tag{I.10-4}$$

$\mathbf{p}$ being the position vector field (I.8-25). $\mathbf{E}_{\mathbf{x}_0}$ is symmetric. If, as we shall assume now, the subbodies of $\mathscr{B}$ that have positive mass are not presently confined to a single plane, $\mathbf{E}_{\mathbf{x}_0}$ is positive. Then $\mathbf{E}_{\mathbf{x}_0}$ has at least one orthonormal triad of proper vectors, the directions of which are called the *principal axes of inertia* of $\mathscr{B}$ with respect to $\mathbf{x}_0$ in χ, and the proper numbers E_k corresponding to them are positive. This statement is ***Segner's Theorem***. The sum of the three latent roots E_k, or tr $\mathbf{E}_{\mathbf{x}_0}$, is the *polar moment of inertia of* $\mathscr{B}$ *about* $\mathbf{x}_0$, and tr $\mathbf{E}_{\mathbf{x}_0} - E_k$, which is positive, is the *moment of inertia*[1] about the kth principal axis through $\mathbf{x}_0$.

If the motion is rigid, the position vector $\mathbf{p}^*(X, t)$ of a body-point X does not change in a rest framing $\oint^*$. The corresponding tensor $\mathbf{E}^*_{\mathbf{x}_0^*}$ is then constant in time. It is determined once and for all by the mass function and by the shape of $\mathscr{B}$ in the rest frame.

Exercise I.10.3. Prove that if $\mathbf{S}$ is a constant tensor field, then

$$\int_{\mathscr{B}} \mathbf{p} \wedge \mathbf{S}\mathbf{p}\, dM = \mathbf{E}_{\mathbf{x}_0}\mathbf{S}^{\mathrm{T}} - \mathbf{S}\mathbf{E}_{\mathbf{x}_0}. \tag{I.10-5}$$

We consider first the case when $\dot{\mathbf{x}}_0 = \dot{\mathbf{x}}_0^* = \mathbf{0}$. Then $\mathbf{c} = \mathbf{0}$ by (3), so substitution of $(1)_2$ into (I.8-2), followed by use of (5), yields the following

Theorem (EULER). *Let a given body* $\mathscr{B}$ *undergo a rigid motion such that in* $\oint$ *one of its body-points remains at rest at the place* $\mathbf{x}_0$. *Then in* $\oint$

$$\mathbf{M}_{\mathbf{x}_0} = -\mathbf{E}_{\mathbf{x}_0}\mathbf{W} - \mathbf{W}\mathbf{E}_{\mathbf{x}_0}. \tag{I.10-6}$$

Equivalently,

$$\mathbf{Q}\mathbf{M}_{\mathbf{x}_0}\mathbf{Q}^{\mathrm{T}} = \mathbf{E}^*_{\mathbf{x}_0^*}\mathbf{A} + \mathbf{A}\mathbf{E}^*_{\mathbf{x}_0^*}, \tag{I.10-7}$$

$\mathbf{E}^*_{\mathbf{x}_0^*}$ *being calculated in a rest frame.*

Exercise I.10.4. For general functions $\mathbf{x}_0(\cdot)$ and $\mathbf{x}_0^*(\cdot)$, in a rigid motion of a body whose mass is M

$$\begin{aligned} \mathbf{M}_{\mathbf{x}_0} &= M\bar{\mathbf{p}} \wedge \mathbf{c} - \mathbf{E}_{\mathbf{x}_0}\mathbf{W} - \mathbf{W}\mathbf{E}_{\mathbf{x}_0}, \\ \mathbf{Q}\mathbf{M}_{\mathbf{x}_0}\mathbf{Q}^{\mathrm{T}} &= M\bar{\mathbf{p}} \wedge (\mathbf{Q}\dot{\mathbf{x}}_0 - \dot{\mathbf{x}}_0^*) + \mathbf{E}^*_{\mathbf{x}_0^*}\mathbf{A} + \mathbf{A}\mathbf{E}^*_{\mathbf{x}_0^*}, \end{aligned} \tag{I.10-8}$$

$\bar{\mathbf{p}}$ and $\bar{\mathbf{p}}^*$ being the position vectors of the center of mass with respect to $\mathbf{x}_0$ in $\oint$ and with respect to $\mathbf{x}_0^*$ in $\oint^*$, respectively.

Even if no body-point remains at rest in $\oint$, we can always choose $\mathbf{x}_0^*$ as the center of mass of $\mathscr{B}$ in a rest frame. Then, in general, $\dot{\mathbf{x}}_0 \neq \mathbf{0}$, but of course $\dot{\mathbf{x}}_0^* = \mathbf{0}$, $\bar{\mathbf{p}}^* = \mathbf{0}$, and $\bar{\mathbf{p}} = \mathbf{0}$, so again (6) and (7) follow. The second of these results is particularly important since $\mathbf{E}^*_{\mathbf{x}_0^*}$, being calculated in the rest frame, does not change in time.

Now regarding (6), we consider a vector $\mathbf{e}$ that lies upon a principal axis of inertia in $\oint$. Then $\mathbf{E}_{\mathbf{x}_0}\mathbf{e} = E_k\mathbf{e}$, and hence $(\mathbf{E}_{\mathbf{x}_0}\mathbf{W} + \mathbf{W}\mathbf{E}_{\mathbf{x}_0})\mathbf{e} = (\mathbf{E}_{\mathbf{x}_0} + E_k\mathbf{1})\mathbf{W}\mathbf{e}$. Since $\mathbf{E}_{\mathbf{x}_0} + E_k\mathbf{1}$ is positive, in order that $(\mathbf{E}_{\mathbf{x}_0} + E_k\mathbf{1})\mathbf{W}\mathbf{e} = \mathbf{0}$

[1] Traditionally the tensor $(\operatorname{tr}\mathbf{E}_{\mathbf{x}_0})\mathbf{1} - \mathbf{E}_{\mathbf{x}_0}$ is called the *tensor of inertia*, and Segner's Theorem is expressed in terms of it.

it is necessary and sufficient that $\mathbf{We} = \mathbf{0}$. Similar reasoning may be applied to $\mathbf{E}^*_{\mathbf{x}_0*}\mathbf{A} + \mathbf{A}\mathbf{E}^*_{\mathbf{x}_0*}$.

Thus we have the following

Theorem. *Let $\mathbf{x}_0(t)$ be either the place occupied in $\mathfrak{f}$ at the time t by the center of mass of $\mathscr{B}$, or the place occupied in $\mathfrak{f}$ by some point of $\mathscr{B}$ that remains at rest in $\mathfrak{f}$. Then any two of the following three properties of a line imply the third:*

1. *It is a principal axis of inertia at $\mathbf{x}_0(t)$.*
2. *It is the axis of spin.*
3. *It is the axis of rotational momentum with respect to $\mathbf{x}_0(t)$.*

Exercise I.10.5. Prove that

$$\dot{\mathbf{W}} = -\mathbf{Q}^{\mathrm{T}}\dot{\mathbf{A}}\mathbf{Q}, \qquad \mathbf{W}^2 = \mathbf{Q}^{\mathrm{T}}\mathbf{A}^2\mathbf{Q}. \tag{I.10-9}$$

We shall now calculate the acceleration field of a rigid motion. Supposing that $\mathfrak{f}^*$ be a rest framing for that motion, we could set $\ddot{\chi}^* = \mathbf{0}$ in (I.9-21) after generalizing it so as to allow $\mathbf{x}_0$ to depend on t, but it is easier to differentiate $(1)_2$ instead. Doing so, we obtain

$$\begin{aligned} \ddot{\chi} &= \dot{\mathbf{c}} + \mathbf{W}(\mathbf{c} - \dot{\mathbf{x}}_0) + (\dot{\mathbf{W}} + \mathbf{W}^2)(\chi - \mathbf{x}_0), \\ &= \ddot{\mathbf{x}}_0 + \mathbf{Q}^{\mathrm{T}}[-\ddot{\mathbf{x}}_0^* + 2\mathbf{A}\dot{\mathbf{x}}_0^* - (\dot{\mathbf{A}} - \mathbf{A}^2)(\chi^* - \mathbf{x}_0^*)], \end{aligned} \tag{I.10-10}$$

the second step being a consequence of (9) and (3).

Again supposing first that one body-point of $\mathscr{B}$ remain fixed at the place $\mathbf{x}_0$ in $\mathfrak{f}$, we calculate the rate of change of rotational momentum with respect to that place. To this end we need only set $\dot{\mathbf{x}}_0 = \dot{\mathbf{x}}_0^* = \ddot{\mathbf{x}}_0 = \ddot{\mathbf{x}}_0^* = \mathbf{0}$ in (10), substitute the result into $(\mathrm{I}.8\text{-}5)_2$, and use (9). Thus we obtain the following

Theorem (EULER). *Let a body $\mathscr{B}$ undergo a rigid motion such that in $\mathfrak{f}$ one of its body-points remains at rest at the place $\mathbf{x}_0$. Then in $\mathfrak{f}$*

$$\begin{aligned} \dot{\mathbf{M}}_{\mathbf{x}_0} &= -\mathbf{E}_{\mathbf{x}_0}\dot{\mathbf{W}} - \dot{\mathbf{W}}\mathbf{E}_{\mathbf{x}_0} + \mathbf{E}_{\mathbf{x}_0}\mathbf{W}^2 - \mathbf{W}^2\mathbf{E}_{\mathbf{x}_0}, \\ \mathbf{Q}\dot{\mathbf{M}}_{\mathbf{x}_0}\mathbf{Q}^{\mathrm{T}} &= \mathbf{E}^*_{\mathbf{x}_0*}\dot{\mathbf{A}} + \dot{\mathbf{A}}\mathbf{E}^*_{\mathbf{x}_0*} + \mathbf{E}^*_{\mathbf{x}_0*}\mathbf{A}^2 - \mathbf{A}^2\mathbf{E}^*_{\mathbf{x}_0*}. \end{aligned} \tag{I.10-11}$$

Exercise I.10.6. Derive (11) by differentiating (7), noting that $\mathbf{E}^*_{\mathbf{x}_0*}$ is constant, and then using $(8)_2$.

An axis of rotation that is constant in time is called a *steady axis of rotation.* If $\mathbf{We} = \mathbf{0}$ and $\dot{\mathbf{e}} = \mathbf{0}$, then $\dot{\mathbf{W}}\mathbf{e} = \mathbf{0}$, so we may apply essentially the same reasoning to $\dot{\mathbf{M}}_{\mathbf{x}_0}$ as we did to $\mathbf{M}_{\mathbf{x}_0}$ and conclude the following

Corollary (EULER). *Let a body $\mathscr{B}$ undergo a rigid motion such that in $\mathfrak{f}$ one of its body-points remains at rest at the place $\mathbf{x}_0$. Then in $\mathfrak{f}$ a steady axis of rotation is an axis of the rate of change of rotational momentum with respect to $\mathbf{x}_0$ if and only if it is a principal axis of inertia at $\mathbf{x}_0$.*

Exercise I.10.7. Show that in the case of rotation about a steady axis $\dot{\mathbf{W}} + \mathbf{W}^2$ has the same nullspace as does $\mathbf{W}$, namely, the axis of spin. Hence complete the proof of the foregoing theorem.

In general, no body-point will remain at rest in $\mathfrak{f}$. A result of simple form may be found even so by taking moments and position vectors with respect to the center of mass $\mathbf{x}_c$ of $\chi(\mathscr{B}, t)$, so that $\bar{\mathbf{p}} = \mathbf{0}$. If we choose $\mathbf{x}_0$ in $(8)_1$ as $\mathbf{x}_c$ and then differentiate the result with respect to t, we obtain

$$\dot{\mathbf{M}}_{\mathbf{x}_c} = -\mathbf{E}_{\mathbf{x}_c}\dot{\mathbf{W}} - \dot{\mathbf{W}}\mathbf{E}_{\mathbf{x}_c} + \mathbf{E}_{\mathbf{x}_c}\mathbf{W}^2 - \mathbf{W}^2\mathbf{E}_{\mathbf{x}_c}. \tag{I.10-12}$$

Returning to use of a fixed place $\mathbf{x}_0$, by substituting (12) into (I.8-29) we prove that

$$\dot{\mathbf{M}}_{\mathbf{x}_0} = \bar{\mathbf{p}} \wedge \dot{\mathbf{m}} - \mathbf{E}_{\mathbf{x}_c}\dot{\mathbf{W}} - \dot{\mathbf{W}}\mathbf{E}_{\mathbf{x}_c} + \mathbf{E}_{\mathbf{x}_c}\mathbf{W}^2 - \mathbf{W}^2\mathbf{E}_{\mathbf{x}_c}. \tag{I.10-13}$$

Exercise I.10.8 (KÖNIG, EULER). Prove that the kinetic energy of a body in rigid motion is given by

$$K = \tfrac{1}{2}M|\mathbf{c}|^2 + M\mathbf{c} \cdot \mathbf{W}\bar{\mathbf{p}} - \tfrac{1}{2}\mathbf{W}^2 \cdot \mathbf{E}_{\mathbf{x}_0}. \tag{I.10-14}$$

If $\mathbf{x}_0(t)$ is taken as the place $\mathbf{x}_c(t)$ occupied by the center of mass of $\mathscr{B}$ at the time t, then the kinetic energy of $\mathscr{B}$ may be decomposed into translational and rotational parts as follows:

$$K = \tfrac{1}{2}M|\dot{\mathbf{x}}_c|^2 - \tfrac{1}{2}\mathbf{A}^2 \cdot \mathbf{E}^*_{\mathbf{x}_0^*}. \tag{I.10-15}$$

The first summand is the kinetic energy of a mass-point whose mass M is that of $\mathscr{B}$ and which moves with the speed of the center of mass of $\mathscr{B}$, while the second term is the kinetic energy that would correspond to the spin and shape of $\mathscr{B}$ if the center of mass of $\mathscr{B}$ were at rest in $\mathfrak{f}$.

11. Frame-Indifference

While the event world $\mathscr{W}$ is the seat of phenomena, we may apprehend these only through the intermediary of a framing, since we always report observations in terms of places $\mathbf{x}$ and times t. A phenomenon, of course, is independent of framing, though a description of it in one framing is generally different from a description of it in another. The same phenomenon is reported differently by different observers. Thus arises the question how to relate statements made in terms of different framings.

First, we may always make a statement with respect to one framing ϕ and then simply translate it into a statement with respect to any other framing ϕ^*. We have seen an example in the case of a motion χ of a body $\mathscr{B}$. If χ is given with respect to ϕ, we *define* χ^* in ϕ^* by (I.9-11), which simply reflects our interpretation of the concepts of motion and change of frame. We may do the same thing with other quantities we regard as intrinsic to the event world. The other principal examples in mechanics, as we shall see later, are mass, force, torque, temperature, internal energy, and calory. We say that such quantities are *frame-indifferent*. We shall discuss the frame-indifference of mass, force, and torque in the next section.

Second, and more commonly, we shall encounter a prescription that delivers a particular function in each framing. The prescription itself is *frame-indifferent* in the sense that it is equally effective in all framings. We have seen examples already, namely, the velocity and the acceleration, which are calculated from the motion by rules that make no mention of framings and hence apply for any choice of ϕ. These particular rules have been stated as (I.9-12). We have then been able to express the velocity and acceleration in ϕ^* in terms of their counterparts in ϕ, with the aid, of course, of the functions $\mathbf{Q}$, $\mathbf{x}_0$, and $\mathbf{x}_0^*$ that specify the change (I.9-4) from ϕ to ϕ^*. The results so obtained have been stated as (I.9-14) and (I.9-21). It is clear from them that the velocity and acceleration as observed by ϕ and ϕ^* are not simply functions defined on the event world $\mathscr{W}$ and then referred to frames, for if they were, under change of frame their values, which are vectors, would have to follow the transformation such a change induces on the translation space $\mathscr{V}$ of $\mathscr{E}$, and this transformation, as we have seen, is (I.9-7). Because in general $\dot{\chi}^* \neq \mathbf{Q}\dot{\chi}$ and $\ddot{\chi}^* \neq \mathbf{Q}\ddot{\chi}$, we say that velocity and acceleration are *not frame-indifferent*. This example makes it clear that *a frame-indifferent prescription or definition leads in general to a quantity that is not frame-indifferent.*

Of course some prescriptions, although stated in terms of a framing, do lead to quantities intrinsic to $\mathscr{W}$. Such quantities we shall call *frame-indifferent*, since in principle they could have been introduced abstractly without use of any framing. Suppose certain prescriptions deliver in ϕ and ϕ^* the scalar fields A and A^*, respectively. If

$$A^*(\mathbf{x}^*, t^*) = A(\mathbf{x}, t) \tag{I.11-1}$$

when $(\mathbf{x}^*, t^*)$ is related to $(\mathbf{x}, t)$ through (I.9-4), we shall say that A and A^* represent a *frame-indifferent scalar*. Loosely, we shall refer to the value of A, which of course is a number assigned to a place and time in ϕ, as being itself a frame-indifferent scalar. Likewise, the vector field $\mathbf{v}$ and the tensor field $\mathbf{T}$ will be called *frame-indifferent* if for all t

$$\begin{aligned}\mathbf{v}^*(\mathbf{x}^*, t^*) &= \mathbf{Q}(t)\mathbf{v}(\mathbf{x}, t),\\ \mathbf{T}^*(\mathbf{x}^*, t^*) &= \mathbf{Q}(t)\mathbf{T}(\mathbf{x}, t)\mathbf{Q}(t)^{\mathrm{T}},\end{aligned} \qquad \text{(I.11-2)}$$

respectively, where again $(\mathbf{x}^*, t^*)$ is related to $(\mathbf{x}, t)$ through (I.9-4), and $\mathbf{Q}(t)$ is the relative orientation of $\mathfrak{f}^*$ and $\mathfrak{f}$ at the time t. The first of these requirements asserts that $\mathbf{v}^*$ and $\mathbf{v}$ are the same "arrow" at the same event as observed in different framings. The second asserts, as we have seen in §I.9, that $\mathbf{T}^*$ and $\mathbf{T}$ are at the same event the same linear transformations of such arrows. For details the reader may refer back to the discussion between (I.9-6) and (I.9-10).

The position vector $\mathbf{p}$ of $\mathbf{x}$ with respect to $\mathbf{x}_0$, defined by (I.8-25), is obviously a frame-indifferent vector. Hence the Euler tensor $\mathbf{E}_{\mathbf{x}_0}$, defined by (I.10-4), is a frame-indifferent tensor, and the principal moments of inertia are frame-indifferent scalars.

Most of the fields we encounter in mechanics are not frame-indifferent. The examples of velocity and acceleration suggest, however, that if we restrict attention to a subgroup $\mathscr{g}$ of changes of frame, we may obtain results of the forms (1) or (2). In such a case we may say that a particular scalar, vector, or tensor is *frame-indifferent in* $\mathscr{g}$. For example, from $(\text{I.9-21})_2$ we see that $\ddot{\chi}^* = \mathbf{Q}\ddot{\chi}$ for all motions if and only if $\ddot{\mathbf{x}}_0^* = \mathbf{0}$ and $\mathbf{A} = \mathbf{0}$, so that $\dot{\mathbf{x}}_0^* = \text{const.}$ and $\mathbf{Q} = \text{const.}$ This subgroup of changes of frame, consisting in those under which the acceleration is frame-indifferent, is called the group of *Galilean transformations*. These transformations interconvert the framings of observers moving at uniform velocities with respect to one another and with no change of relative orientation in time.

The term "Galilean" is merely traditional and should not be regarded as referring to anything in the work of the baroque savant GALILEO.

Exercise I.11.1. Prove that the rate of change $\dot{\mathbf{m}}$ of the linear momentum of a body $\mathscr{B}$ in a motion χ is frame-indifferent in the subgroup of Galilean transformations.

The class of all framings obtainable from a given one by Galilean transformations constitute the *Galilean class* of that framing.

Specifically, from (I.9-14) we see that $\dot{\chi}^* = \mathbf{Q}\dot{\chi}$ for all motions if and only if $\dot{\mathbf{x}}_0^* = \mathbf{0}$ and $\mathbf{A} = \mathbf{0}$. Thus $\mathbf{x}_0$ and $\mathbf{Q}$ are constants. This subgroup of the Galilean group of changes of frame, consisting in those under which the velocity is frame-indifferent, is called the group of *constant rigid transformations.* These transformations interconvert the framings of observers at rest with respect to one another. The class of all framings obtainable from a given one by rigid transformations is the *rigid class* of that frame. In §I.9 a rigid motion was defined as one whose velocity field vanishes in some $\mathfrak{f}^*$. We now see that the velocity field of a rigid motion is independent of place in all framings belonging to the rigid class determined by $\mathfrak{f}^*$, and only in such framings. In particular, all rest framings for a rigid motion are obtained from any given one by changes of frame in which $\mathbf{x}_0^* = \text{const.}$, $\mathbf{Q} = \text{const.}$ As is plain from the concept of rigid motions, these framings may be obtained from one another by time-independent translations and rotations. These also constitute a subgroup, the *rest class* of the given rigid motion.

Exercise I.11.2. Prove that the gradient of a frame-indifferent scalar is a frame-indifferent vector; that the proper numbers, trace, and determinant of a frame-indifferent tensor are frame-indifferent scalars; that the proper vectors of such a tensor are frame-indifferent vectors; that the scalar product of two frame-indifferent vectors is a frame-indifferent scalar; and that the tensor product and exterior product of frame-indifferent vectors are frame-indifferent tensors.

Exercise I.11.3. Prove that an oriented unit normal to a surface is a frame-indifferent vector.

At the beginning of this section we remarked that prescription of a quantity in one particular $\mathfrak{f}$ can always be extended trivially to form the definition of a corresponding frame-indifferent quantity. So as to illustrate this fact, we now consider the acceleration $\ddot{\mathbf{x}}$ in $\mathfrak{f}$. If

$$\boldsymbol{\alpha}_{\mathfrak{f}} \equiv \ddot{\boldsymbol{\chi}}^* - \ddot{\mathbf{x}}_0^* - 2\mathbf{A}(\dot{\boldsymbol{\chi}}^* - \dot{\mathbf{x}}_0^*) - (\dot{\mathbf{A}} - \mathbf{A}^2)(\boldsymbol{\chi}^* - \mathbf{x}_0^*), \qquad \text{(I.11-3)}$$

$\ddot{\boldsymbol{\chi}}^*$ and $\dot{\boldsymbol{\chi}}^*$ being the acceleration and the velocity in $\mathfrak{f}^*$, and $\mathbf{A}$ being the spin of $\mathfrak{f}$ with respect to $\mathfrak{f}^*$, then by (I.9-21) we recognize $\boldsymbol{\alpha}_{\mathfrak{f}}$ as being that frame-indifferent vector field over $\mathscr{B}$ which in $\mathfrak{f}$ is the acceleration field of $\mathscr{B}$. Of course, it is the acceleration field in all framings in the Galilean class of $\mathfrak{f}$. The frame-indifferent vector field $\boldsymbol{\alpha}_{\mathfrak{f}}$ is of central importance in dynamics.

12. Axioms of Mechanics

Mechanics relates the motions of bodies to the masses assigned to them and the forces which act on them. Bodies are encountered only in their shapes. Masses and forces, therefore, can be correlated with experience in nature only when they are assigned to the shapes of bodies. Indeed, the value of the mass of a body is a real number, and we may simply transport that number to the shape $\chi(\mathscr{B}, t)$. However, as we have seen in §I.9, different observers generally see $\mathscr{B}$ in different shapes, related to one another by a change of frame (I.9-4). By our assumption that the value of the mass is a scalar on $\mathscr{B}$, we have already implied such independence: *Mass is frame-indifferent.* We may state this fact formally as

Axiom A1.

$$M^* = M, \tag{I.12-1}$$

the notation being that used in §I.11.

We dignify A1 by the title "axiom" since such it would have to be if we had chosen to describe everything in terms of frames from the start.

Since a force is a vector in $\mathscr{V}$, the translation space of $\mathscr{E}$, assignment of forces presumes the previous assignment of a frame of reference.[1] If forces are to have primary meaning, the transport of them to the shapes of bodies must be independent of the observer. The forces acting upon the shapes of $\mathscr{B}$ in ϕ and ϕ^* at the corresponding times t and t^* should be related by the transformation the change of frame from ϕ to ϕ^* induces in $\mathscr{V}$. In other words, *we require that the forces be frame-indifferent.* Formally, we lay down

Axiom A2.

$$\mathbf{f}^* = \mathbf{Q}\mathbf{f}, \tag{I.12-2}$$

the notation being again that of §I.11.

The forces $\mathbf{f}$ and $\mathbf{f}^*$ generally depend upon the time, which is dropped from the notation.

Axiom A1 is part of the assumption commonly called "the principle of conservation of mass"; the other part, which asserts that the mass of a body is the same in all shapes of the body, is implied by our Axiom M1 in §4, according to which mass is assigned to bodies with no mention of any shapes they may assume. Axiom A2, until recently, was left to be inferred from the context and hence was not given a name.

[1] It would be possible but not useful to think of $\mathscr{V}$ as just some vector space given *a priori*, not necessarily the translation space of the instantaneous space $\mathscr{E}$. It would then be difficult to motivate the frame-indifference of forces.

Without exception, the various traditional ways of presenting the foundations of mechanics leave the concept of force in the shadows of intuition. Some even foster the illusion that force is a derived concept, the existence of which follows from some mysterious legerdemain with potential functions and variational principles and magic δs. Assumptions must be made about forces in these treatments, since nothing comes from nothing, but the assumptions are tacit, if not concealed. Modern fundamental thought in mechanics has reverted to the viewpoint of NEWTON and EULER: Forces are basic, *a priori* concepts in mechanics. While NEWTON and EULER left forces, as they did many other things, largely unformalized, today we apply to mechanics the requirement of HILBERT, now universally accepted in the rest of mathematics: An object which enters a mathematical structure must be described by explicit, formal axioms specifying mathematical properties which make it possible to prove theorems. If one such axiomatic basis suffices, so do infinitely many others. The one we adopt in this book is close to the ideas used informally and successfully be engineers for over a century, but of course other axiomatic bases are equally admissible.

It will be noticed that Axioms A1 and A2 require that mass and force as observed in $\mathfrak{f}$ and in $\mathfrak{f}^*$ be assigned the same units, just as the change of frame (I.9-4) leaves the units of length and time unchanged. Of course, a fully general formulation, while allowing it to be *possible* that different observers use the same units, *i.e.*, to choose the same metrics in $\mathscr{E}$, $\mathscr{R}$, and $\mathscr{V}$, would not require them to do so. The generality so obtained is merely apparent and is not worth the complication it introduces into the mathematics at this level. It may be achieved, if desired, by simply allowing free change of units afterward in all frames, once the requirements of frame-indifference have been satisfied, if they can be, by one choice of units.

Exercise I.12.1. Show that Axiom A2 implies that $\int_{\mathscr{A}} (\chi - \mathbf{x}_0) \otimes d\mathbf{f}_{\mathscr{B}^e}$ is frame-indifferent. Hence, in particular, the resultant torque is frame-indifferent.

In a more general system of mechanics allowing for couples as well as the moments of forces, an additional axiom is needed: The torques are frame-indifferent.

In §I.5 we have remarked that a linear combination $A\mathbf{f}_1 + B\mathbf{f}_2$ of two systems of forces $\mathbf{f}_1$ and $\mathbf{f}_2$ is a system of forces. If, as is natural, we require the scalar coefficients A and B to be frame-indifferent, then Axiom A2 is satisfied also by $A\mathbf{f}_1 + B\mathbf{f}_2$. Thus, even after the imposition of Axiom A2, a linear combination of two systems of forces is a system of forces. Conversely, if $\mathbf{f}$ and $\mathbf{g}$ are systems of forces, the trivial decomposition $\mathbf{f} = \mathbf{g} + (\mathbf{f} - \mathbf{g})$ allows us to regard $\mathbf{f}$ as the sum of $\mathbf{g}$ and another system of forces. To justify this decomposition, we cannot take for

$\mathbf{g}$ simply any function that satisfies the axioms of forces listed in §I.5. Rather, we must be sure that $\mathbf{g}$ is frame-indifferent, since Axiom A2 requires that *all forces be frame-indifferent*.

We are now in a position to impose requirements relating forces to motions, or, in looser terms, to state the effects of forces in producing motions. Specifically, we lay down

NOLL's Axiom. *For every assignment of forces to bodies, the working of a system of forces acting on each body is frame-indifferent, no matter what be the motion.*

Formally, in the notations (I.8-7) and (I.11-1),

Axiom A3.

$$W^* = W \qquad \forall \mathscr{B} \in \overline{\Omega}, \quad \forall \chi. \tag{I.12-3}$$

On the assumption that A2 is satisfied, we can prove that NOLL's Axiom is a necessary and sufficient condition that the resultant force and torque on each body $\mathscr{B}$ shall vanish. Indeed, by applying (I.11-1) and (I.9-13) to the definition (I.8-7) we see that, for given $\mathscr{B}$ and χ,

$$\begin{aligned} W^* - W &= \int_{\mathscr{B}} (\dot{\chi}^* \cdot d\mathbf{f}^*_{\mathscr{B}^e} - \dot{\chi} \cdot d\mathbf{f}_{\mathscr{B}^e}), \\ &= \int_{\mathscr{B}} [\dot{\mathbf{x}}_0^* + \dot{\mathbf{Q}}(\chi - \mathbf{x}_0) + \mathbf{Q}\dot{\chi}] \cdot \mathbf{Q}\, d\mathbf{f}_{\mathscr{B}^e} - \int_{\mathscr{B}} \dot{\chi} \cdot d\mathbf{f}_{\mathscr{B}^e}, \\ &= \mathbf{Q}^{\mathrm{T}}\dot{\mathbf{x}}_0^* \cdot \int_{\mathscr{B}} d\mathbf{f}_{\mathscr{B}^e} - \mathbf{Q}^{\mathrm{T}}\dot{\mathbf{Q}} \cdot \int_{\mathscr{B}} (\chi - \mathbf{x}_0) \otimes d\mathbf{f}_{\mathscr{B}^e}, \\ &= \mathbf{Q}^{\mathrm{T}}\dot{\mathbf{x}}_0^* \cdot \mathbf{f}(\mathscr{B}, \mathscr{B}^e) - \tfrac{1}{2}\mathbf{Q}^{\mathrm{T}}\dot{\mathbf{Q}} \cdot \mathbf{F}(\mathscr{B}; \mathscr{B}^e)_{\mathbf{x}_0}. \end{aligned} \tag{I.12-4}$$

By Axiom A3 the right-hand side of this equation must vanish for all choices of the functions $\mathbf{Q}$ and $\dot{\mathbf{x}}_0^*$. We consider a particular time t and choose $\mathbf{Q}$ such that $\dot{\mathbf{Q}}(t) = \mathbf{0}$. Since $\mathbf{Q}(t)^{\mathrm{T}}\dot{\mathbf{x}}_0^*(t)$ may be any vector whatever, Axiom A3 requires that

$$\mathbf{f}(\mathscr{B}, \mathscr{B}^e) = \mathbf{0}. \tag{I.12-5}$$

This being so, Axiom A3 again applied to (4) shows that in the space of skew tensors, $\mathbf{F}(\mathscr{B}, \mathscr{B}^e)_{\mathbf{x}_0}$ must be perpendicular to every tensor of the form $\mathbf{Q}(t)^{\mathrm{T}}\dot{\mathbf{Q}}(t)$, if $\mathbf{Q}(t)$ is an orthogonal tensor. If $\mathbf{W}$ is a skew tensor, and if $\mathbf{Q}(t) \equiv e^{(t-t_0)\mathbf{W}}$, then $\mathbf{Q}(t_0) = \mathbf{1}$ and $\dot{\mathbf{Q}}(t_0) = \mathbf{W}$, so $\mathbf{Q}(t_0)^{\mathrm{T}}\dot{\mathbf{Q}}(t_0) = \mathbf{W}$. Thus the skew tensor $\mathbf{F}(\mathscr{B}, \mathscr{B}^e)_{\mathbf{x}_0}$ must be perpendicular to every skew tensor. Therefore

$$\mathbf{F}(\mathscr{B}, \mathscr{B}^e)_{\mathbf{x}_0} = \mathbf{0}. \tag{I.12-6}$$

Conversely, (5) and (6) suffice for the truth of Axiom A3, it being presumed always that Axiom A2 holds. Thus we have established the following

Theorem (NOLL). *The working is frame-indifferent if and only if the system of forces and the system of torques are both balanced.*

The reader accustomed to the usual treatments of mechanics needs to be reminded that here forces of all kinds are included. The common and useful separation of forces into "applied" forces and "inertial" forces will be made in the following section.

As a consequence of NOLL's theorem here, NOLL's corollary in §I.5, and the counterpart for torques mentioned in §I.8, Axioms A2 and A3 imply the

Corollary (Principle of Action and Reaction). *For each pair of separate bodies $\mathscr{B}$ and $\mathscr{C}$*

$$\begin{aligned} \mathbf{f}(\mathscr{B}, \mathscr{C}) &= -\mathbf{f}(\mathscr{C}, \mathscr{B}), \\ \mathbf{F}(\mathscr{B}, \mathscr{C})_{\mathbf{x}_0} &= -\mathbf{F}(\mathscr{C}, \mathscr{B})_{\mathbf{x}_0}. \end{aligned} \qquad \text{(I.12-7)}$$

While, as we have seen in §I.8, the special assumptions of analytical dynamics, once the system of forces is assumed balanced, reduce the balance of torques to the hypothesis of central forces, in more general and typical universes of mechanics the balance of torques is independent of the balance of forces. The proof of NOLL's theorem makes it clear that the balance of forces expresses the invariance of the working under translations, while the balance of torques expresses the invariance of the working under rotations. Since rotations and translations may be chosen independently in a change of frame, no relation between the two principles can be expected except in degenerate cases.

We have made the existence of a rest framing ϕ^* the definition of a rigid motion of $\mathscr{B}$ (§I.10). In a rest frame, directly from the definition (I.8-7) we see that $W^* = 0$. By Axiom A3, therefore, $W = 0$ in any framing ϕ: *The working of any system of forces vanishes in a rigid motion.* This is the ***work theorem*** of the dynamics of rigid motions. Thus for a rigid motion the value of the quantity whose frame-indifference NOLL's Axiom asserts is in fact 0.

Any motion of a single mass-point is rigid. Thus we may obtain again the trivial result (I.8-12). Our earlier proof assumed the system of forces to be balanced, which NOLL's theorem ensures.

Work theorems similar to that just stated hold in some other special branches of mechanics also, but by no means in all of them. For example, if in analytical dynamics we consider the body consisting in X_1 and X_2, for it the working does not generally vanish. Likewise, if $\mathscr{B}$ is the join of two parts, each of which is in rigid motion, W does not generally vanish unless both parts have the same spin.

13. The Axioms of Inertia. Euler's Laws of Motion

Thus far we have considered a framework on which models of all mechanical occurrences may be constructed: all the bodies in the universe, set in motion through the entire event world. By its nature, human experience can never use with profit, let alone test the worth of so embracing a picture, for human experience is limited to a portion of the event world and to those bodies which have occupied that portion within a limited period of time. This subset of the universe may be a small one, this interval of time a short one; at most, the former represents all bodies whose existence has so far been seen or inferred by man, and the latter, the total length of time through which human experience is known to have existed or can be shrewdly extrapolated. Whatever be the limitation chosen, some limitation there must be, for otherwise we could not isolate a class of putative phenomena from all the rest so as to form models for experiments or for the future course of nature.

On the other hand, we cannot simply disregard the existence of all bodies but those in the subcollection or *great system* Σ in Ω that we choose to isolate for attention, since such further bodies as may exist will generally exert forces upon those we do consider. The idea of "isolation" requires merely that the forces among members of the excluded set of bodies, and the consequent motions of those bodies, need not be known. If $\mathscr{B} \in \Sigma$, and if we denote by Σ^e the join of all the bodies exterior to Σ, then we consider $\mathbf{f}(\mathscr{B}, \Sigma^e)$ and disregard whatever forces the parts of Σ^e may exert upon each other and whatever motions those parts may undergo.

We may also limit the event world $\mathscr{W}$ and the space of instants, but in classical mechanics it is not usual to do so.

In classical mechanics in its most general form, the great system Σ is characterized by two *axioms of inertia*.

Axiom I1. *There is a framing such that if* $\mathbf{m}(\mathscr{B}, \chi)$ *is constant over an open interval of time, then in that interval* $\mathbf{f}(\mathscr{B}, \Sigma^e) = \mathbf{0}$, *and conversely.* Equivalently, by (I.8-28), *there is a framing such that the center of mass* $\bar{\mathbf{p}}$ *of* $\mathscr{B}$ *moves along a straight line at uniform speed in that framing if and only if* Σ^e *exerts no force on* $\mathscr{B}$.

When such a framing is used, the frame $\mathscr{E} \times \mathscr{R}$ itself is called an *inertial frame*; according to the First Axiom of Inertia, *such a frame exists*.

The First Axiom of Inertia, while it asserts the existence of a particular framing, is itself a frame-indifferent statement in that the condition it lays down restricts but does not depend upon the assignment of a frame to the event world. Moreover, it does not depend upon what system of forces is being considered. Axiom A2 asserts that all forces are frame-indifferent. Therefore, no matter what be the function **f**, so long as it satisfies the axioms imposed on systems of forces, the force exerted by Σ^e on $\mathscr{B}$ vanishes in one framing if and only if it vanishes in all framings.

We can express the First Axiom of Inertia in another way. The exterior $\mathscr{B}^e$ of $\mathscr{B}$ may be decomposed into two separate parts: Σ^e and the join of all bodies of Σ separate from $\mathscr{B}$:

$$\mathscr{B}^e = \mathscr{B}^e_\Sigma \vee \Sigma^e. \tag{I.13-1}$$

Here $\mathscr{B}^e_\Sigma$ is the part of the exterior of $\mathscr{B}$ which is also part of the great system Σ, while Σ^e, of course, is the exterior of the great system. If $\mathscr{A} \prec \mathscr{B}$, by Axiom F3 in §I.5

$$\begin{aligned} \mathbf{f}(\mathscr{A}, \mathscr{B}^e) &= \mathbf{f}(\mathscr{A}, \mathscr{B}^e_\Sigma) + \mathbf{f}(\mathscr{A}, \Sigma^e), \\ d\mathbf{f}_{\mathscr{B}^e} &= d\mathbf{f}_{\mathscr{B}_\Sigma{}^e} + d\mathbf{f}_{\Sigma^e}; \end{aligned} \tag{I.13-2}$$

the second equation refers to the vector-valued measures which the three systems of forces define over $\mathscr{B}$ in virtue of Axiom F4 in §I.5. By (I.8-8) we have a similar decomposition for the torque with respect to $\mathbf{x}_0$:

$$\begin{aligned} \mathbf{F}(\mathscr{A}, \mathscr{B}^e)_{\mathbf{x}_0} &= \mathbf{F}(\mathscr{A}, \mathscr{B}^e_\Sigma)_{\mathbf{x}_0} + \mathbf{F}(\mathscr{A}, \Sigma^e)_{\mathbf{x}_0}, \\ (d\mathbf{F}_{\mathscr{B}^e})_{\mathbf{x}_0} &= (d\mathbf{F}_{\mathscr{B}_\Sigma{}^e})_{\mathbf{x}_0} + (d\mathbf{F}_{\Sigma^e})_{\mathbf{x}_0}. \end{aligned} \tag{I.13-3}$$

In particular, the resultant force and resultant torque have such decompositions:

$$\begin{aligned} \mathbf{f}(\mathscr{B}, \mathscr{B}^e) &= \mathbf{f}(\mathscr{B}, \mathscr{B}^e_\Sigma) + \mathbf{f}(\mathscr{B}, \Sigma^e), \\ \mathbf{F}(\mathscr{B}, \mathscr{B}^e)_{\mathbf{x}_0} &= \mathbf{F}(\mathscr{B}, \mathscr{B}^e_\Sigma)_{\mathbf{x}_0} + \mathbf{F}(\mathscr{B}, \Sigma^e)_{\mathbf{x}_0}. \end{aligned} \tag{I.13-4}$$

In each case the first term on the right-hand side, since it depends only on the bodies within the great system Σ, is accessible in principle to observation and measurement. We call these terms the *applied force* on $\mathscr{B}$ and the *applied torque* on $\mathscr{B}$, respectively, and we denote the corresponding functions of $\mathscr{B}$ by $\mathbf{f}^a$ and $\mathbf{F}^a_{\mathbf{x}_0}$:

$$\mathbf{f}^a(\mathscr{B}) \equiv \mathbf{f}(\mathscr{B}, \mathscr{B}^e_\Sigma), \qquad \mathbf{F}^a(\mathscr{B})_{\mathbf{x}_0} \equiv \mathbf{F}(\mathscr{B}, \mathscr{B}^e_\Sigma)_{\mathbf{x}_0}. \tag{I.13-5}$$

While Axiom I1 may be imposed independently of the general axioms of mechanics laid down in §I.12, we shall of course wish to adopt those axioms also. Then in virtue of NOLL's theorem in §I.12 the left-hand sides of $(4)_1$ and $(4)_2$ vanish, so (4) becomes

$$\begin{aligned} \mathbf{f}^a(\mathscr{B}) &= -\mathbf{f}(\mathscr{B}, \Sigma^e), \\ \mathbf{F}^a(\mathscr{B})_{\mathbf{x}_0} &= -\mathbf{F}(\mathscr{B}, \Sigma^e)_{\mathbf{x}_0}. \end{aligned} \tag{I.13-6}$$

Accordingly, we may express Axiom I1 in the following forms, provided we grant Axioms A1–A3 in §I.12:

1. *There is a framing in which the linear momentum of $\mathscr{B}$ is constant if and only if no applied force acts on $\mathscr{B}$.*
2. *There is a framing in which the center of mass of $\mathscr{B}$ moves along a straight line at uniform speed if and only if no applied force acts on $\mathscr{B}$.*

NEWTON set forth in 1687 three Laws of Motion. The first of these was, "Every body perseveres in its state of rest or of uniform motion straight ahead, unless it be compelled to change that state by forces impressed upon it." In the generality maintained in modern mechanics, this axiom is not always valid, for a body may be subject to internal or external constraints not expressed in terms of a system of forces. For example, a rigid body subject to no applied force spins about some axis through its center of mass; its parts, which also are bodies, move in such a way that their centers of mass describe circles about that axis. NEWTON himself did not specify any mathematical properties of bodies or forces, so his intentions must be inferred by the reader, and in the course of time different readers have read different meanings into his words. Our Axiom I1 may be regarded as including one interpretation of NEWTON's First Law.

A Galilean class was defined in §I.11 as the set of all framings with respect to transformations among which the acceleration is frame-indifferent. If the acceleration of a certain body-point vanishes in one framing, it vanishes in all framings belonging to the same Galilean class. In view of (I.8-5), then, the linear momentum of a body is constant in one framing if and only if it is constant in all framings of the same Galilean class. Thus, finally, *the class of inertial frames is a Galilean class.* Accordingly, Axiom I1 requires that if the system of forces $\mathbf{f}$ be such that $\mathbf{f}(\mathscr{B}, \Sigma^e) = \mathbf{0}$ in $\mathfrak{f}$, then $\mathbf{f}^*(\mathscr{B}, \Sigma^e) = \mathbf{0}$ in every $\mathfrak{f}^*$ belonging to the Galilean class containing $\mathfrak{f}$. It imposes no restriction at all upon bodies $\mathscr{B}$ and forces $\mathbf{f}$ such that $\mathbf{f}(\mathscr{B}, \Sigma^e) \neq \mathbf{0}$.

According to astronomers, certain of the most distant stars seem to be nearly at rest with respect to one another. It is customary to interpret the class of inertial frames in the theory as being those that are obtained by uniform translation of one in which those "fixed stars" are stationary. The theory itself, however, merely assumes that there are inertial frames and does not enter into the question of how they should be interpreted in nature.

Once a frame satisfying Axiom I1 is given, we may ask what forces are exerted upon a body $\mathscr{B}$ experiencing general motion with respect to it. These forces are restricted by the following conditions:

1. Since $\mathbf{f}$ is a function of pairs of separate bodies, $\mathbf{f}(\mathscr{B}, \Sigma^e)$ should depend upon the motions of bodies at most through the motion of $\mathscr{B}$ and the motion of Σ^e.
2. Since we know nothing about the nature of Σ^e or its motion, $\mathbf{f}(\mathscr{B}, \Sigma^e)$ should depend upon $\mathscr{B}$ and its motion alone.
3. For consistency with Axiom I1, $\mathbf{f}(\mathscr{B}, \Sigma^e)$ should vanish if $\mathbf{m}(\mathscr{B}, \chi) =$ const.

Classical mechanics rests upon what seems to be the simplest assumption consistent with these three requirements, namely,

Axiom I2 (EULER). *In an inertial frame*

$$\mathbf{f}(\mathscr{B}, \Sigma^e) = -\dot{\mathbf{m}}(\mathscr{B}; \chi). \tag{I.13-7}$$

Up to now, the units of length, time, mass, and force have been independent, and Axiom I1 does not require there to be any relation among them, since it merely asserts that a certain force vanishes when a certain acceleration vanishes. Before it becomes legitimate even to state Axiom I2, we must assume that *forces can be specified in mechanical units*—in particular, that the dimensions of force are the dimensions of (mass) × (acceleration), which are (mass)(length)(time)$^{-2}$.

The origin of this assumption seems not to have been any particular experiment or observation but rather the fact that at first only special forces, namely, *weights*, were recognized. Weight was seen in time to be proportional to mass, and indeed in the beginning force, weight, and mass seem to have been confused often. That the units of force are of the special kind required in order that we be allowed even to consider Axiom I2 as a possible assumption in a theory of natural phenomena, should be accessible to test by experiment.[1] While no specific experiment seems ever to have been proposed, let alone effected, so as to test this assumption, it seems to be universally accepted.

Axiom I2 is consistent with Axiom F4 in §I.5, since, as shown by $(\text{I.8-5})_1$, the rate of change of linear momentum of a part of $\mathscr{B}$ is the value of a measure over $\mathscr{B}$. Specifically, for smooth motions $(\text{I.8-5})_1$ enables us to express (7) in the form

$$\begin{aligned} d\mathbf{f}_{\Sigma^e} &= -\ddot{\chi}\, dM, \\ (d\mathbf{F}_{\Sigma^e})_{\mathbf{x}_0} &= -(\chi - \mathbf{x}_0) \wedge \ddot{\chi}\, dM. \end{aligned} \tag{I.13-8}$$

[1] The question is strictly parallel to that underlying the "first law of thermodynamics", which requires that flow of heat may be measured in mechanical units.

The second assertion follows from the first because we have assumed (I.8-8). In a more general system of mechanics, we should have to lay down $(8)_2$, or some other axiom, independently of $(8)_1$.

The forces and torques given by (8) are called *inertial.* Providing the framing $\mathfrak{f}$ be an inertial one, these forces and torques are those exerted upon the bodies of the great system Σ by the bodies, whatever they may be, that are outside Σ. When we choose instead to use a general framing $\mathfrak{f}^*$, we think of the unknown motions of the exterior Σ^e as being subjected to the same change from the framing $\mathfrak{f}$ to the framing $\mathfrak{f}^*$ as are the motions of Σ. Therefore, the second axiom of inertia, while it refers to a particular class of framings, is itself a frame-indifferent statement. While we follow tradition in stating it as we have, in terms of an inertial frame, we need not do so. Axiom A2 of §I.12 asserts that *all forces are frame-indifferent.* Thus the quantity on the left-hand side of $(8)_1$ is frame-indifferent. Accordingly, a frame-indifferent statement that reduces to $(8)_1$ in an inertial frame is

$$d\mathbf{f}(\mathscr{B}, \Sigma^e) = -\boldsymbol{\alpha}_{\mathfrak{f}}\, dM, \qquad \text{(I.13-9)}$$

$\boldsymbol{\alpha}_{\mathfrak{f}}$ being that frame-indifferent vector field over $\mathscr{B}$ which in the inertial frame $\mathfrak{f}$ reduces to $\ddot{\chi}$. We have already calculated $\boldsymbol{\alpha}_{\mathfrak{f}}$ and recorded it in (I.11-3).

In the remainder of this book we shall follow the tradition of mechanics in assuming tacitly that *the frame used is an inertial one,* so (8) holds.

Our use of an inertial frame rests on more than respect for tradition. An essential feature of classical mechanics is *the existence of special frames in which the relation between forces and the motions they produce is especially simple.* Since we have these felicitous frames, it would be simply foolish not to use them. When for purposes of interpretation in a particular case we need to employ some frame that is not inertial, as for example in problems referred to a rotating earth, we formulate the laws of mechanics first in an inertial frame and then transform them to the other frame of interest. Such is the traditional approach, which derives from CLAIRAUT and EULER. In replacing (8) by (9) we formulate that approach in general terms.

Recalling that the general axioms of mechanics imply (6), from Axiom I2 we see that

$$\mathbf{f}^a(\mathscr{B}) = \dot{\mathbf{m}}(\mathscr{B}; \chi), \qquad \mathbf{F}^a(\mathscr{B})_{\mathbf{x}_0} = \dot{\mathbf{M}}(\mathscr{B}, \chi)_{\mathbf{x}_0}, \qquad \text{(I.13-10)}$$

where to obtain the second statement we have used $(8)_2$. *That is, the applied force on $\mathscr{B}$ equals the rate of change of the linear momentum of $\mathscr{B}$ in an inertial frame, and the applied torque equals the rate of change of rotational momentum of $\mathscr{B}$ in that frame,* both torque and rotational momentum being taken with respect to a place $\mathbf{x}_0$ that is stationary in the inertial frame. These two statements are ***Euler's Laws of Motion.*** The formal treatment in the rest of this book is based upon them rather than upon the more general ideas from which we have developed them. We shall generally write them in the shorter notation

$$\mathbf{f}^a = \dot{\mathbf{m}}, \qquad \mathbf{F}^a = \dot{\mathbf{M}}. \tag{I.13-11}$$

If for a given body

$$\mathbf{f}^a = \mathbf{0}, \qquad \mathbf{F}^a = \mathbf{0}, \tag{I.13-11A}$$

that body is *isolated*. From Euler's Laws (11) we see that *the linear and rotational momenta of a body remain constant if and only if that body is isolated.*

We have seen that in a general framing $(8)_1$ must be replaced by (9). The corresponding replacement in $(8)_2$ has to be treated with care, since in it $\mathbf{x}_0$ is a fixed place in an inertial frame. The resulting general forms of EULER's Laws (10) are

$$\begin{aligned} \mathbf{f}^a(\mathscr{B}) &= \int_{\mathscr{B}} \boldsymbol{\alpha}_{\mathfrak{f}}\, dM \\ \mathbf{F}^a(\mathscr{B})_{\mathbf{y}_0} &= \int_{\mathscr{B}} (\boldsymbol{\chi} - \mathbf{x}_0) \wedge \boldsymbol{\alpha}_{\mathfrak{f}}\, dM. \end{aligned} \tag{I.13-12}$$

Here $\mathbf{y}_0$ is a fixed place in an inertial frame $\mathfrak{f}$, and $\mathbf{x}_0$ is the corresponding place in the general frame $\mathfrak{f}^*$. Only if the frame is inertial are the right-hand sides of (12) equal to the rates of change of linear momentum and rotational momentum, respectively.

Returning to use of an inertial frame, as we shall do henceforth in this book, we note from (I.8-28) that EULER's First Law $(11)_1$ can be written in terms of the motion of the center of mass $\bar{\mathbf{p}}$ of $\mathscr{B}$:

$$\mathbf{f}^a = M\ddot{\bar{\mathbf{p}}}. \tag{I.13-13}$$

Thus the applied force on a body equals the mass of that body times the acceleration of its center of mass in an inertial frame.

This last is one of the oldest of the commonly accepted principles of mechanics, used again and again, with or without explicit statement, in the eighteenth century. It is sometimes regarded as expressing the Second Law of NEWTON: "The change of motion is proportional to the impressed motive force, and it is made in the direction of the right line along which that force is impressed."

The point $\mathbf{x}_0$ with respect to which torques and rotational momenta entering EULER's Second Law are calculated is a fixed point in an inertial frame. We may use EULER's two laws together so as to calculate the effect of the applied loads upon the rotational momentum with respect to the center of mass $\mathbf{x}_c$. In (I.8-9) we replace $\mathscr{B}^e$ by $\mathscr{B}^e_\Sigma$. Then EULER's Second Law $(10)_2$ makes the left-hand side of the result equal the left-hand side of (I.8-29). If in the former we take $\mathbf{x}_c$ for $\mathbf{x}_1$, by use of EULER's First Law $(10)_1$ we conclude at once that $(10)_2$ holds with $\mathbf{x}_c$ replacing $\mathbf{x}_0$. *That is, the applied torque on $\mathscr{B}$ equals the rate of change of rotational momentum of $\mathscr{B}$ in an inertial frame when both torque and rotational momentum are taken with respect to the center of mass.*

In analytical dynamics the "environment" X_0 is generally considered to have two separate parts, one inside the system Σ and the other being Σ^e:

$$X_0 = X_e \vee \Sigma^e, \tag{I.13-14}$$

say, so that

$$\begin{aligned} \mathbf{f}^e_k = \mathbf{f}(X_k, X_0) &= \mathbf{f}(X_k, X_e) + \mathbf{f}(X_k, \Sigma^e), \\ &= \mathbf{f}^0_k - M_k \ddot{\mathbf{x}}_k, \end{aligned} \tag{I.13-15}$$

and

$$\mathbf{f}^0_k \equiv \mathbf{f}(X_k, X_e), \qquad \ddot{\mathbf{x}}_k \equiv \ddot{\chi}(X_k, \cdot). \tag{I.13-16}$$

The force $\mathbf{f}^0_k$ is called the *external* or *extrinsic* applied force acting upon X_k. In terms of it and the mutual forces $\mathbf{f}_{kq}$, EULER's First Law $(10)_1$ assumes the form

$$\mathbf{f}^0_k + \sum_{q=1}^{n}{}' \mathbf{f}_{kq} = M_k \ddot{\mathbf{x}}_k, \qquad k = 1, 2, \ldots, n, \tag{I.13-17}$$

as may be seen also from putting (15) into $(\text{I.5-24})_1$. Equations of this form are often called "Newtonian", though they occur nowhere in the writings of NEWTON.

Exercise I.13.1 (NOLL). Show that the axioms of inertia when applied to analytical dynamics do not alter the requirement (I.5-22) and NOLL's theorem at the end of §I.8. Thus in analytical dynamics EULER's Second Law is *equivalent*, the first being presumed imposed already, to the statement that the mutual forces are central.

Moreover, for the entire system of n mass-points

$$\begin{aligned} \sum_{k=1}^{n} \mathbf{f}^0_k &= \frac{d}{dt} \sum_{k=1}^{n} M_k \dot{\mathbf{x}}_k = M\ddot{\bar{\mathbf{p}}}, \\ \sum_{k=1}^{n} (\mathbf{x}_k - \mathbf{x}_0) \wedge \mathbf{f}^0_k &= \frac{d}{dt} \sum_{k=1}^{n} (\mathbf{x}_k - \mathbf{x}_0) \wedge M_k \dot{\mathbf{x}}_k; \end{aligned} \tag{I.13-18}$$

the second relation, while its form suggests the principle of rotational momentum, is a simple consequence of (17) and (I.8-23). These are the *theorems of linear and rotational momentum* of analytical dynamics.

Comparison of $(18)_1$ with (13) shows that *in an inertial frame, the motion of the center of mass of a body $\mathscr{B}$ is the same as that of a mass-point having the same mass as $\mathscr{B}$, located at the center of mass of $\mathscr{B}$, and subject to the* *resultant applied force on $\mathscr{B}$*. Thus if we are satisfied with knowing no more about the motion of a body than the motion of its center of mass, and if we can determine the applied force on that body, we need enter no more deeply into mechanics than the level of analytical dynamics. As HAMEL wrote in 1909, "what is understood in practice as the mechanics of points is neither more nor less than the theorem on the center of gravity." This fact goes far to explain the pragmatic success of analytical dynamics. In particular, use of it does not require that the body $\mathscr{B}$ really occupy no more than a discrete set of points in space, but only that our curiosity be slaked by determining the motions of such a set of points. The standard example here is furnished by the sun and its planets and comets. It is a typical example in that whether or not analytical dynamics be sufficient to describe its motion depends on how far we choose to inquire into it. For certain problems or in certain refined cases we need to take account of the spins and even the shapes of the bodies, and then analytical dynamics, as embodied in (17), (I.5-22), and (I.8-23), no longer suffices.

From (13) we see that the motion of the center of mass of any body is determined, to within arbitrarily assigned position and velocity at some one time, if the resultant force on that body is a known function of time. For the case of a rigid motion still more can be said. Consider first the case when one body-point $\mathbf{x}_0$ of a body $\mathscr{B}$ in rigid motion remains at rest in an inertial frame. Substitution of $(\text{I}.10\text{-}11)_2$ into $(11)_2$ then yields ***Euler's Differential Equation*** for a rigid motion:

$$\mathbf{Q}\mathbf{F}^{\mathrm{a}}_{\mathbf{x}_0}\mathbf{Q}^{T} = \mathbf{F}^{\mathrm{a}*}_{\mathbf{x}_0^*} = \mathbf{E}^*_{\mathbf{x}_0^*}\dot{\mathbf{A}} + \dot{\mathbf{A}}\mathbf{E}^*_{\mathbf{x}_0^*} + \mathbf{E}^*_{\mathbf{x}_0^*}\mathbf{A}^2 - \mathbf{A}^2\mathbf{E}^*_{\mathbf{x}_0^*}. \qquad (\text{I}.13\text{-}19)$$

Here $\mathbf{F}^{\mathrm{a}*}_{\mathbf{x}_0^*}$ is the applied torque in the rest frame with respect to the stationary place $\mathbf{x}_0^*$ occupied in that frame by the body-point that remains at rest at the one place $\mathbf{x}_0$ in an inertial frame. Since $\mathbf{E}^*_{\mathbf{x}_0^*}$ is a known, constant tensor, (19) is a differential equation of first order for the spin $\mathbf{A}$ of the rest framing ϕ^* with respect to the inertial framing ϕ, on the presumption that the resultant torque $\mathbf{F}^{\mathrm{a}*}_{\mathbf{x}_0^*}$ be known.

Even if no body-point remains at rest in an inertial frame, we may appeal to the italicized theorem at the top of p. 61 and so by use of (I.10-12) conclude that

$$\mathbf{F}^{\mathrm{a}}_{\mathbf{x}_c} = -\mathbf{E}_{\mathbf{x}_c}\dot{\mathbf{W}} - \dot{\mathbf{W}}\mathbf{E}_{\mathbf{x}_c} + \mathbf{E}_{\mathbf{x}_c}\mathbf{W}^2 - \mathbf{W}^2\mathbf{E}_{\mathbf{x}_c}, \qquad (\text{I}.13\text{-}20)$$

or, equivalently,

$$\mathbf{F}^{\mathrm{a}*}_{\mathbf{x}_c^*} = \mathbf{E}^*_{\mathbf{x}_c^*}\dot{\mathbf{A}} + \dot{\mathbf{A}}\mathbf{E}^*_{\mathbf{x}_c^*} + \mathbf{E}^*_{\mathbf{x}_c^*}\mathbf{A}^2 - \mathbf{A}^2\mathbf{E}^*_{\mathbf{x}_c^*}, \qquad (\text{I}.13\text{-}21)$$

$\mathbf{x}_c^*$ being the place occupied by the center of mass in the rest framing ϕ^*. This result is of the same form as (19) and can be interpreted similarly.

If $\mathbf{F}^{\mathrm{a}*}_{\mathbf{x}_0^*}$ is a continuous function of time, there is a unique solution $\mathbf{A}$ of (19) corresponding to any given initial value $\mathbf{A}(t_0)$, and if that initial value is skew, so is $\mathbf{A}(t)$ for all t, as the reader will easily verify. A theorem given in §I.9 states conditions under which the spin $\mathbf{A}$ determines the relative orientation $\mathbf{Q}$. Similar reasoning may be applied to (21).

Summarizing all these results, we have the following

Theorem. *Let a body $\mathscr{B}$ be in rigid motion, and let $\mathbf{E}^*_{\mathbf{x}_0^*}$ be its Euler tensor in a rest frame with respect to the place $\mathbf{x}_0^*$. Suppose that either:*

A. *the place $\mathbf{x}_0^*$ in ϕ^* is occupied by a body-point that remains at rest at the place $\mathbf{x}_0$ in the inertial framing ϕ, or*
B. *the place $\mathbf{x}_0^*$ in ϕ^* is occupied by the center of mass of $\mathscr{B}$.*

In Case B, suppose that the place $\mathbf{x}_0(t)$ occupied in the inertial framing ϕ by the center of mass of $\mathscr{B}$ be known, e.g. *by integration of* (13).

Then the assignment of the initial orientation $\mathbf{Q}(t_0)$ of ϕ^* with respect to ϕ determines a unique rest framing ϕ^* and hence a unique rigid motion of $\mathscr{B}$.

Roughly, if one body-point of $\mathscr{B}$ remains at rest in an inertial frame, or if the motion of the center of mass of $\mathscr{B}$ with respect to an inertial frame is known, a rigid motion of $\mathscr{B}$ is determined by an assigned resultant torque, to within inessential constants.

This theorem enables us to refine, if we so desire, the bare skeleton of mechanics furnished by analytical dynamics. If we are content to regard the motion of a body as rigid, we may calculate that motion from the resultant torque, once the existence of a fixed point or the motion of the center of mass has been determined. For this purpose we need to know about the body itself only its Euler tensor $\mathbf{E}^*_{\mathbf{x}_0^*}$ with respect to an appropriate place $\mathbf{x}_0^*$ in a rest frame.

As we noticed above, to apply the mechanics of mass-points we need not assume that the shape of $\mathscr{B}$ be a single place; rather, we must simply be content with determining the motion of the center of mass of $\mathscr{B}$, leaving unknown such motion relative to that center as the remaining points of $\mathscr{B}$ may have. Likewise, to apply the theory of rigid motions, we need not assume that the body $\mathscr{B}$ be susceptible *only* of such motions; rather, we must simply remain content with specifying some one shape of $\mathscr{B}$ and supposing that in *some* frame, that shape shall remain unchanged. In rough terms, analytical dynamics and the theory of rigid motions determine certain *aspects* of the motions of all bodies, whether or not they be mass-points or rigid bodies.

If we cast back one look at the purely kinematical corollary on steady rotations at the end of §I.10, by use of the axioms of inertia we may now obtain from it a major proposition of dynamics. An *axis of free rotation* is a line whose direction is steady in a rest frame and about which a body subject to no resultant torque with respect to some point on that axis may spin. Such an axis is necessarily a steady axis of rotation and, of course, an axis of rotational momentum. Thus we have the following

Theorem (EULER). *The axes of free rotation through the center of mass of a body or through the place of a body-point which is at rest in an inertial frame are the principal axes of inertia.*

In particular, *a body of a given shape cannot spin freely about any line that is not one of its axes of inertia.* Since $\mathbf{E}^*_{\mathbf{x}_0^*}$ is symmetric, there are either exactly three such axes, which are orthogonal to one another, or infinitely many. In the latter case, either every line is a principal axis of inertia, or the principal axes of inertia consist in one certain line and all lines in a certain plane perpendicular to it.

Exercise I.13.2 (EULER). Let $\mathbf{e}$, $\mathbf{f}$, $\mathbf{g}$ be an orthonormal triad of proper vectors of $\mathbf{E}^*_{\mathbf{x}_0^*}$ in a rest frame, and suppose $\mathscr{B}$ be in rotation about the principal axis of inertia defined by $\mathbf{e}$, so that

$$\mathbf{A} = \omega \mathbf{f} \wedge \mathbf{g}. \tag{I.13-22}$$

Then $\omega = |\mathbf{A}|/\sqrt{2}$. Let $\mathbf{x}_0^* = \mathbf{x}_0 =$ a place on the axis. Prove that EULER's equation (21) reduces to

$$\mathbf{F}^{a*}_{\mathbf{x}_0^*} = F \mathbf{f} \wedge \mathbf{g}, \tag{I.13-23}$$

where

$$F = I\dot{\omega} \qquad \text{and} \qquad I = E_2 + E_3, \tag{I.13-24}$$

E_2 and E_3 being the proper numbers of the Euler tensor $\mathbf{E}^*_{\mathbf{x}_0^*}$ corresponding to the proper vectors $\mathbf{f}$ and $\mathbf{g}$.

14. Energy

In §I.12 we have imposed the requirement that the working W be frame-indifferent. In an inertial frame, the working has an especially striking interpretation. Namely, if we substitute $(\text{I.13-8})_1$ into (I.13-2) and then substitute the result into the definition (I.8-7) of W, by comparison with $(\text{I.8-5})_3$ we see that

$$W = P - \dot{K}, \tag{I.14-1}$$

P being the *power*, namely, the working of the forces exerted on $\mathscr{B}$ by the exterior bodies in the great system Σ alone:

$$P = \int_{\mathscr{B}} \dot{\chi} \cdot d\mathbf{f}(\mathscr{B}, \mathscr{B}^e_\Sigma), \tag{I.14-2}$$

and K being the kinetic energy of $\mathscr{B}$. We have eased the writing by leaving arguments such as $\mathscr{B}$ and χ unwritten. The result (1) asserts that the working W is the power of the forces exerted upon $\mathscr{B}$ by the exterior of $\mathscr{B}$ in the great system Σ, less the rate of increase of the kinetic energy of $\mathscr{B}$, in an inertial frame. We may say equally that the working of the inertial forces is $-\dot{K}$.

If in an inertial frame all work done is converted into kinetic energy, $P = \dot{K}$, so that

$$W = 0, \tag{I.14-3}$$

and the term *mechanically perfect* is applied. That term may refer to the body, to the system of forces, or to the motion, whichever of these we choose to regard as being restricted by the statement. The condition (3) is frame-indifferent, so it may be imposed on all bodies, all motions, or all systems of forces, in any combination we please. In §I.12 we have proved that a rigid motion of any body and all motions of a single mass-point are mechanically perfect. This statement is a consequence of NOLL's Axiom in §I.12 and does not require the Axioms of Inertia. These latter, however, enable us to interpret the result as stating that in an inertial frame, the working of the forces on $\mathscr{B}$ is balanced by increase of the kinetic energy of $\mathscr{B}$.

More generally, the exercise of forces need not give rise to motion but may in whole or part be consumed in production of heat, and, *vice versa*, heating a body may set it in motion, as we may see from the example of compressing a gas by a piston, or allowing the gas by expanding to move such a piston and so cool itself by working. Also there are circumstances in which a body may be heated or cooled with no consequent effects recognizable as being motions. This type of heating suggests the introduction of a *system of heatings* $Q(\mathscr{B}, \mathscr{C})$, a scalar-valued function of pairs of separate bodies which satisfies axioms of just the same form as F1–F4 in §I.5, and a scalar-valued function $E(\mathscr{B})$, the *internal energy* of $\mathscr{B}$. Both these quantities depend, in general, on the time, which we omit from the notation. The *resultant heating* is $Q(\mathscr{B}, \mathscr{B}^e)$, and the idea that the heat flowing into or out of $\mathscr{B}$ is stored within $\mathscr{B}$ or taken from it is embodied in the equation

$$\dot{E}(\mathscr{B}) = Q(\mathscr{B}, \mathscr{B}^e). \tag{I.14-4}$$

Circumstances in which this equation holds are called *energetically perfect.*

Exercise I.14.1. On the assumption that (4) holds, show that $\dot{E}$ is an additive set function if and only if $Q(\mathscr{B}, \mathscr{C}) = -Q(\mathscr{C}, \mathscr{B})$ for all pairs of separate bodies $\mathscr{B}$ and $\mathscr{C}$.

Since E and Q are defined over bodies and pairs of bodies, respectively, they may be transferred to the shapes of bodies. Since there is no basis for assigning preference to one framing rather than another in considerations of energy and heat, we assume that both E and Q are frame-indifferent scalars. Specifically, we lay down in analogy to Axioms A1–A3 in §I.12

Axiom E1.

$$E^* = E, \qquad Q^* = Q. \tag{I.14-5}$$

The occasional connection between heating and the action of forces, mentioned above, shows that the conditions (3) and (4) cannot be general. It suggests that W, $\dot{E}$, and Q may be related, but it does not dictate any particular relation. The units assigned to Q and hence determined for E by (4) were specified originally in terms of conditions in which forces and motions were absent. These units are called "thermal", and they are still in wide use today. Decisive experiments by JOULE and others showed that, despite the dissimilarity of the original concepts of heating from those of mechanics, *heating may be measured in units of working*. This is the *Principle of Equivalence of Heat and Work*. It suggests that we may consider heating Q and working W as co-operating to produce energy, and this further assumption is called the ***Balance of Energy***,[1] or, sometimes, "the first law of thermodynamics":

Axiom E2.

$$\dot{E} = W + Q. \tag{I.14-6}$$

The arguments $\mathscr{B}$, $\mathscr{B}^e$, χ, etc., are omitted for ease of writing.

The scalar function that the axiom of balance of energy asserts to vanish is frame-indifferent since $\dot{E}$, W, and Q are frame-indifferent.

The concept of inertial frame and *a fortiori* the axioms of inertia play no part in the theory of energy. We have followed the historical order in introducing energy only after the exposition of mechanics, but in truth a logical order from the general to the particular would introduce Axiom E2 immediately after Axiom A3 in §I.12.

By comparing Axiom E2 with (4) and (3) we see that "*mechanically perfect*" *and* "*energetically perfect*" *are equivalent concepts.*

[1] A probing analysis of this axiom, its basis, and its implications is given in the paper by GURTIN & WILLIAMS cited in the references at the end of this chapter.

Theories of mechanically and energetically perfect motions or bodies are untypical of general mechanics, since they permit us to study and determine the effects of forces without specifying or even mentioning heating, or the effects of heating without specifying or even mentioning forces. Examples of the former kind are furnished by any motion of a mass-point and by the rigid motion of any body, since as we have seen in §I.12, in both these cases $W = 0$ always. An example of the latter is furnished by the classical theory of the conduction of heat, which assumes at its very start that $\dot{E} = Q$. If in that theory it is assumed that all bodies are at rest, obviously $W = 0$, but in fact the theory in its classical form is consistent with Axiom E2 *only* when $W = 0$.

Axiom E2 makes no use of the Axioms of Inertia and may be imposed independently of them. In an inertial frame we have the relation (1) between the power P of the forces within the great system Σ and the kinetic energy K, so Axiom E2 yields

$$\dot{K} + \dot{E} = P + Q. \tag{I.14-7}$$

This consequence of the Axioms of Inertia and the Balance of Energy will be our starting point when we come to consider heating and energy, which we shall do in Volume 3.

The sum $K + E$ is called the *total energy* of $\mathscr{B}$ in its actual shape $\chi_\Omega(\mathscr{B}, t)$. Thus (7) states that *in an inertial frame, the sum of the heating of* $\mathscr{B}$ *and the power of the forces within the great system acting on* $\mathscr{B}$ *is the rate of increase of the total energy of* $\mathscr{B}$.

Internal energy must not be confused with what is called "potential energy", which exists only for certain special kinds of systems of forces. This term has somewhat different meanings in different special theories. For illustration we shall select analytical dynamics. First, from the appropriate specialization of (I.8-5) we see by use of (I.13-17) that

$$\dot{K} = \sum_{k=1}^{n} \dot{\mathbf{x}}_k \cdot \left(\mathbf{f}_k^0 + \sum_{q=1}^{n}{}' \mathbf{f}_{kq}\right). \tag{I.14-8}$$

Second, we impose the restrictive assumption that the extrinsic applied forces $\mathbf{f}_k^0$ and the mutual forces $\mathbf{f}_{kq}$ be derivable from frame-indifferent scalar *potential functions* $V_0(\mathbf{x})$ and $V_{\mathrm{m}}(\mathbf{v})$, as follows:

$$\begin{aligned} \mathbf{f}_k^0 &= -\nabla V_0(\mathbf{x})\Big|_{\mathbf{x}=\mathbf{x}_k}, \\ \mathbf{f}_{kq} &= -\nabla V_{\mathrm{m}}(\mathbf{v})\Big|_{\mathbf{v}=\mathbf{x}_q-\mathbf{x}_k}, \end{aligned} \tag{I.14-9}$$

where $\mathbf{x}_k$, as usual, is the position of X_k at the time t in the motion χ, and all indices run from 1 to n. We suppose also that $V_{\mathrm{m}}(0) = 0$, $\nabla V_{\mathrm{m}}(0) = \mathbf{0}$, so that $\mathbf{f}_{kk} = \mathbf{0}$. For a reason to be seen presently, such a system of forces is called *conservative*.

Exercise I.14.2. Show that under change of frame $V_{\mathrm{m}}(\mathbf{v}^*) = V_{\mathrm{m}}(\mathbf{v})$ if and only if $V_{\mathrm{m}}(\mathbf{v}) = V_{\mathrm{m}}(|\mathbf{v}|)$, and hence the mutual forces are central. Interpret this requirement in terms of frame-indifference. Prove that if under change of frame $V_0(\mathbf{x}^*, \mathbf{x}_0^*) = V_0(\mathbf{x}, \mathbf{x}_0)$, $\mathbf{x}_0$ being a fixed place, then $\mathbf{f}_k$ is parallel to $\mathbf{x}_k - \mathbf{x}_0$, and conversely. If $\mathbf{n}$ is a frame-indifferent unit vector, prove that $V_0(\mathbf{x}^* - \mathbf{x}_0^*, \mathbf{n}^*) = V_0(\mathbf{x} - \mathbf{x}_0, \mathbf{n})$ if and only if $\mathbf{f}_k$ is the sum of a vector parallel to $\mathbf{x} - \mathbf{x}_0$ and a vector parallel to $\mathbf{n}$.

In terms of a motion χ we define the following functions of time only:

$$V_k(t;\chi) \equiv V_0(\mathbf{x}_k),$$
$$V_{kq}(t;\chi) \equiv \begin{cases} V_{\mathrm{m}}(|\mathbf{x}_k - \mathbf{x}_q|), & k \neq q, \\ 0, & k = q. \end{cases} \tag{I.14-10}$$

Hence

$$\begin{aligned} \dot{V}_k &= -\mathbf{f}_k^0 \cdot \dot{\mathbf{x}}_k, \\ \dot{V}_{kq} &= \dot{V}_{qk} = -\mathbf{f}_{qk} \cdot \dot{\mathbf{x}}_k - \mathbf{f}_{kq} \cdot \dot{\mathbf{x}}_q; \end{aligned} \tag{I.14-11}$$

to obtain the second result we have used (I.5-22). The *potential energy* V of a system of forces of the form (9) is the following function of n places $\mathbf{y}_k$ and $\frac{1}{2}n(n-1)$ vectors $\mathbf{v}_{lm}$, $l > m$, $m = 1, 2, \ldots, n-1$:

$$V(\{\mathbf{y}_k\}, \{\mathbf{v}_{lm}\}) \equiv \sum_{k=1}^{n} V_0(\mathbf{y}_k) + \sum_{\substack{k,\,q=1 \\ k \geqq q}}^{n} V_{\mathrm{m}}(\mathbf{v}_{kq}); \tag{I.14-12}$$

for formal convenience we have set $V_{\mathrm{m}}(0) \equiv 0$. In a particular motion χ, if we set $\mathbf{y}_k \equiv \mathbf{x}_k$ and $\mathbf{v}_{kq} \equiv \mathbf{x}_k - \mathbf{x}_q$, each summand in (12) defines a function of t by (10), and we thus obtain such a function for the motion of the body $\mathscr{B}$ comprising all n mass-points:

$$V(\mathscr{B}; t; \chi) = \sum_{k=1}^{n} V_k(t;\chi) + \sum_{\substack{k,\,q=1 \\ k \geqq q}}^{n} V_{kq}(t;\chi). \tag{I.14-13}$$

If we differentiate this formula with respect to t and substitute (11) into the result, we see that

$$\dot{V} = -\sum_{k=1}^{n} \dot{\mathbf{x}}_k \cdot \left(\mathbf{f}_k^0 + \sum_{q=1}^{n} \mathbf{f}_{kq}\right). \tag{I.14-14}$$

Comparison with (8) yields the ***Energy Theorem:*** *If the system of forces acting upon the universe of analytical dynamics is conservative, then*

$$K + V = \text{const.} \tag{I.14-15}$$

That is, the sum of the kinetic and potential energies of the whole system has a constant value in any particular motion.

Exercise I.14.3. State and prove a corresponding theorem for a rigid motion.

The quantity $K + V$ is sometimes called the "total energy" of the system, but in this book we reserve that term for $K + E$. If a potential energy exists, we do not attempt to include it in E. The internal energy E enters a *general* law of mechanics and hence belongs in its basic structure. In some special problems, indeed, we may avoid use or even mention of E, but it is always there, on demand. The potential energy V, on the contrary, exists only in special cases. It is an ancillary quantity; useful though it is in the description of important special systems and in the solution of important special problems, it does not enter the conceptual structure of mechanics as a whole.

General References

W. NOLL, "The foundations of classical mechanics in the light of recent advances in continuum mechanics," pp. 266–281 of *The Axiomatic Method, with Special Reference to Geometry and Physics* (Colloquium at Stanford, 1957), Amsterdam, North-Holland Publ., 1959. Reprinted in W. NOLL, *The Foundations of Mechanics and Thermodynamics*, New York, Heidelberg, and Berlin, Springer-Verlag, 1974.

W. NOLL, "La mécanique classique, basée sur un axiome d'objectivité," pp. 47–56 of *La Méthode Axiomatique dans les Mécaniques Classiques et Modernes* (Colloque International à Paris, 1959), Paris, Gauthier-Villars, 1963. Reprinted along with the preceding.

W. NOLL, "Euclidean geometry and Minkowskian chronometry," *American Mathematical Monthly* **71**, 129–144 (1964). Reprinted along with the preceding.

W. NOLL, "Lectures on the Foundations of Continuum Mechanics and Thermodynamics," *Archive for Rational Mechanics and Analysis* **52**, 62–92 (1973). Reprinted along with the preceding.

M. E. GURTIN & W. O. WILLIAMS, "On the first law of thermodynamics," *Archive for Rational Mechanics and Analysis* **42**, 77–92 (1971).

Chapter II

Kinematics

These theorems render the forms of motion... at least approachable in concept.

HELMHOLTZ
On the integrals of the
hydrodynamical equations that
correspond to vortex motion
Journal für die Reine und Angewandte Mathematik **55**, 25–55 (1858)

The great clarity which geometrical investigation lends to the study of the dynamics of solids leads us to expect significant success in hydrodynamics through a study of the kinematics of deformable systems.

ZHUKOVSKI
Кинематика Жидкаго Тѣла (1876)

The theory of these general phenomena of motion in continuous media has a yet unbounded scope of development. Nevertheless, it is necessary to approach them entirely without prejudice....

JAUMANN
Introduction to *Die Grundlagen der Bewegungslehre von einem modernen Standpunkte aus*
Leipzig (1905)

1. Bodies, Placements, Motions

In §I.4 we have agreed that by the term *body* $\mathscr{B}$ we shall mean the closure of an open set in some measure space Ω over which a non-negative measure M, called *mass*, is defined, and that M can be extended to a Borel measure over the σ-algebra of Borel sets in Ω. The elements X of $\mathscr{B}$ are called *body-points*. In continuum mechanics we assume that $\mathscr{B}$ is in fact a homeomorph of the closure of a regular region[1] of $\mathscr{E}$. In §I.7 we have defined a *motion* χ of $\mathscr{B}$, namely, a mapping of the body-points comprised by $\mathscr{B}$ onto a region of a 3-dimensional Euclidean space $\mathscr{E}$ at the time t:

$$\mathbf{x} = \chi(X, t), \qquad X \in \mathscr{B}, \quad t \in \mathscr{R}. \tag{II.1-1}$$

For each t the map $\chi(\cdot, t)$ is a *placement* of the body-points of $\mathscr{B}$; the place $\mathbf{x}$ is *occupied* by the body-point X at the time t in the motion χ. The range of the placement is the *shape* assumed by $\mathscr{B}$ at the time t. It is the closure of an open set in $\mathscr{E}$. When regarding t as the present time, we shall call the shape of $\mathscr{B}$ its *present shape*.

Without fear of confusion we may write $\chi(\mathscr{B}, t)$ for the shape of $\mathscr{B}$ at the time t, thus using the symbol χ in two different though closely related senses: as a mapping of body-points onto places and as a mapping of the bodies they constitute onto regions of space. For the latter sense we used the symbol χ_Ω in §I.7.

While in physical experience bodies are available to us only in some shape or other, the shapes are not to be confused with the bodies themselves. In analytical dynamics (§I.3, Example 1) the bodies are discrete and hence stand in one-to-one correspondence with the numbers 1, 2, ..., n. Nobody ever confuses the sixth body with the number 6, or with the place the sixth body happens to occupy at some time. The number 6 is merely a label

[1] A "regular region" is defined in §IV.9 of O. D. KELLOGG's *Foundations of Potential Theory*, Berlin, Springer-Verlag, 1929, variously reprinted. An example of a regular region is the closure of an open set whose boundary is the union of a finite number of surfaces with continuous normal field, joined together at vertices or regular arcs. A "regular arc" is a continuously differentiable image of a closed interval. The boundary of a regular region may contain a finite number of edges and conical points.

In §I.4 we have explained why a more general concept of body, while presently lacking a proper mathematical expression, would be useful in mechanics. The restriction to images of closures of regular regions is only for want of a better way, so as to allow us to apply the divergence theorem to the shapes of bodies, providing the fields to which we wish to apply it be reasonably smooth. *Cf*. also the special case discussed after Example 3 in §I.3.

Extensions of the divergence theorem, in regard to the fields and the regions, are discussed by KELLOGG, *op. cit*. §§IV.10–IV.11.

attached to the body, and other labels would do just as well. Similarly, in continuum mechanics a body may assume infinitely many different shapes.

We shall refer to the subbodies $\mathscr{P}$ of a given body $\mathscr{B}$ as the *parts* of $\mathscr{B}$, as in §I.2.

In this book we assume that the shapes of a body at various times are bounded regions of space. With some technical detail it is possible to include also bodies filling infinite regions. We shall sometimes describe motions of such bodies, as for example in the case of flow of a fluid filling all of space, but in the mathematical treatment we shall confine attention to some part $\mathscr{P}$ whose shapes in some finite interval of time remain within a bounded region, or we shall carry out a limit process with such parts. Unless the contrary is stated explicitly, the term "body" henceforth will be taken to refer only to such parts.

We have assumed also in §I.7 that χ is differentiable as often as need be with respect to t, and we have defined the *velocity* $\mathbf{v}$, the *acceleration* $\mathbf{a}$, and the nth *velocity* ${}_n\mathbf{v}$, $n \geqq 1$, as the values of the successive time derivatives:

$$\begin{aligned} \mathbf{v} &= \dot{\chi}(X, t), \\ \mathbf{a} &= \ddot{\chi}(X, t), \quad \ldots, \\ {}_n\mathbf{v} &= \overset{(n)}{\chi}(X, t). \end{aligned} \qquad \text{(I.7-7)}_r$$

For most of our analysis it will suffice to assume that two time derivatives exist; for some of it, one. These derivatives, for each fixed X and t, are vectors in $\mathscr{V}$, the translation space of $\mathscr{E}$. The nth velocity field $\overset{(n)}{\chi}$ is a time-dependent vector field over $\mathscr{B}$.

In continuum mechanics, to which we henceforth address ourselves, the motion χ is assumed to be invertible at each t. Specifically, for each t the function $\chi(\cdot, t)$: $\mathscr{B} \to \mathscr{E}$ has an inverse $\chi^{-1}(\cdot, t)$, defined over the shape $\chi(\mathscr{B}, t)$:

$$X = \chi^{-1}(\mathbf{x}, t) \qquad \forall \mathbf{x} \in \chi(\mathscr{B}, t), \quad \forall t \in \mathscr{R}. \qquad \text{(II.1-2)}$$

Moreover, we shall always assume that both χ and χ^{-1} are continuous functions of their arguments. Thus the shape of a neighborhood of a body-point is always an open set in a Euclidean 3-dimensional space, and any open set in the shape of a body is the image of some neighborhood in $\mathscr{B}$ of some body-point that instantaneously occupies a place in that shape.

In §§I.4 and I.5 we have introduced mass $M(\cdot)$ and resultant force $\mathbf{f}(\cdot, \mathscr{B}^e)$, and we have assumed that both of them are Borel measures over $\mathscr{B}$. Since χ is continuous, M and $\mathbf{f}$ can be regarded also as Borel measures over the collection of Borel sets of the shape of $\mathscr{B}$ at the time t.

As explained in Example 2 of §I.3, the operations $\curlyvee$ and $\curlywedge$ in the Boolean lattice of bodies are defined in continuum mechanics as follows:

$$\mathscr{B} \curlyvee \mathscr{C} \equiv \mathscr{B} \cup \mathscr{C}, \tag{II.1-3}$$

$$\mathscr{B} \curlywedge \mathscr{C} \equiv \overline{\mathring{\mathscr{B}} \cap \mathring{\mathscr{C}}}, \tag{I.3-1)$_r$}$$

in which the superimposed ° denotes "interior" and the superimposed bar denotes "closure". Equivalently

$$\begin{aligned} \chi(\mathscr{B} \curlyvee \mathscr{C}) &= \chi(\mathscr{B}) \cup \chi(\mathscr{C}), \\ \chi(\mathscr{B} \curlywedge \mathscr{C}) &= \overline{\chi(\mathring{\mathscr{B}}) \cap \chi(\mathring{\mathscr{C}})}. \end{aligned} \tag{II.1-4}$$

2. Mass-Density

Since in continuum mechanics $\mathscr{B}$ is the closure of an open set, it contains infinitely many distinct body-points X. However, the assignment of mass M is left arbitrary so far and might be discrete, or partially so. Of primary interest in continuum mechanics are masses which are absolutely continuous functions of volume, which is defined as a non-negative Borel measure[1] on the Euclidean space $\mathscr{E}$. To assume that M is absolutely continuous is to assume that if a part assumes a shape having sufficiently small volume, then that part has arbitrarily small mass. Thus, formally, concentrated masses are excluded, and analytical dynamics will not emerge directly as a special case of continuum mechanics (though the two are always related through the connection established at the end of §I.13).

Let $\boldsymbol{\sigma}$ be a placement of $\mathscr{B}$. By the Radón–Nikodym Theorem,[2] the mass of any massy part $\mathscr{P}$ of $\mathscr{B}$ may be expressed as the Lebesgue integral of a non-negative *mass-density* $\rho_{\boldsymbol{\sigma}}$ over the shape of $\mathscr{P}$:

$$M(\mathscr{P}) = \int_{\boldsymbol{\sigma}(\mathscr{P})} \rho_{\boldsymbol{\sigma}} \, dV. \tag{II.2-1}$$

The density $\rho_{\boldsymbol{\sigma}}$ exists and is unique almost everywhere in $\boldsymbol{\sigma}(\mathscr{P})$. Clearly it depends upon $\boldsymbol{\sigma}$.

The existence of a mass-density expresses a relation between the body $\mathscr{B}$ and such shapes as it may assume. At almost every place $\mathbf{x}$ in $\boldsymbol{\sigma}(\mathscr{P})$ the density is the ultimate ratio of mass to volume in the following sense: If $\mathscr{P}_k$

[1] The field of Lebesgue-measurable sets contains certain null sets that are not Borel sets, but these seem not to be of interest in continuum mechanics, so Borel measure is used in most axiomatic studies.

[2] For example, Theorem 6.9 of W. RUDIN's *Real and Complex Analysis*, New York, McGraw-Hill, 1966.

is a suitably chosen sequence of parts such that $\mathscr{P}_{k+1} \subset \mathscr{P}_k$, that all the $\mathscr{P}_k$ have but the single body-point $\boldsymbol{\sigma}^{-1}(\mathbf{x})$ in common, and that $V(\boldsymbol{\sigma}(\mathscr{P}_k)) \to 0$ as $k \to \infty$, then

$$\rho_{\boldsymbol{\sigma}}(\mathbf{x}) = \lim_{k \to \infty} \frac{M(\mathscr{P}_k)}{V(\boldsymbol{\sigma}(\mathscr{P}_k))}. \tag{II.2-2}$$

In all its shapes a part $\mathscr{P}$ has the same mass $M(\mathscr{P})$. We have made this assumption plain by assigning masses directly to the massy parts of $\mathscr{B}$. To each shape of $\mathscr{B}$ we may apply (1). Thus

$$M(\mathscr{P}) = \int_{\boldsymbol{\sigma}_1(\mathscr{P})} \rho_{\boldsymbol{\sigma}_1}\, dV = \int_{\boldsymbol{\sigma}_2(\mathscr{P})} \rho_{\boldsymbol{\sigma}_2}\, dV. \tag{II.2-3}$$

Since both $\boldsymbol{\sigma}_1$ and $\boldsymbol{\sigma}_2$ are homeomorphisms of $\mathscr{B}$ with parts of $\mathscr{E}$, there is a homeomorphism $\boldsymbol{\lambda}$ that maps $\boldsymbol{\sigma}_1(\mathscr{P})$ onto $\boldsymbol{\sigma}_2(\mathscr{P})$. In fact $\boldsymbol{\lambda} = \boldsymbol{\sigma}_2 \circ \boldsymbol{\sigma}_1^{-1}$. We shall assume that $\boldsymbol{\sigma}_2 \circ \boldsymbol{\sigma}_1^{-1}$ is continuously differentiable for all placements $\boldsymbol{\sigma}_2$ and $\boldsymbol{\sigma}_1$. Then not only is $\boldsymbol{\lambda}$ continuously differentiable, but also, as we see by interchanging the roles of $\boldsymbol{\sigma}_1$ and $\boldsymbol{\sigma}_2$, so is $\boldsymbol{\lambda}^{-1}$. Then the chain rule of differential calculus holds; thus $(\nabla\boldsymbol{\lambda})(\nabla\boldsymbol{\lambda}^{-1}) = \mathbf{1}$, so $(\det \nabla\boldsymbol{\lambda}) \times (\det \nabla\boldsymbol{\lambda}^{-1}) = 1$. Therefore, neither $\det \nabla\boldsymbol{\lambda}$ nor $\det \nabla\boldsymbol{\lambda}^{-1}$ can vanish. If we write J for the absolute value of the Jacobian determinant of $\boldsymbol{\lambda}$:

$$J \equiv |\det \nabla\boldsymbol{\lambda}|, \tag{II.2-4}$$

then

$$J > 0. \tag{II.2-5}$$

A theorem of integral calculus[1] yields the relation

$$\int_{\boldsymbol{\sigma}_1(\mathscr{P})} \rho_{\boldsymbol{\sigma}_1}\, dV = \int_{\boldsymbol{\sigma}_1(\mathscr{P})} \rho_{\boldsymbol{\sigma}_2} J\, dV \tag{II.2-6}$$

for each massy part $\mathscr{P}$. Hence follows an equation relating the two densities almost everywhere:

$$\rho_{\boldsymbol{\sigma}_2}(\boldsymbol{\lambda}(\mathbf{X}))J(\mathbf{X}) = \rho_{\boldsymbol{\sigma}_1}(\mathbf{X}), \qquad \mathbf{X} \in \boldsymbol{\sigma}_1(\mathscr{P}). \tag{II.2-7}$$

[1] The formula (6), inferred by a formal or pictorial argument, derives from EULER's researches on hydrodynamics in the middle of the eighteenth century. A clear and simple statement and proof within the theory of Riemann integration is given in Theorems 3-13 and 3-14 of M. SPIVAK, *Calculus on Manifolds*, New York, Benjamin, 1965; for Lebesgue integrals, in Theorem 8.26 of W. RUDIN, *Real and Complex Analysis*, New York, McGraw-Hill. In both cases the integrand is merely assumed integrable, and $\boldsymbol{\lambda}$ is assumed to be a bijective, differentiable map of an open set of $\mathscr{E}_n$ into $\mathscr{E}_n$. For the former theorem, $\boldsymbol{\lambda}$ is assumed continuously differentiable; for the latter, $\boldsymbol{\lambda}^{-1}$ is assumed continuous, and Range $\boldsymbol{\lambda}$ is assumed open and bounded. In both cases J may vanish on a set of measure 0.

Thus the mass-density field over one shape of $\mathscr{B}$ determines the mass-density fields over all others.

In particular, these results hold for the shapes $\mathscr{B}$ assumes when it undergoes a motion.

As in §I.4, integration with respect to mass is defined on the massy parts $\mathscr{P}$ of a body $\mathscr{B}$. In continuum mechanics the assumption that M is an absolutely continuous function of V enables us to replace all integrals so defined by others which are taken over regions of $\mathscr{E}$. Thus

$$\int_{\mathscr{P}} f\, dM = \int_{\chi(\mathscr{P},\, t)} \rho f\, dV. \tag{II.2-8}$$

On the left-hand side, f stands for $f(X, t)$, while on the right-hand side, f stands for $f(\chi^{-1}(\mathbf{x}, t), t)$. Alternatively, we may start with the right-hand side and regard f as standing for $f(\mathbf{x}, t)$, so that on the left-hand side f stands for $f(\chi(X, t), t)$. This abbreviated notation, which is common in continuum mechanics, will be developed further in §II.6.

In this book we shall always assume not only that mass is an absolutely continuous function of volume but even that mass is ultimately bounded by volume: For any placement of $\mathscr{B}$ there is a constant K such that if $V(\mathscr{P})$ is sufficiently small, then

$$M(\mathscr{P}) \leqq KV(\mathscr{P}). \tag{II.2-9}$$

Equivalently, ρ_σ is essentially bounded. Then any additive set function that is bounded with respect to volume is bounded also with respect to mass, and conversely. In passages where the manipulations of differential calculus are brought to bear, we shall presume the still stronger assumption that ρ_σ is a continuously differentiable function of its arguments at all places and times we may choose to consider.

3. Reference Placement. Transplacement

Often it is convenient to select the placement of $\mathscr{B}$ at some one time t in some putative motion, not necessarily the motion χ being studied, and refer everything concerning $\mathscr{B}$ and its motion to that placement. We denote by $\mathbf{X}$ the place given to the body-point X by the particular placement $\boldsymbol{\kappa}$:

$$\mathbf{X} = \boldsymbol{\kappa}(X). \tag{II.3-1}$$

Since $\boldsymbol{\kappa}$ is invertible, by assumption,

$$X = \boldsymbol{\kappa}^{-1}(\mathbf{X}), \tag{II.3-2}$$

and both $\boldsymbol{\kappa}$ and $\boldsymbol{\kappa}^{-1}$ are continuous. Hence the motion (II.1-1) may be written in the form

$$\mathbf{x} = \chi(\boldsymbol{\kappa}^{-1}(\mathbf{X}), t) \equiv \chi_{\boldsymbol{\kappa}}(\mathbf{X}, t). \tag{II.3-3}$$

In the description furnished by this equation, the motion is expressed as a mapping $\chi_{\boldsymbol{\kappa}}$ of the *reference shape* $\boldsymbol{\kappa}_{\Omega}(\mathscr{B})$ onto the actual shapes $\chi(\mathscr{B}, t)$ as t progresses. Thus the motion is visualized as mapping parts of space onto parts of space. A reference shape is introduced so as to allow us to employ immediately the apparatus of Euclidean geometry. The mapping $\chi_{\boldsymbol{\kappa}}$ is the *transplacement* of the body-points of $\mathscr{B}$ from their reference places $\mathbf{X}$ into their actual places $\mathbf{x}$ at the time t.

We shall call the mapping $\boldsymbol{\kappa}$ itself the *reference placement* of the body-points of $\mathscr{B}$. With no fear of confusion we may drop the subscript Ω and write $\boldsymbol{\kappa}(\mathscr{B})$ for the reference shape of $\mathscr{B}$, just as we have already written $\chi(\mathscr{B}, t)$ for $\chi_{\Omega}(\mathscr{B}, t)$. As in ordinary language a body is "deformed" when its shape changes, we shall say that $\chi_{\boldsymbol{\kappa}}$ *deforms* the reference shape $\boldsymbol{\kappa}(\mathscr{B})$ into the actual shape $\chi(\mathscr{B}, t)$. While "strain" is commonly used to denote deformation or some aspect of it, in this book we do not introduce the term in any precise meaning but shall use the word descriptively from time to time.

The choice of reference placement, like the choice of a co-ordinate system, is arbitrary. The reference placement, which may be any smooth mapping of $\mathscr{B}$ into $\mathscr{E}$, need not be the value of the motion $\chi(\mathscr{B}, \cdot)$ at some time t.

For each different $\boldsymbol{\kappa}$, a different function $\chi_{\boldsymbol{\kappa}}$ for the same motion χ is defined by (3). Thus one motion of the body is represented by infinitely many different mappings of parts of space in the course of time, one for each choice of $\boldsymbol{\kappa}$. For some choice of $\boldsymbol{\kappa}$ we may get a particularly simple description, just as in geometry one choice of co-ordinates may lead to a simple equation for a particular figure while another may not, but the reference placement itself has nothing to do with such motions as it may be used to describe, just as the co-ordinate system has nothing to do with geometrical figures themselves. A reference placement is introduced so as to allow the use of mathematical apparatus familiar in other contexts. Again there is an analogy to co-ordinate geometry, where co-ordinates are introduced, not because they are natural or germane to geometry, but because they allow the familiar apparatus of algebra to be applied at once.

4. Descriptions of Motion

There are four methods of describing the motion of a body: the material, the referential, the spatial, and the relative. Because of our hypotheses of smoothness, all are equivalent.

In the *material description* we deal directly with the body-points X. This description corresponds to the only one used in analytical dynamics, where we always speak of the first, second, ..., n^{th} masses. To be precise, there we should say, "the mass-point X_q whose mass is M_q", but commonly this expression is abbreviated to "the mass q" or "the body M_q", *etc.* In continuum mechanics every body $\mathscr{B}$ comprises infinitely many body-points X. The material description employs as independent variables X and t, the body-point and the time. While the material description is the most natural in concept, it was not mentioned in continuum mechanics until a few decades ago and is still used little. For some time the term "material description" was used to denote another and older description often confused with it, the description to which we turn next.

The *referential description* employs some assigned reference placement $\boldsymbol{\kappa}$. Thus it describes the motion χ by means of the transplacement $\chi_{\boldsymbol{\kappa}}$. We must always bear in mind that the choice of $\boldsymbol{\kappa}$ is ours, that $\boldsymbol{\kappa}(\mathscr{B})$ is merely some shape that $\mathscr{B}$ has occupied or might occupy, and that it must be possible always to state hypotheses and equations in forms valid for any choice of $\boldsymbol{\kappa}$, although for one choice of $\boldsymbol{\kappa}$ the corresponding transplacement $\chi_{\boldsymbol{\kappa}}$ may show the important properties of some particular motion far more easily than the transplacements corresponding to other choices of $\boldsymbol{\kappa}$. Any motion of a body has infinitely many different referential descriptions, equally valid.

For the purposes of this book, and for most purposes in mechanics, the material description and the referential description may be confused, at least locally, as they long have been. To see that they are in principle different, and that the referential description may not always suffice, we need only consider the two-body problem of analytical dynamics. No one would find it convenient to use as labels for the first and second mass-points the places they occupied at some particular time. If that time were one at which the two mass-points collided, such names would not distinguish those two bodies. Since analytical dynamics always envisions the chance that collisions may occur, the distinction between material and referential descriptions is not a matter of purism or mere abstraction, and in fact nobody has ever used the referential description in analysis of the motions of discrete systems. The referential description is useful only for systems in which it is convenient to use a place as a name for an element of an abstract manifold. Such naming is indeed convenient in continuum mechanics.

In the mid-eighteenth century EULER introduced the description that hydrodynamicists still call "Lagrangean". This is a particular referential description, in which the Cartesian co-ordinates of the position $\mathbf{X}$ of the body-point X at the time $t = 0$ are used as a label for that body-point. It was recognized that such labelling by initial co-ordinates was arbitrary; writers on the foundations of hydrodynamics have often mentioned that the essential results must be and are independent of the choice of the initial time, and some have remarked that the parameters of any triple system of surfaces moving with the material would do just as well. The referential description, taking $\mathbf{X}$ and t as independent variables, includes all these possibilities. Some form of it is always used in classical elasticity theory, and the best studies of the foundations of classical hydrodynamics from EULER's day to the present have employed it almost without fail. It is the description commonly used in modern works on continuum mechanics, and we shall use it in this book.

In view of (II.3-2), any function $F(X, t)$ may be replaced by a function $F_{\boldsymbol{\kappa}}(\mathbf{X}, t)$ that has the same value at corresponding arguments X and $\mathbf{X}$, for given $\boldsymbol{\kappa}$:

$$F(X, t) = F(\boldsymbol{\kappa}^{-1}(\mathbf{X}), t) \equiv F_{\boldsymbol{\kappa}}(\mathbf{X}, t). \qquad \text{(II.4-1)}$$

In (II.3-3) we have already encountered a special case. Moreover,

$$\partial_t F = \partial_t F_{\boldsymbol{\kappa}}. \qquad \text{(II.4-2)}$$

We shall employ a superimposed dot to denote also time derivatives of functions of the *referential variables* $\mathbf{X}$ and t. Thus by differentiating (II.3-3) and using the definitions (I.7-7) and (II.3-2) we see that for each choice of $\boldsymbol{\kappa}$

$$\dot{\chi} = \dot{\chi}_{\boldsymbol{\kappa}}, \qquad \ddot{\chi} = \ddot{\chi}_{\boldsymbol{\kappa}}, \qquad \dots, \qquad \overset{(n)}{\chi} = \overset{(n)}{\chi}_{\boldsymbol{\kappa}}, \qquad \text{(II.4-3)}$$

the arguments of the functions on the left-hand sides being X and t, those of the functions on the right-hand sides being $\mathbf{X}$ and t.

In the *spatial description*, attention is focused on the present placement of the body. This description, which was introduced by DANIEL BERNOULLI and D'ALEMBERT, is called "Eulerian" by the hydrodynamicists. The place $\mathbf{x}$ and the time t are taken as independent variables. In view of (II.1-1), any function $F(X, t)$ may be replaced by a function of the *spatial variables*, $\mathbf{x}$ and t, that has the same value at corresponding arguments X and $\mathbf{x}$:

$$F(X, t) = F[\chi^{-1}(\mathbf{x}, t), t] \equiv f(\mathbf{x}, t). \qquad \text{(II.4-4)}$$

The function f, moreover, is unique. Thus, while there are infinitely many referential descriptions of a given motion, there is only one spatial description, just as there is only one material description. With the spatial description, we watch what is occurring in a fixed region of space as time goes on. This description seems perfectly suited to studies of fluids, where often a rapidly deforming mass comes from no one knows where and goes no one knows whither, so that we may prefer to consider what happens here and now before our eyes. However convenient kinematically, the spatial description is awkward for questions of principle in mechanics, since in fact the laws of dynamics refer to what is suffered by the body, not by the region of space the body momentarily occupies. Some relations which are obvious and easy to derive in the material or referential descriptions seem to require some contorted reasoning if approached by the strictly spatial standpoint sometimes adopted by specialists in applied hydrodynamics.

According to (4), the value of any function of the body-points of $\mathscr{B}$ at the time t is given also by a field defined over the actual shape $\chi(\mathscr{B}, t)$. In this way, for example, we obtain from (I.7-7) the *velocity field* $\dot{\mathbf{x}}$, the *acceleration field* $\ddot{\mathbf{x}}$, and the nth *velocity field* $\overset{(n)}{\mathbf{x}}$:

$$\mathbf{v} = \dot{\mathbf{x}}(\mathbf{x}, t), \qquad \mathbf{a} = \ddot{\mathbf{x}}(\mathbf{x}, t), \qquad \ldots, \qquad {}_n\mathbf{v} = \overset{(n)}{\mathbf{x}}(\mathbf{x}, t). \tag{II.4-5}$$

The fields $\overset{(n)}{\mathbf{x}}$ and $\overset{(n)}{\chi}$ have the common value ${}_n\mathbf{v}$ at arguments related through the motion:

$$\begin{aligned} {}_n\mathbf{v} &= \overset{(n)}{\mathbf{x}}(\chi(X, t), t) = \overset{(n)}{\chi}(X, t), \qquad X \in \mathscr{B} \\ &= \overset{(n)}{\mathbf{x}}(\mathbf{x}, t) = \overset{(n)}{\chi}(\chi^{-1}(\mathbf{x}, t), t), \qquad \mathbf{x} \in \chi_\Omega(\mathscr{B}, t). \end{aligned} \tag{II.4-6}$$

In §I.11 we have calculated the frame-indifferent field $\boldsymbol{\alpha}_{\mathfrak{f}}$ that reduces in the inertial framing $\mathfrak{f}$ to the acceleration. While as given by the right-hand side of (I.11-3) this field is defined over the body $\mathscr{B}$, of course we may convert it into a field over the actual shape $\chi^*(\mathscr{B}, t)$ in the general framing $\mathfrak{f}^*$. Calling that field $\mathbf{a}_{\mathfrak{f}}$, we calculate it as follows from (I.11-3):

$$\mathbf{a}_{\mathfrak{f}} = \ddot{\mathbf{x}}^* - \ddot{\mathbf{x}}_0^* - 2\mathbf{A}(\dot{\mathbf{x}}^* - \dot{\mathbf{x}}_0^*) - (\dot{\mathbf{A}} - \mathbf{A}^2)(\mathbf{x}^* - \mathbf{x}_0^*); \tag{II.4-7}$$

here $\mathbf{x}^*$, $\dot{\mathbf{x}}^*$, and $\ddot{\mathbf{x}}^*$ are the spatial fields of place, velocity, and acceleration over the actual shape $\chi^*(\mathscr{B}, t)$ in the framing $\mathfrak{f}^*$, and $\mathbf{A}$ is the spin of the inertial framing $\mathfrak{f}$ with respect to $\mathfrak{f}^*$.

The fourth common description, the relative one, we shall develop in §II.8.

5. Local Deformation

The gradient of the transplacement χ_κ is called the *local deformation*[1] $\mathbf{F}$:

$$\mathbf{F} \equiv \mathbf{F}_\kappa(\mathbf{X}, t) \equiv \nabla\chi_\kappa(\mathbf{X}, t). \tag{II.5-1}$$

It is the linear approximation to the mapping (II.3-3). More precisely, we should call it the gradient of the transplacement of $\boldsymbol{\kappa}$ into χ, but when, as is usual, a single reference placement $\boldsymbol{\kappa}$ is laid down once and for all, no confusion should result from failure to remind ourselves that the very concepts of transplacement and local deformation presume use of a reference placement. If, as we may, we select independently co-ordinates X^α and x^m in the reference shape and the actual shape, respectively, so that the motion (II.3-3) assumes the co-ordinate form

$$x^m = \chi_\kappa^m(X^1, X^2, X^3, t), \qquad m = 1, 2, 3, \tag{II.5-2}$$

then the components of $\mathbf{F}$ are simply the nine partial derivatives of the functions χ_κ^m with respect to the X^α, *viz*

$$F_\alpha^m = x^m{}_{,\alpha} = \partial_{X^\alpha}\chi_\kappa^m(X^1, X^2, X^3, t), \qquad m = 1, 2, 3, \quad \alpha = 1, 2, 3. \tag{II.5-3}$$

Throughout this book, such co-ordinates as are introduced will be general ones, unless the contrary is stated. Only rarely shall we invoke co-ordinate representations so as to prove results of a general character, but there would be no loss in logical strictness were we to write out everything, as the older authors on continuum mechanics did, in rectangular Cartesian co-ordinates. In practice, abstract notations are easier to understand and more efficient to manipulate, once they be grown familiar, and proofs using them are easier to follow, except in special cases where a special choice of components is virtually dictated by the problem itself. In such cases, as we shall see later, those special components are often curvilinear, sometimes even anholonomic. That is, while any reasonably simple problem refers to certain particular directions and hence suggests use of a particular basis, that basis need not be the natural basis of any co-ordinate system.[2] Thus it is to our advantage to express all the principles of our science directly

[1] The reader ought not confuse $\mathbf{F}$ or $\mathbf{F}_\kappa$ with the torque $\mathbf{F}_{\mathbf{x}_0}$ of a system of forces with respect to $\mathbf{x}_0$.

[2] Components of a vector or tensor field may be defined by any field of bases. When the bases are not normal to any families of surfaces, they are called "anholonomic". Anholonomic components, while they serve perfectly well to define or specify vector and tensor fields, are not components with respect to any co-ordinate system.

in terms of concepts, without the complicating intermediary of co-ordinate systems.

In §II.2 we have introduced the mass-density $\rho_{\boldsymbol{\sigma}}$ that corresponds to the placement $\boldsymbol{\sigma}(\mathscr{B})$. Henceforth we shall write simply ρ for ρ_{χ}; thus ρ is the mass density field over the actual shape $\boldsymbol{\chi}(\mathscr{B}, t)$. Choosing for $\boldsymbol{\sigma}_1$ the reference placement $\boldsymbol{\kappa}$, from (II.2-7) we obtain

$$\rho J = \rho_{\boldsymbol{\kappa}}, \tag{II.5-4}$$

on the understanding that when the argument of $\rho_{\boldsymbol{\kappa}}$ is $\mathbf{X}$, the arguments of ρ and J are $\boldsymbol{\chi}_{\boldsymbol{\kappa}}(\mathbf{X}, t)$ and t, and that

$$J \equiv |\det \mathbf{F}|. \tag{II.5-5}$$

Henceforth J will be used in the sense just defined rather than in the more general one expressed by (II.2-4). The relation (4) is EULER's *referential equation for the mass-density.*

The reader will recall that

$$J > 0. \tag*{(II.2-5)$_\mathrm{r}$}$$

While (4) is often called "the Lagrangean equation of continuity", that name is misleading, since if the deformation is smooth enough, (4) holds, but if the deformation is not differentiable, let alone not continuous, J cannot be defined at all, so (4) cannot even be stated, let alone used. Obviously (4) is neither more nor less than a formula that delivers the actual density ρ, once the local deformation $\mathbf{F}$ and the reference density $\rho_{\boldsymbol{\kappa}}$ be known.

In the older literature (4) is sometimes related to an "axiom of impenetrability", according to which two distinct body-points never come to occupy the same place, and thus no body enters into the shape of another body at the same time (*cf.* §I.7). In truth, however, a formal condition such as (4) does not express that axiom but rather presumes that some such axiom has been laid down already.

Exercise II.5.1 (EULER). By using the formula for differentiating a determinant, prove that

$$\operatorname{div} \dot{\mathbf{x}} = \dot{J}/J; \tag{II.5-6}$$

here the superimposed dot on the right-hand side denotes the time derivative, and $\operatorname{div} \dot{\mathbf{x}}$ is the divergence of the velocity field (II.4-5)$_1$.

In (6), as in (4), the field on the right-hand side is a referential one, while that on the left-hand side is a spatial one. Both results assert that fields of these two kinds have at each time t the same values at the places $\mathbf{X}$ and $\mathbf{x}$, respectively, selected so as to correspond to each other through the referential description (II.3-3) of the motion. It leads to less awkward statements if in such cases we simply presume that any referential field is replaced by the corresponding spatial one. For example, if we differentiate (4) with respect to time and then use (6), we obtain D'ALEMBERT and EULER's *spatial equation for the density*:

$$\dot{\rho} + \rho \operatorname{div} \dot{\mathbf{x}} = 0, \tag{II.5-7}$$

in which, following the convention just stated, we have written ρ and $\dot{\rho}$ for the spatial fields that are defined by ρ and $\dot{\rho}$ as fields over $\boldsymbol{\kappa}(\mathscr{B})$ at the time t. This equation has exactly the same meaning as (4), which, conversely, may be gotten from it by use of (6) followed by integration.

Exercise II.5.2 (LAGRANGE). Regarding (7) as a first-order differential equation for ρ in the spatial description, integrate it by the method of characteristics so as to obtain (4).

Exercise II.5.3 (D'ALEMBERT, EULER). A motion of $\mathscr{B}$ is called *isochoric* if the volume $V(\chi(\mathscr{P}, t))$ of the shape of each part $\mathscr{P}$ of $\mathscr{B}$ remains constant in time. Show that any one of the following three conditions is necessary and sufficient for isochoric motion:

1. $\operatorname{div} \dot{\mathbf{x}} = 0.$ $\qquad$ (II.5-8)$_1$
2. There is a reference placement $\boldsymbol{\kappa}$ such that

$$\rho = \rho_{\boldsymbol{\kappa}}. \tag{II.5-8$_2$}$$

3. There is a reference placement $\boldsymbol{\kappa}$ such that

$$J = 1. \tag{II.5-8$_3$}$$

Exercise II.5.4 (D'ALEMBERT). If the velocity field $\dot{\mathbf{x}}$ is everywhere parallel to a single plane, show that the general solution of $(8)_1$ is given in terms of a single-valued *stream function* q by

$$\dot{\mathbf{x}} = (\nabla q)^{\perp}, \tag{II.5-9}$$

∇ denoting the gradient operator in the plane and $\perp$ denoting rotation counter-clockwise through a right angle about the normal to the plane. Show that $\dot{\mathbf{x}}(\mathbf{x}, t)$ is tangent to the curve $q(\cdot, t) = \text{const.}$ through $\mathbf{x}$ at each fixed t.

Exercise II.5.5. Given a vector field $\mathbf{v}$ defined on $\chi(\mathscr{B}, t)$, let it be desired to find a vector field $\mathbf{v}_{\boldsymbol{\kappa}}$ such that

$$\int_{\mathscr{S}} \mathbf{v}_{\boldsymbol{\kappa}} \cdot \mathbf{n}_{\boldsymbol{\kappa}} \, dA = \int_{\chi_{\boldsymbol{\kappa}}(\mathscr{S}, t)} \mathbf{v} \cdot \mathbf{n} \, dA \tag{II.5-10}$$

for any surface $\mathscr{S}$ in $\boldsymbol{\kappa}(\mathscr{B})$. Show that

$$\mathbf{v}_{\boldsymbol{\kappa}} = J\mathbf{F}^{-1}\mathbf{v}. \tag{II.5-11}$$

When a motion is described in terms of a velocity field, it is often called a *flow*.

A function of place alone is called a *steady* function in the context of the spatial description. For example, a velocity field $\dot{\mathbf{x}}$ which is independent of t is called a *steady flow*. Other flows are called *unsteady*. A steady flow may or may not have a steady density.

A motion that is steady in one framing generally fails to be steady in another. The property of steadiness is not even a Galilean invariant. It is not a simple matter to determine whether a given motion that is not steady in the framing in which it is defined be steady in some other framing.

6. Material Time Rates and Gradients in the Spatial Description. Material Surfaces. Kinematic Boundaries

In continuum mechanics the need to distinguish a vast number of quantities often deprives us of the clarity gained by using for a function a symbol different from that for its value, as logically we ought to do. If two functions of different variables have the same value and if *both* are denoted by that value, when we come to effect some functional operation it is not clear which function is intended. The distinction, which of course is essential, is traditionally made by introducing different symbols for the differential operators. Henceforth

$$\dot{f} \qquad \text{and} \qquad \operatorname{Grad} f$$

shall denote the partial time derivative and the gradient of the function $G(\mathbf{X}, t)$ such that

$$f = G(\mathbf{X}, t), \tag{II.6-1}$$

while

$$f' \qquad \text{and} \qquad \operatorname{grad} f$$

shall denote the partial time derivative and the gradient of the function $g(\mathbf{x}, t)$ that has the same value as G, namely,

$$f = g(\mathbf{x}, t) = G(\chi_{\kappa}^{-1}(\mathbf{x}, t), t), \tag{II.6-2}$$

by (II.3-3). If we apply the chain rule to the equation $G(\mathbf{X}, t) = g(\chi_{\kappa}(\mathbf{X}, t), t)$ and then denote by f *both* functions G and g, we obtain the classical formulae of EULER:

$$\begin{aligned} \dot{f} &= f' + (\operatorname{grad} f) \cdot \dot{\mathbf{x}}, \\ \dot{\mathbf{f}} &= \mathbf{f}' + (\operatorname{grad} \mathbf{f})\dot{\mathbf{x}}, \end{aligned} \tag{II.6-3}$$

where f is scalar-valued and $\mathbf{f}$ is vector-valued, and an analogous rule holds for tensor-valued functions. In particular, the acceleration field $\ddot{\mathbf{x}}$ is calculated from the velocity field $\dot{\mathbf{x}}$ by the D'ALEMBERT–EULER formula

$$\ddot{\mathbf{x}} = \dot{\mathbf{x}}' + (\text{grad}\ \dot{\mathbf{x}})\dot{\mathbf{x}}. \tag{II.6-4}$$

The dot operator as defined by (3) is sometimes called the *material derivative*.[1] We have already agreed to use the dot to denote the time derivative in the material and referential descriptions, and the definition (3) has been framed so as to render the two usages consistent with each other.

Likewise,

$$\text{Grad}\, f = \mathbf{F}^{\mathrm{T}}\ \text{grad}\, f. \tag{II.6-5}$$

The notations div and Div shall stand for the divergence formed from grad and Grad, respectively.

We have already introduced a special case of these conventions in (II.5-6) and (II.5-7). For example, by $(3)_1$ the latter equation may be written explicitly in the forms

$$\rho' + (\text{grad}\ \rho)\cdot\dot{\mathbf{x}} + \rho\ \text{div}\ \dot{\mathbf{x}} = 0, \qquad \rho' + \text{div}(\rho\dot{\mathbf{x}}) = 0. \tag{II.6-6}$$

For a motion with steady density $(6)_2$ reduces to $\text{div}(\rho\dot{\mathbf{x}}) = 0$.

In §II.2 we have shown how to convert integration with respect to mass on $\mathscr{B}$ into integration with respect to volume on $\mathscr{E}$. If $f(X, t)$ is continuously differentiable with respect to t, the theorem on differentiation of an integral with respect to a parameter assures us that

$$\frac{d}{dt}\int_{\mathscr{P}} f\, dM = \int_{\mathscr{P}} \partial_t f\, dM. \tag{II.6-7}$$

We may now use (II.2-8) to convert the right-hand side into an integral over $\chi(\mathscr{P}, t)$. According to the convention of notation just established the function of $\mathbf{x}$ and t whose value is $\partial_t f(X, t)$ at $\chi^{-1}(\mathbf{x}, t)$ is denoted by $\dot{f}$. Thus

$$\frac{d}{dt}\int_{\mathscr{P}} f\, dM = \int_{\chi(\mathscr{P}, t)} \rho\dot{f}\, dV. \tag{II.6-8}$$

More generally, if Ψ denotes a tensor field of any order,

$$\frac{d}{dt}\int_{\mathscr{P}} \Psi\, dM = \frac{d}{dt}\int_{\chi(\mathscr{P}, t)} \rho\Psi\, dV = \int_{\chi(\mathscr{P}, t)} \rho\dot{\Psi}\, dV, \tag{II.6-9}$$

[1] The material derivative is only one of many time rates that may be calculated on the basis of a given time-dependent field such as a velocity field. Others are introduced below by (II.13-4) and (II.13-7). The problem is discussed from a general point of view by H. BOLDER, "Deformation of tensor fields described by time-dependent mappings," *Archive for Rational Mechanics and Analysis* **35**, 321–341 (1969).

and $\dot{\Psi}$ is to be calculated by an appropriate rule of the type (3). (The central expression, which involves an undefined operation d/dt, is to be regarded only as a suggestive way of writing the left-hand expression.) The commutation formula (9) is used so often in continuum mechanics that it is taken for granted without special reference. It expresses the time-rate of change of the integral of Ψ over a body $\mathscr{B}$ as that body moves through space, in terms of an integral over the present shape $\chi(\mathscr{B}, t)$ of $\mathscr{B}$.

Exercise II.6.1. By simple rearrangement of (9), supplemented by use of (II.5-7), prove the *Reynolds Transport Theorem*: For a given part $\mathscr{P}$ of $\mathscr{B}$,

$$\left(\int_{\chi(\mathscr{P}, t)} \mathfrak{A}\, dV\right)^{\cdot} = \left(\int_{\chi(\mathscr{P}, t)} \mathfrak{A}\, dV\right)' + \int_{\partial\chi(\mathscr{P}, t)} \mathfrak{A} \otimes \dot{\mathbf{x}}\mathbf{n}\, dA, \qquad \text{(II.6-10)}$$

the notations being defined as follows:

$$\left(\int_{\chi(\mathscr{P}, t)} \mathfrak{A}\, dV\right)^{\cdot} \equiv \frac{d}{dt}\int_{\mathscr{P}} \frac{\mathfrak{A}}{\rho(\chi(X, t), t)}\, dM,$$
$$\left(\int_{\chi(\mathscr{P}, t)} \mathfrak{A}\, dV\right)' \equiv \int_{\chi(\mathscr{P}, t)} \mathfrak{A}'\, dV. \qquad \text{(II.6-11)}$$

Interpret the result and apply it so as to calculate for $\mathscr{P}$ in χ the rates of change $\dot{\mathbf{m}}$ and $\dot{\mathbf{M}}_{\mathbf{x}_0}$:

$$\dot{\mathbf{m}} = \mathbf{m}' + \int_{\partial\chi(\mathscr{P}, t)} \rho\dot{\mathbf{x}} \otimes \dot{\mathbf{x}}\mathbf{n}\, dA,$$
$$\dot{\mathbf{M}}_{\mathbf{x}_0} = \mathbf{M}'_{\mathbf{x}_0} + \int_{\partial\chi(\mathscr{P}, t)} (\mathbf{x} - \mathbf{x}_0) \wedge \rho\dot{\mathbf{x}} \otimes \dot{\mathbf{x}}\mathbf{n}\, dA. \qquad \text{(II.6-12)}$$

A stationary surface $\mathscr{S}_{\boldsymbol{\kappa}}$ in the reference shape $\boldsymbol{\kappa}(\mathscr{B})$ is described by an equation of the form $f(\mathbf{X}) = 0$, and hence

$$\dot{f} = 0. \qquad \text{(II.6-13)}$$

Conversely, if (13) is satisfied by a function $f(\mathbf{X}, t)$, then in fact the surface $f = 0$ is a stationary surface in the reference shape, provided of course that $\mathbf{X} \in \boldsymbol{\kappa}(\mathscr{B})$. At the time t, the body-points that make up $\mathscr{S}_{\boldsymbol{\kappa}}$ constitute a certain surface $\mathscr{S}$ in the actual shape assumed by $\mathscr{B}$ in its motion. These surfaces are the successive forms of a single *material surface*. In accord with the convention we have established, we write f also for the function of $\mathbf{x}$ and t whose value at $\chi_{\boldsymbol{\kappa}}(\mathbf{X}, t)$ is $f(\mathbf{X})$, and so in order for the locus $f = 0$ to represent a material surface we have the necessary and sufficient condition

(13), where now the dot operation is defined by $(3)_1$. Thus in the spatial description this requirement becomes EULER's condition:

$$f' + (\operatorname{grad} f) \cdot \dot{\mathbf{x}} = 0. \qquad \text{(II.6-14)}$$

If $\mathbf{n}$ is the oriented unit normal to the surface $f = 0$, where of course f now stands for the function such that $f(\mathbf{x}, t) = 0$ is the locus of $\mathscr{S}$, then (14) may be written alternatively in the form

$$S_n = \mathbf{n} \cdot \dot{\mathbf{x}}, \qquad \text{(II.6-15)}$$

provided S_n, which is called the *speed of displacement* of $\mathscr{S}$, be the speed at which that surface advances in the direction normal to itself in space:

$$S_n = \frac{-f'}{|\operatorname{grad} f|}. \qquad \text{(II.6-16)}$$

EULER's condition (14) thus asserts that the speed of displacement of $\mathscr{S}$ at $(\mathbf{x}, t)$ is just the same as the speed at which the body-point now occupying $(\mathbf{x}, t)$ is moving in the direction normal to $\mathscr{S}$.

Exercise II.6.2. Let a surface $\mathscr{S}$ have parametric representation $\mathbf{x} = \mathbf{g}(\mathbf{A}, t)$, where the parameter $\mathbf{A}$ is a 2-dimensional vector. Regarding $\mathbf{A}$ as permanently denoting a particular point on $\mathscr{S}$ as $\mathscr{S}$ moves, calculate its velocity $\mathbf{u}$ and show that $\mathbf{n} \cdot \mathbf{u} = S_n$. Show likewise that if $\mathscr{S}$ is represented by some other spatial equation, say $h(\mathbf{x}, t) = 0$, the same field S_n is obtained in this way. Thus justify the name "speed of displacement".

Exercise II.6.3 (LAGRANGE). Using the spatial description only, regard (14) as a partial differential equation for f, integrate it by the method of characteristics, and interpret the result.

A *kinematic boundary* is a surface which separates permanently two parts of $\mathscr{B}$, one of them being possibly the null body. Thus a kinematic boundary is a material surface, and conversely. The special term "boundary" is introduced so as to distinguish particular material surfaces, usually assigned in advance like a wall or at least given some special role such as a surface separating two parts having different properties. The simplest case is a *stationary wall*, a surface $f(\mathbf{x}) = \text{const}$. In order for such surface to be material and hence a possible kinematic boundary for a given motion of $\mathscr{B}$, by (15) we have the following necessary and sufficient condition relating the unit normal $\mathbf{n}$ to the velocity:

$$\mathbf{n} \cdot \dot{\mathbf{x}} = 0. \qquad \text{(II.6-17)}$$

That is, *the velocity field is tangent to the wall*, as is obvious.

In some cases a stronger kinematic condition is imposed, that of *adherence*. The body is then constrained to move with the kinematic boundary. If the places on the wall have an assigned velocity $\mathbf{v}$, then

$$\dot{\mathbf{x}} = \mathbf{v}. \tag{II.6-18}$$

In the case of a stationary wall this condition becomes

$$\dot{\mathbf{x}} = \mathbf{0}. \tag{II.6-19}$$

Exercise II.6.4. Let the surface $\mathscr{S}$ whose equation is $g(\mathbf{x}, t) = 0$ in $\chi(\mathscr{B}, t)$ be the image of the surface $\mathscr{S}_\kappa$ whose equation is $G(\mathbf{X}, t) = 0$ in the reference shape $\boldsymbol{\kappa}(\mathscr{B})$. (Note that $\mathscr{S}_\kappa$, in contradistinction with the material surfaces discussed above, generally moves with respect to $\boldsymbol{\kappa}(\mathscr{B})$.) With the conventions of notation set at the beginning of this section, show that the oriented unit normals $\mathbf{n}_\kappa$ and $\mathbf{n}$ to these two surfaces are related by

$$\mathbf{n}_\kappa = \frac{|\operatorname{grad} f|}{|\operatorname{Grad} f|}\mathbf{F}^{\mathsf{T}}\mathbf{n}; \tag{II.6-20}$$

that the speed of advance S_κ of the surface $\mathscr{S}_\kappa$ in the direction normal to itself in $\boldsymbol{\kappa}(\mathscr{B})$ is given by

$$S_\kappa = -\frac{\dot{f}}{|\operatorname{Grad} f|}; \tag{II.6-21}$$

and that

$$S_\kappa = \frac{|\operatorname{grad} f|}{|\operatorname{Grad} f|}(S_n - \mathbf{n}\cdot\dot{\mathbf{x}}). \tag{II.6-22}$$

The speed S_κ given by (21) is called the *speed of propagation* of the surface $\mathscr{S}_\kappa$ in $\boldsymbol{\kappa}(\mathscr{B})$. It is the normal speed of advance of $\mathscr{S}_\kappa$ in $\boldsymbol{\kappa}(\mathscr{B})$. Its reciprocal, S_κ^{-1}, is the *slowness* of $\mathscr{S}_\kappa$, and the vector $S_\kappa^{-1}\mathbf{n}_\kappa$ is the *slowness vector* of that surface.

When, as we may, we take the actual shape as being also $\boldsymbol{\kappa}(\mathscr{B})$, the corresponding speed of propagation is denoted by S and called the *intrinsic speed of propagation* of $\mathscr{S}$. At $(\mathbf{x}, t)$ it is the speed at which the surface is advancing in the direction normal to itself and relative to the velocity of the body-point instantaneously situate upon it. The *intrinsic slowness vector* of $\mathscr{S}$ is $S^{-1}\mathbf{n}$.

Obviously, the intrinsic speed of propagation of $\mathscr{S}$ is related as follows to the speed of displacement of $\mathscr{S}$:

$$S = S_n - \mathbf{n}\cdot\dot{\mathbf{x}}, \tag{II.6-23}$$

and of course this formula is a special case of (22). Finally, comparison with (22) shows that

$$S = \frac{|\operatorname{Grad} f|}{|\operatorname{grad} f|}S_\kappa. \tag{II.6-24}$$

7. Change of Reference Placement

Let the same motion (II.1-1) be described alternatively by transplacements χ_{κ_1} and χ_{κ_2} with respect to two different reference placements, $\boldsymbol{\kappa}_1$ and $\boldsymbol{\kappa}_2$:

$$\begin{aligned} \chi_{\kappa_1}&: \boldsymbol{\kappa}_1(\mathscr{B}) \to \chi(\mathscr{B}, t), \\ \chi_{\kappa_2}&: \boldsymbol{\kappa}_2(\mathscr{B}) \to \chi(\mathscr{B}, t). \end{aligned} \tag{II.7-1}$$

The local deformations $\mathbf{F}_1$ and $\mathbf{F}_2$ at $(\mathbf{X}, t)$ are of course generally different. Let $\mathbf{X}_1$ and $\mathbf{X}_2$ denote the places occupied by the body-point X in $\boldsymbol{\kappa}_1$ and $\boldsymbol{\kappa}_2$:

$$\mathbf{X}_1 = \boldsymbol{\kappa}_1(X), \qquad \mathbf{X}_2 = \boldsymbol{\kappa}_2(X). \tag{II.7-2}$$

Thus

$$\mathbf{X}_2 = \boldsymbol{\kappa}_2 \circ \boldsymbol{\kappa}_1^{-1}(\mathbf{X}_1) \equiv \boldsymbol{\lambda}(\mathbf{X}_1), \tag{II.7-3}$$

say. The transplacement from $\boldsymbol{\kappa}_1$ to χ can be effected in two ways: either straight off by use of χ_{κ_1} itself, or by using $\boldsymbol{\lambda}$ to get to $\boldsymbol{\kappa}_2$ and then using χ_{κ_2} to get to χ. Thus

$$\chi_{\kappa_1} = \chi_{\kappa_2} \circ \boldsymbol{\lambda}. \tag{II.7-4}$$

Because this relation holds among the three mappings, we see that their linear approximations, the gradients, are related in the same way:

$$\mathbf{F}_{\kappa_1} = \mathbf{F}_{\kappa_2}\mathbf{P}, \qquad \mathbf{P} \equiv \nabla\boldsymbol{\lambda}. \tag{II.7-5}$$

Of course, the relation (5) may be written as a "chain rule" of differential calculus:

$$\partial_{X^\alpha} \chi^m_{\kappa_1} = (\partial_{X^A} \chi^m_{\kappa_2}) \partial_{X^\alpha} \lambda^A; \tag{II.7-6}$$

in this notation (X^α) are the co-ordinates of the place occupied by X in $\boldsymbol{\kappa}_1$, (X^A) are the co-ordinates of the place occupied by X in $\boldsymbol{\kappa}_2$, and the summation convention is followed.

8. Present Placement as Reference

To serve as a reference, a placement need only be a differentiable homeomorphism on $\mathscr{B}$. So far, we have employed a reference placement independent of time, but we could just as well use a varying one. Thus one motion may be described in terms of any other. The only varying placement often useful as a reference placement is the present one. If we take the present placement as reference, we describe the past and future as they seem to an observer fixed to the body-point X now at the place $\mathbf{x}$. The corresponding description is called *relative*.

To see how such a description is constructed, consider places that are values of the motion of X at the two times t and τ:

$$\begin{aligned} \boldsymbol{\xi} &= \chi(X, \tau), \\ \mathbf{x} &= \chi(X, t). \end{aligned} \tag{II.8-1}$$

That is, $\boldsymbol{\xi}$ is the place occupied at the time τ by the body-point that at the time t occupies $\mathbf{x}$:

$$\begin{aligned} \boldsymbol{\xi} &= \chi(\chi^{-1}(\mathbf{x}, t), \tau), \\ &\equiv \chi_t(\mathbf{x}, \tau), \end{aligned} \tag{II.8-2}$$

say. The function χ_t just defined is called the *relative transplacement*.

Sometimes we shall wish to calculate the relative transplacement when the motion is given to us only through the spatial description of the velocity field:

$$\mathbf{v} = \dot{\mathbf{x}}(\mathbf{x}, t). \tag{II.4-5$_r$}$$

By $(1)_1$

$$\partial_\tau \boldsymbol{\xi} = \dot{\mathbf{x}}(\boldsymbol{\xi}, \tau). \tag{II.8-3}$$

Since the right-hand side is a given function, we thus have a differential equation to integrate. The initial condition to be satisfied by the integral $\boldsymbol{\xi} = \chi_t(\mathbf{x}, \tau)$ is

$$\boldsymbol{\xi}|_{\tau=t} = \chi_t(\mathbf{x}, t) = \mathbf{x}. \tag{II.8-4}$$

When the motion is described by (2), we shall use a subscript t to denote quantities derived from the relative transplacement χ_t. Thus $\mathbf{F}_t$, defined by

$$\mathbf{F}_t \equiv \mathbf{F}_t(\tau) \equiv \operatorname{grad} \chi_t, \tag{II.8-5}$$

is the *local relative deformation*. Of course,

$$\mathbf{F}_t(t) = \mathbf{1}. \tag{II.8-6}$$

By (II.7-5), at $\mathbf{X}$

$$\mathbf{F}(\tau) = \mathbf{F}_t(\tau)\mathbf{F}(t). \tag{II.8-7}$$

As the fixed reference placement with respect to which $\mathbf{F}(\tau)$ and $\mathbf{F}(t)$ are taken we may select the placement of the body at the time t'. Then (7) yields

$$\mathbf{F}_{t'}(\tau) = \mathbf{F}_t(\tau)\mathbf{F}_{t'}(t), \tag{II.8-8}$$

a formula which, like (II.7-5), expresses a chain rule of differential calculus.

9. Stretch and Rotation

Since the transplacement χ_κ is invertible, so is its gradient $\mathbf{F}$, and the polar decomposition theorem of CAUCHY[1] yields two unique expressions for $\mathbf{F}$ in terms of an orthogonal tensor $\mathbf{R}$ and positive symmetric tensors $\mathbf{U}$ and $\mathbf{V}$:

$$\mathbf{F} = \mathbf{RU} = \mathbf{VR}. \tag{II.9-1}$$

$\mathbf{R}$ is orthogonal but need not be proper-orthogonal: $\mathbf{RR}^\mathsf{T} = \mathbf{1}$, so $\det \mathbf{R} = +1$ or -1, and $\det \mathbf{R}$ maintains either the one value or the other for all $\mathbf{X}$ and t, by continuity. Thus $\det \mathbf{U} = \det \mathbf{V} = |\det \mathbf{F}| = J$. $\mathbf{R}$ is called the *rotation tensor*[2]; $\mathbf{U}$ and $\mathbf{V}$, which of course satisfy the obvious relation

$$\mathbf{V} = \mathbf{RUR}^\mathsf{T}, \tag{II.9-2}$$

are called the *right* and *left stretch tensors*, respectively. These tensors, like $\mathbf{F}$ itself, are to be interpreted as comparing aspects of the present shape of $\mathscr{B}$ with their counterparts in the reference shape. Just how they do so, we shall now proceed to show.

First, since $\mathbf{U}$ is symmetric, it has at least one orthogonal triad of principal axes; the members of any such triad are called *principal axes of strain at* $\mathbf{X}$ in the reference shape $\boldsymbol{\kappa}(\mathscr{B})$. Likewise, $\mathbf{V}$ has an orthogonal triad of principal axes which are called *principal axes of strain at* $\mathbf{x}$ in the present shape $\chi(\mathscr{B}, t)$. By (2), $\mathbf{U}$ and $\mathbf{V}$ have their proper numbers in

[1] This theorem is proved in any book on linear algebra, *e.g.* in §83 of P. R. HALMOS, *Finite-Dimensional Vector Spaces*, 2nd ed., Princeton, Toronto, and London, Van Nostrand, 1958. It was discovered and proved by CAUCHY in the present context.

[2] To reconcile the term with the definition, we could have imposed from the start the requirement that only reference placements such that $\det \mathbf{F} > 0$ be allowed, which would have implied that $\det \mathbf{R} = 1$ and have made $\mathbf{R}$ a rotation in the usual sense of that term. Since there is no reason to do so other than the convenience of language, we take advantage of that convenience without imposing the restriction. That is, in the text above we leave to the reader such trivial changes of wording as may be needed to include the case when $-\mathbf{R}$ rather than $\mathbf{R}$ is proper-orthogonal.

common. Indeed, if $\mathbf{e}_k$ is a proper vector of $\mathbf{U}$ corresponding to the proper number v_k, then

$$\mathbf{U}\mathbf{e}_k = v_k \mathbf{e}_k, \tag{II.9-3}$$

so by (1) and (2)

$$\mathbf{V}(\mathbf{R}\mathbf{e}_k) = (\mathbf{R}\mathbf{U}\mathbf{R}^{\mathrm{T}})(\mathbf{R}\mathbf{e}_k) = v_k(\mathbf{R}\mathbf{e}_k). \tag{II.9-4}$$

Thus the rotation $\mathbf{R}$ carries principal axes of strain at $\mathbf{X}$ into principal axes of strain at $\mathbf{x}$. (Since $\mathbf{R}$ is unique but the principal axes of strain need not be, we cannot always use this property as the definition of $\mathbf{R}$.) If $\mathbf{e}_k$ points along the kth principal axis of strain at $\mathbf{X}$ in $\boldsymbol{\kappa}(\mathscr{B})$, then v_k is the ratio of the length of the image $\mathbf{F}\mathbf{e}_k$ in $\chi(\mathscr{B}, t)$ to the length of the original $\mathbf{e}_k$. Thus the v_k are called the *principal stretches*. Since $\mathbf{U}$ and $\mathbf{V}$ are positive, $v_k > 0$. When $\mathbf{R} = \mathbf{1}$, the local deformation is called a *pure stretch* at $\mathbf{X}$, t. In a pure stretch, $\mathbf{U} = \mathbf{V}$; the principal axes of strain at $\mathbf{X}$ and $\mathbf{x}$ coincide; and we may visualize the local deformation as being effected by stretching elements along those axes in the ratios v_1, v_2, v_3. If $\mathbf{U} = \mathbf{V} = \mathbf{1}$, the local deformation is called a *rotation* at $\mathbf{X}$, t. CAUCHY's decomposition tells us that the local deformation may be obtained by effecting a pure stretch with principal stretches v_k along three suitable mutually orthogonal directions $\mathbf{e}_k$, followed by a rotation of those directions, or by performing the same rotation first and then effecting the same stretches along the resulting directions.

The *right* and *left Cauchy–Green tensors* $\mathbf{C}$ and $\mathbf{B}$ are defined as follows:

$$\begin{aligned} \mathbf{C} &\equiv \mathbf{U}^2 = \mathbf{F}^{\mathrm{T}}\mathbf{F}, \\ \mathbf{B} &\equiv \mathbf{V}^2 = \mathbf{F}\mathbf{F}^{\mathrm{T}} = \mathbf{R}\mathbf{C}\mathbf{R}^{\mathrm{T}}. \end{aligned} \tag{II.9-5}$$

While the fundamental decomposition (1) plays the major part in the proof of general theorems, calculation of $\mathbf{U}$, $\mathbf{V}$, and $\mathbf{R}$ from $\mathbf{F}$ in special cases may be awkward, since irrational operations are usually required. $\mathbf{C}$ and $\mathbf{B}$, however, are calculated by mere multiplication of $\mathbf{F}$ and $\mathbf{F}^{\mathrm{T}}$. *E.g.*, if g_{km} and $g^{\alpha\beta}$ are the covariant and contravariant metric components in arbitrarily selected co-ordinate systems in space and in the reference shape, respectively, components of $\mathbf{C}$ and $\mathbf{B}$ are[1]

$$\begin{aligned} C_{\alpha\beta} &= F^k_{\alpha} F^m_{\beta} g_{km}, \\ B^{km} &= F^k_{\alpha} F^m_{\beta} g^{\alpha\beta}, \end{aligned} \tag{II.9-6}$$

[1] If both systems of co-ordinates (x^k) and (X^α) are rectangular Cartesian, (6) follows at once from (5) and (II.5-3). To derive (6) in general co-ordinates it suffices to observe that $(6)_1$ and $(6)_2$ are tensorial equations that reduce in Cartesian co-ordinates to the equations already demonstrated in the case when those co-ordinates are used.

where $F^k_\alpha = x^k{}_{,\alpha} \equiv \partial_{X^\alpha} \chi^k_\kappa(X^1, X^2, X^3, t)$. The proper numbers of $\mathbf{C}$ and $\mathbf{B}$ are the squares v_i^2 of the principal stretches. The *principal invariants* of $\mathbf{C}$ and $\mathbf{B}$ are given by

$$I \equiv \operatorname{tr} \mathbf{B} = \operatorname{tr} \mathbf{C} = v_1^2 + v_2^2 + v_3^2,$$
$$II \equiv \tfrac{1}{2}[(\operatorname{tr} \mathbf{B})^2 - \operatorname{tr} \mathbf{B}^2] = \tfrac{1}{2}[(\operatorname{tr} \mathbf{C})^2 - \operatorname{tr} \mathbf{C}^2] = v_1^2 v_2^2 + v_2^2 v_3^2 + v_3^2 v_1^2, \quad \text{(II.9-7)}$$
$$III \equiv \det \mathbf{B} = \det \mathbf{C} = J^2 = v_1^2 v_2^2 v_3^2.$$

Any symmetric function of v_1, v_2, and v_3 is equal to a function of I, II, and III.

The formulae obtained so far in this section apply to any invertible tensor, making no use of the fact that $\mathbf{F}$ is the gradient of χ_κ. Also often elsewhere in continuum mechanics the relation between the values of $\mathbf{F}$ at different arguments $\mathbf{X}$ is of no importance, but usually it must be taken into account. One such example is furnished by the chain rule (II.8-7). For another, we note from $(7)_7$ and (II.5-4) that

$$\rho_\kappa/\rho = \det \mathbf{U} = \det \mathbf{V} = \sqrt{III}. \quad \text{(II.9-8)}$$

Another example is furnished by the following exercise.

Exercise II.9.1 (MICHAL). Prove that if $\kappa(\mathscr{B})$ is connected, a transplacement whose local deformation is orthogonal at each point is either a rigid rotation or the product of one by a central inversion. Denoting two transplacements by χ_κ and $\bar{\chi}_\kappa$, show that for $\bar{\chi}_\kappa \circ \chi_\kappa^{-1}$ to preserve the distances between points it is necessary and sufficient that $\bar{\mathbf{U}} = \mathbf{U}$.

If we begin with the local relative deformation $\mathbf{F}_t$, defined by (II.8-5), and apply to it the polar decomposition theorem, we obtain the *relative rotation tensor* $\mathbf{R}_t$, the *relative stretch tensors* $\mathbf{U}_t$ and $\mathbf{V}_t$ and the *relative Cauchy–Green tensors* $\mathbf{C}_t$ and $\mathbf{B}_t$:

$$\mathbf{F}_t = \mathbf{R}_t \mathbf{U}_t = \mathbf{V}_t \mathbf{R}_t, \qquad \mathbf{C}_t = \mathbf{U}_t^2, \qquad \mathbf{B}_t = \mathbf{V}_t^2. \quad \text{(II.9-9)}$$

Exercise II.9.2. Prove from (II.8-7) that

$$\mathbf{C}(\tau) = \mathbf{F}(t)^{\mathsf{T}} \mathbf{C}_t(\tau) \mathbf{F}(t). \quad \text{(II.9-10)}$$

When a transplacement is laid down for study, it is a trivial matter to calculate from it the tensors $\mathbf{B}$ and $\mathbf{C}$. We consider here two examples, both of which will be useful later in connection with specific materials. In a *simple shear* each member of a family of parallel planes is moved a distance proportional to its distance from a certain member of the family, and in a direction lying in that plane. If we let the particular plane be $X_1 = 0$, and if we let the direction of the shear be that of the co-ordinate X_2, then a simple shear is given in the rectangular co-ordinate system X_1, X_2, X_3 by the following components of transplacement:

$$\begin{aligned} x_1 &= X_1, \\ x_2 &= X_2 + KX_1, \\ x_3 &= X_3. \end{aligned} \tag{II.9-11}$$

The constant K is called the *amount* of shear.

Since

$$[\mathbf{F}] = \begin{Vmatrix} 1 & 0 & 0 \\ K & 1 & 0 \\ 0 & 0 & 1 \end{Vmatrix}, \tag{II.9-12}$$

it follows that

$$\begin{aligned} [\mathbf{B}] = [\mathbf{F}\mathbf{F}^{\mathrm{T}}] &= \begin{Vmatrix} 1 & K & 0 \\ K & 1+K^2 & 0 \\ 0 & 0 & 1 \end{Vmatrix}, \\ [\mathbf{B}^{-1}] &= \begin{Vmatrix} 1+K^2 & -K & 0 \\ -K & 1 & 0 \\ 0 & 0 & 1 \end{Vmatrix}, \end{aligned} \tag{II.9-13}$$

$$I = \operatorname{tr} \mathbf{B} = 3 + K^2 = II = \operatorname{tr} \mathbf{B}^{-1}, \qquad III = 1.$$

Exercise II.9.3 (KELVIN & TAIT). In simple shear the principal stretches are expressed as follows in terms of the amount of shear:

$$\begin{aligned} v_1^2 &= 1 + \tfrac{1}{2}K^2 + K\sqrt{1 + \tfrac{1}{4}K^2}, \\ v_2^2 &= 1 + \tfrac{1}{2}K^2 - K\sqrt{1 + \tfrac{1}{4}K^2} = \frac{1}{v_1^2}, \\ v_3 &= 1. \end{aligned} \tag{II.9-14}$$

Show that the angle θ through which the principal axes of strain in $\boldsymbol{\kappa}(\mathscr{B})$ are rotated so as to become the principal axes of strain in $\chi(\mathscr{B})$ is given by $\tan\theta = \frac{1}{2}K$.

An example illustrating the use of curvilinear co-ordinate systems is given by the following components of transplacement in cylindrical polar co-ordinates:

$$r = \sqrt{AR^2 + B}, \qquad \theta = \Theta + DZ, \qquad z = FZ, \qquad AF = 1, \tag{II.9-15}$$

A, B, D, and F being constants. The cylinders $R = \text{const.}$ are mapped into the cylinders $r = \text{const.}$, and choice of the constants A and B allows an arbitrary expansion or contraction as well as an eversion of these cylinders.

At the same time, there is a stretch F in the direction of the axis of the cylinders, so adjusted as to make the transplacement isochoric. Finally the planes $Z = \text{const.}$ are rotated about the axis through angles proportional to their distance from the particular plane $Z = 0$. Thus a torsion of amount D/F is superimposed upon the isochoric expansion or contraction of the cylinders.

Exercise II.9.4. Using $(6)_2$, show that

$$\|B^{km}\| = \begin{Vmatrix} A^2R^2/r^2 & 0 & 0 \\ 0 & R^{-2} + D^2 & DF \\ 0 & DF & F^2 \end{Vmatrix}. \tag{II.9-16}$$

To calculate $(B^{-1})_{km}$, either invert the matrix (16) or first prove and then use the formula

$$(B^{-1})_{km} = X^{\alpha}{}_{,k} X^{\beta}{}_{,m} g_{\alpha\beta} \tag{II.9-17}$$

so as to obtain

$$\|(B^{-1})_{km}\| = \begin{Vmatrix} r^2/(A^2R^2) & 0 & 0 \\ 0 & R^2 & -ADR^2 \\ 0 & -ADR^2 & A^2(1 + D^2R^2) \end{Vmatrix},$$

$$I = \operatorname{tr} \mathbf{B} = g_{km} B^{km} = \frac{A^2R^2}{r^2} + r^2\left(\frac{1}{R^2} + D^2\right) + F^2, \tag{II.9-18}$$

$$II = \operatorname{tr} \mathbf{B}^{-1} = g^{km}(B^{-1})_{km} = \frac{r^2}{A^2R^2} + \frac{R^2}{r^2} + A^2(1 + D^2R^2),$$

$$III = 1.$$

Exercise II.9.5. In *simple torsion* $A = F = 1$, $B = 0$. By comparing (16) with $(13)_1$, show that simple torsion may be regarded as effecting on each cylinder $R = \text{const.}$, when cut along a generator and developed onto a plane, a simple shear of amount DR.

Exercise II.9.6. Show that, in the notation used in §II.6,

$$|\operatorname{Grad} f|^2 = \operatorname{grad} f \cdot \mathbf{B} \operatorname{grad} f, \tag{II.9-19}$$

and hence (II.6-20) can be written in the form

$$\mathbf{n}_\kappa = \frac{1}{\sqrt{\mathbf{n} \cdot \mathbf{Bn}}} \mathbf{F}^{\mathrm{T}}\mathbf{n}; \tag{II.9-20}$$

likewise, (II.6-22) becomes

$$S_\kappa = \frac{1}{\sqrt{\mathbf{n} \cdot \mathbf{Bn}}} (S_n - \mathbf{n} \cdot \dot{\mathbf{x}}). \tag{II.9-21}$$

10. Histories

Let Ψ denote a function of time whose value is a scalar, a vector, or a tensor. We shall often wish to consider the restriction of Ψ to present and past times only. For convenience, if t is the present time, we shall represent the past time t' by the positive quantity $s \equiv t - t'$. The *history of* Ψ *up to time* t is denoted by Ψ^t, the value of which is $\Psi^t(s)$:

$$\Psi^t(s) \equiv \Psi(t - s), \qquad t \text{ fixed}, \quad s \geqq 0. \qquad \text{(II.10-1)}$$

For each t the history Ψ^t is defined on $[0, \infty[$. The history Ψ^t, as its name suggests, is the portion of a function of all time which corresponds to the present and past times only. Histories turn out to be of major importance in mechanics because it is the present and past that determine the future.

In this notation $\mathbf{C}_t^t$, for example, is the history of the relative right Cauchy–Green tensor $\mathbf{C}_t$ up to time t.

11. Stretching and Spin

For the instantaneous time derivative of a tensor defined from the relative transplacement, for example $\mathbf{F}_t$, we introduce the notation[1]

$$\dot{\mathbf{F}}_t(t) \equiv \partial_u \mathbf{F}_t(u)\Big|_{u=t} = -\partial_s \mathbf{F}_t^t(s)\Big|_{s=0}, \qquad \text{(II.11-1)}$$

$\mathbf{X}$ being held constant. Set

$$\begin{aligned} \mathbf{G} &\equiv \dot{\mathbf{F}}_t(t), \\ \mathbf{D} &\equiv \dot{\mathbf{U}}_t(t) = \dot{\mathbf{V}}_t(t), \\ \mathbf{W} &\equiv \dot{\mathbf{R}}_t(t). \end{aligned} \qquad \text{(II.11-2)}$$

$\mathbf{D}$, which is called the *stretching*, is the rate of change of the stretch at the place of X in the shape at time $t + \varepsilon$ with respect to that at time t, in the limit as $\varepsilon \to 0$. Likewise, $\mathbf{W}$, which is called the *spin*, is the ultimate rate of change of the rotation at $\mathbf{x}$ from the present shape to one the body had just before or will have just afterward. Since $\mathbf{U}_t$ is symmetric, so is $\mathbf{D}$, being its derivative with respect to a parameter:

$$\mathbf{D}^{\mathrm{T}} = \mathbf{D}, \qquad \text{(II.11-3)}$$

[1] This notation could not be confused with the material derivative introduced in §II.6, since $\mathbf{F}_t(t) = \mathbf{1}$ and $\dot{\mathbf{1}} = \mathbf{0}$.

but $\mathbf{D}$, unlike $\mathbf{U}_t$, generally fails to be positive. Since $\mathbf{D}(\mathbf{x}, t)$ is symmetric, its proper numbers are real, and it has at least one orthogonal triad of proper vectors. The latent roots of $\mathbf{D}(\mathbf{x}, t)$ are called the *principal stretchings* $d_k(\mathbf{x}, t)$, $k = 1, 2, 3$; the directions of a corresponding orthogonal triad are called *principal axes of stretching*.

If we differentiate the relation $\mathbf{R}_t(u)\mathbf{R}_t(u)^{\mathrm{T}} = \mathbf{1}$ with respect to u, put $u = t$, and use $(2)_4$, we find that $\mathbf{W}$ is skew:

$$\mathbf{W}^{\mathrm{T}} + \mathbf{W} = \mathbf{0}. \tag{II.11-4}$$

From its definition $(2)_1$, $\mathbf{G}$ is the ultimate rate of change of $\mathbf{F}_t$, but that is not all, for by differentiating (II.8-7) with respect to τ and then putting $\tau = t$ we obtain

$$\mathbf{G} = \dot{\mathbf{F}}\mathbf{F}^{-1}. \tag{II.11-5}$$

Differentiation of (II.5-1) with respect to t yields

$$\dot{\mathbf{F}} = \text{Grad } \dot{\boldsymbol{\chi}}_{\boldsymbol{\kappa}} = (\text{grad } \dot{\mathbf{x}})\mathbf{F}; \tag{II.11-6}$$

in view of (II.5-3), the last step follows by the chain rule of differential calculus. Substitution into (5) yields

$$\mathbf{G} = \text{grad } \dot{\mathbf{x}}. \tag{II.11-7}$$

We have shown that the tensor $\mathbf{G}$, which we defined by $(2)_1$, is in fact the spatial *velocity gradient*.

If we differentiate the polar decomposition $(\text{II.9-9})_1$ with respect to τ and then put $\tau = t$, we find that

$$\mathbf{G} = \mathbf{D} + \mathbf{W}. \tag{II.11-8}$$

This result, showing that $\mathbf{D}$ and $\mathbf{W}$ are the symmetric and skew parts of the velocity gradient, expresses the fundamental ***Euler–Cauchy–Stokes Decomposition*** of the instantaneous motion at $\mathbf{x}$, t into the sum of a pure stretching along three mutually orthogonal axes, a rigid spin of those axes, and a translation. The nature of the spin is clarified below by Exercise II.13-4.

Of course, we could have defined $\mathbf{G}$ by (7) as the velocity gradient and $\mathbf{W}$ and $\mathbf{D}$ by (8) as the symmetric and skew parts of $\mathbf{G}$. We should then have had to prove $(2)_{2,4}$ as theorems so as to interpret $\mathbf{G}$, $\mathbf{W}$, and $\mathbf{D}$ kinematically. Writers on hydrodynamics usually prefer the argument in this order.

Motions in which $\mathbf{W} = \mathbf{0}$ are called *irrotational*. They form the main subject of study in classical hydrodynamics. Motions in which $\mathbf{W} \neq \mathbf{0}$ are called *rotational*.

Since $\mathbf{W}$ is skew, it may be represented to within a convention of sign by the axial vector curl $\dot{\mathbf{x}}$, which is called the "vorticity" in hydrodynamics.[1] Nowadays it seems more convenient not to introduce this vector but instead to use the tensor $\mathbf{W}$. We shall use the letter W to denote the magnitude of curl $\dot{\mathbf{x}}$:

$$W \equiv |\text{curl } \dot{\mathbf{x}}| = \sqrt{2}|\mathbf{W}|. \tag{II.11-9}$$

Another important scalar is the *expansion* E, defined as follows:

$$E \equiv \dot{J}/J = \text{div } \dot{\mathbf{x}} = \text{tr } \mathbf{G} = \text{tr } \mathbf{D} = -\dot{\rho}/\rho; \tag{II.11-10}$$

the third and fourth expressions follow by use of (7) and (8), and the last by use of (II.5-7). E is the local rate of increase of volume of a material region, referred to unit volume. We remark again that a necessary and sufficient condition for isochoric motion is $E = 0$.

The velocity field of a rigid motion is given by (I.10-1). Taking the gradient of that equation yields $\mathbf{G} = \mathbf{W}$. Thus in a rigid motion $\mathbf{D} = \mathbf{0}$, and the spin as defined by $(2)_4$ for a general motion reduces in the case of a rigid motion to the constant field having as its value what we have called in §I.10 the spin of that motion. In this sense we may regard the spin field as a generalization of the spin of a rigid motion—in rough language, a "local angular velocity".

Exercise II.11.1 (EULER). Prove the converse. Namely, regarding $\mathbf{D} = \mathbf{0}$ as a differential equation for the spatial velocity field in a connected open set, by integrating it obtain (I.10-1).

These results establish the following theorem: *The condition* $\mathbf{D} = \mathbf{0}$ *in a region at an instant is necessary and sufficient that the motion be rigid in that region at that instant.* In view of the interpretation of $\mathbf{D}$ given just after its definition, the theorem is really obvious.

[1] Since we shall not use 3-dimensional vector analysis in this book, we need only list a few formulae to help the reader compare statements with those in some other works. For the algebra of the Gibbsian cross-product, such a list is provided in §AII.15. The curl of a vector field $\mathbf{v}$ is defined as follows in terms of the components of the gradient $\nabla\mathbf{v}$:

$$\text{curl } \mathbf{v} \equiv -(\nabla\mathbf{v})_{\times},$$

so that

$$(\text{curl } \mathbf{v})_3 = v_{2,1} - v_{1,2}, \qquad \textit{etc.}$$

If $\mathbf{W}$ is the spin and if $\mathbf{w} \equiv \text{curl } \dot{\mathbf{x}}$, then $\mathbf{w} = -\mathbf{W}_{\times}$. For any skew tensor $\mathbf{W}$

$$2\mathbf{W}\mathbf{u} = \mathbf{w} \times \mathbf{u}, \qquad \mathbf{W} \cdot (\mathbf{u} \wedge \mathbf{v}) = \mathbf{w} \cdot (\mathbf{u} \times \mathbf{v}).$$

Also

$$\text{div } (\mathbf{u} \wedge \mathbf{v}) = \text{curl } (\mathbf{u} \times \mathbf{v}).$$

Clearly the spin $\mathbf{W}$ is generally something quite different from $\dot{\mathbf{R}}$, the time-rate of the rotation tensor, as the following two examples show.

In a *simple shearing*, the Cartesian velocity components are

$$\dot{x}_1 = 0, \qquad \dot{x}_2 = \kappa x_1, \qquad \dot{x}_3 = 0, \tag{II.11-11}$$

and each body-point moves ahead along a straight line parallel to the x_2-axis, yet unless $\kappa = 0$, the motion is rotational. In a *simple vortex*, the cylindrical polar contravariant velocity components are

$$\dot{r} = 0, \qquad \dot{\theta} = \omega(r), \qquad \dot{z} = 0, \tag{II.11-12}$$

and each body-point rotates steadily about the axis on a circle $r = \text{const.}$, yet in the case when $\omega(r) = Kr^{-2}$, the motion is irrotational.

The definitions (II.9-1) and $(2)_4$ make the different kinematic meanings of $\mathbf{R}$ and $\mathbf{W}$ clear and suggest that both tensors will be useful in the description and classification of motions.

Likewise, the stretching $\mathbf{D}$ is not generally the rate of change $\dot{\mathbf{U}}$ of the stretch $\mathbf{U}$.

Further enlightenment of the difference between stretch and stretching and between rotation and spin is furnished by the following exercise.

Exercise II.11.2 (E. & F. COSSERAT, COLEMAN & TRUESDELL). Prove that

$$\begin{aligned} \dot{\mathbf{C}} &= 2\mathbf{F}^{\mathrm{T}}\mathbf{D}\mathbf{F}, \\ \mathbf{W} &= \dot{\mathbf{R}}\mathbf{R}^{\mathrm{T}} + \tfrac{1}{2}\mathbf{R}(\dot{\mathbf{U}}\mathbf{U}^{-1} - \mathbf{U}^{-1}\dot{\mathbf{U}})\mathbf{R}^{\mathrm{T}}, \\ \mathbf{D} &= \tfrac{1}{2}\mathbf{R}(\dot{\mathbf{U}}\mathbf{U}^{-1} + \mathbf{U}^{-1}\dot{\mathbf{U}})\mathbf{R}^{\mathrm{T}}, \end{aligned} \tag{II.11-13}$$

where $\mathbf{R}$ and $\mathbf{U}$ have their usual meanings as the rotation and right stretch tensors with respect to a fixed reference placement. Prove also that $\dot{\mathbf{B}}|_{\mathbf{F}=\mathbf{1}} = 2\mathbf{D}$.

Various higher rates of change of stretch and rotation may be defined.

Exercise II.11.3. Including and generalizing (1) and $(2)_1$, set

$$\mathbf{G}_n \equiv \overset{(n)}{\mathbf{F}}_t(t) \equiv \partial_u^n \mathbf{F}_t(u)\Big|_{u=t}, \qquad n = 1, 2, \ldots. \tag{II.11-14}$$

By differentiating (II.8-7) n times with respect to τ and setting $\tau = t$ show that

$$\overset{(n)}{\mathbf{F}}\mathbf{F}^{-1} = \mathbf{G}_n, \tag{II.11-15}$$

and hence by the chain rule show that

$$\mathbf{G}_n = \operatorname{grad} \overset{(n)}{\mathbf{x}}. \tag{II.11-16}$$

The most useful higher rates are the *Rivlin–Ericksen tensors* $\mathbf{A}_n$. They are defined as follows in terms of a notation like (14):

$$\mathbf{A}_n \equiv \overset{(n)}{\mathbf{C}}_t(t). \tag{II.11-17}$$

In particular, $\mathbf{A}_1 = 2\mathbf{D}$.

Exercise II.11.4 (DUPONT, RIVLIN & ERICKSEN). Prove that

$$\mathbf{A}_n = \mathbf{G}_n + \mathbf{G}_n^{\mathrm{T}} + \sum_{j=1}^{n-1} \binom{n}{j} \mathbf{G}_j^{\mathrm{T}} \mathbf{G}_{n-j} \tag{II.11-18}$$

and

$$\mathbf{A}_{n+1} = \dot{\mathbf{A}}_n + \mathbf{A}_n \mathbf{G} + (\mathbf{A}_n \mathbf{G})^{\mathrm{T}}. \tag{II.11-19}$$

Exercise II.11.5 (RIVLIN). By differentiating the relation $\det \mathbf{C}_t(u) = 1$ repeatedly with respect to u and then putting $u = t$, show that in an isochoric motion

$$\begin{aligned} \operatorname{tr} \mathbf{A}_1 &= 0, \\ \operatorname{tr} \mathbf{A}_2 &= \operatorname{tr} \mathbf{A}_1^2, \\ \operatorname{tr} \mathbf{A}_3 &= -2 \operatorname{tr} \mathbf{A}_1^3 + 3 \operatorname{tr} (\mathbf{A}_2 \mathbf{A}_1), \end{aligned} \tag{II.11-20}$$

and in general $\operatorname{tr} \mathbf{A}_n$ is a linear combination of traces of products formed from $\mathbf{A}_1, \mathbf{A}_2, \ldots, \mathbf{A}_{n-1}$.

We now consider restrictions imposed upon $\mathbf{W}$ by boundaries. To do so, we first appeal to KELVIN's transformation of a line integral into a surface integral ("STOKES's theorem"), applied to the velocity field $\dot{\mathbf{x}}$. In this context we may state the result as follows, always assuming that $\dim \mathscr{E} = 3$. Let a surface $\mathscr{S}$ be given by a mapping $\mathbf{x} = \mathbf{f}(a, b)$ on a domain of parameters (a, b), and let $\partial_a \mathbf{x}$ and $\partial_b \mathbf{x}$ denote the partial derivatives of that mapping. Let $\mathscr{A}$ be a subsurface of $\mathscr{S}$, and let $\int_{\mathscr{A}}$ denote integration over the subdomain of the parameters that corresponds to $\mathscr{A}$. Let $\int_{\partial\mathscr{A}}$ denote a line integral over the curve bounding the subsurface $\mathscr{A}$. Then, with an appropriate convention of sign,

$$\int_{\partial\mathscr{A}} \dot{\mathbf{x}} \cdot d\mathbf{x} = \int_{\mathscr{A}} \mathbf{W} \cdot (\partial_a \mathbf{x} \wedge \partial_b \mathbf{x})\, da\, db. \tag{II.11-21}$$

If $\mathscr{S}$ is a surface normal to the velocity field, the left-hand side of (21) vanishes for every subsurface $\mathscr{A}$. If $\mathbf{W}$ and $\partial_a \mathbf{x} \wedge \partial_b \mathbf{x}$ are continuous, then everywhere on $\mathscr{S}$

$$\mathbf{W} \cdot (\partial_a \mathbf{x} \wedge \partial_b \mathbf{x}) = 0. \tag{II.11-22}$$

We have proved the following theorem: *At a point on a surface normal to the velocity field, either* $\mathbf{W} = \mathbf{0}$ *or the axis of* $\mathbf{W}$ *lies in the tangent plane.* Therefore, if $\mathbf{n}$ is the unit normal field to $\mathscr{S}$, we can choose an orthonormal basis field $\{\mathbf{n}, \mathbf{e}, \mathbf{f}\}$ such that $\mathbf{e}$ lies in the nullspace of $\mathbf{W}$ and

$$\mathbf{W} = \tfrac{1}{2}W\mathbf{n} \wedge \mathbf{f}. \tag{II.11-23}$$

Hence

$$\mathbf{W}\mathbf{n} = -\tfrac{1}{2}W\mathbf{f}, \qquad \mathbf{W}\mathbf{f} = \tfrac{1}{2}W\mathbf{n}. \tag{II.11-24}$$

so

$$\mathbf{W} = -\mathbf{n} \otimes \mathbf{W}\mathbf{n} + \mathbf{W}\mathbf{n} \otimes \mathbf{n}. \tag{II.11-25}$$

If there are surfaces everywhere normal to $\dot{\mathbf{x}}$, the field $\dot{\mathbf{x}}$ is called *complex-lamellar*. By definition, there is then a scalar field A such that $\dot{\mathbf{x}} = A\mathbf{n}$. The reasoning given above in regard to a single surface normal to the velocity field shows that a *velocity field is complex-lamellar if and only if at each place the velocity lies in the nullspace of the spin there.* If $\mathbf{W} \neq \mathbf{0}$, then the field of axes of spin is normal to the velocity field.

Exercise II.11.6 (BERKER, CASWELL, TRUESDELL). From the interpretation of the gradient in terms of the directional derivative, show that if $\mathbf{k}$ is any vector in the tangent plane at the place $\mathbf{x}$ on a stationary wall to which the body adheres, then at $\mathbf{x}$

$$\mathbf{G}\mathbf{k} = \mathbf{0}. \tag{II.11-26}$$

Hence show that at $\mathbf{x}$

$$\mathbf{D} = E\mathbf{n} \otimes \mathbf{n} + \mathbf{n} \otimes \mathbf{W}\mathbf{n} + \mathbf{W}\mathbf{n} \otimes \mathbf{n}, \tag{II.11-27}$$

and hence that the principal stretchings are given by

$$\begin{aligned} 2D_1 &= E + \sqrt{E^2 + W^2} \geqq 0, \\ D_2 &= D_{(\mathbf{e})} = 0, \\ 2D_3 &= E - \sqrt{E^2 + W^2} \leqq 0. \end{aligned} \tag{II.11-28}$$

Exercise II.11.7 (CAUCHY). If

$$2\mathbf{W}_2 \equiv \operatorname{grad} \ddot{\mathbf{x}} - (\operatorname{grad} \ddot{\mathbf{x}})^{\mathsf{T}}, \tag{II.11-29}$$

prove that

$$(\mathbf{F}^{\mathsf{T}}\mathbf{W}\mathbf{F})^{\cdot} = \mathbf{F}^{\mathsf{T}}\mathbf{W}_2\mathbf{F}. \tag{II.11-30}$$

Hence a necessary and sufficient condition that

$$\mathbf{W}_2 = \mathbf{0} \tag{II.11-31}$$

along the path of a body-point is

$$\mathbf{F}^T\mathbf{W}\mathbf{F} = \mathbf{W}_\kappa, \tag{II.11-32}$$

$\mathbf{W}_\kappa$ being the value of $\mathbf{W}$ for that body-point at the place it would occupy, were it to assume its place in the reference shape. In particular, (31) is satisfied by an irrotational flow.

Exercise II.11.8 (LAGRANGE (imperfectly), CAUCHY). Prove that if (31) holds, then a body-point is presently in irrotational motion if and only if it always has been so and always will be so.

The condition (31) is of central importance in classical fluid dynamics. There it is applied in a region, not merely along the path of one body-point. It is called the ***D'Alembert–Euler condition***. We shall learn further consequences of it in §§II.13 and VI.6. For the time being we remark only that according to a familiar theorem on lamellar fields, in a simply connected region the field $\ddot{\mathbf{x}}$ satisfies (31) if and only if there is an *acceleration-potential* P_2:

$$\ddot{\mathbf{x}} = -\text{grad } P_2. \tag{II.11-33}$$

Exercise II.11.9 (D'ALEMBERT, EULER, BELTRAMI). Prove that

$$\mathbf{W}_2 = \dot{\mathbf{W}} + \mathbf{D}\mathbf{W} + \mathbf{W}\mathbf{D}. \tag{II.11-34}$$

Exercise II.11.10 (APPELL). Assuming that $\dim \mathscr{E} = 3$, prove that

$$(\tfrac{1}{2}|J\mathbf{W}|^2)^{\cdot} = J^2(\mathbf{W}\cdot\mathbf{W}_2 + |\mathbf{W}|^2\mathbf{n}\cdot\mathbf{D}\mathbf{n}), \tag{II.11-35}$$

$\mathbf{n}$ being any unit vector in the nullspace of $\mathbf{W}$. Hence show that W satisfies the differential equation

$$(JW)^{\cdot} = JW\mathbf{n}\cdot\mathbf{D}\mathbf{n} \tag{II.11-36}$$

if and only if

$$\mathbf{W}\cdot\mathbf{W}_2 = 0. \tag{II.11-37}$$

Exercise II.11.11. Using (I.9-22), prove that in a steady rigid motion at angular speed ω

$$P_2 = \tfrac{1}{2}\omega^2 r^2 + h, \tag{II.11-38}$$

r being the distance from the axis of spin, and h being a function of t only.

12. Homogeneous Transplacement

The transplacement χ_κ of the reference placement κ is said to be *homogeneous* if the body-points occupying each straight line segment in $\kappa(\mathscr{B})$ are carried into some straight-line segment in $\chi(\mathscr{B}, t)$. By a theorem of geometry, any such transplacement χ_κ must be affine at each time t. Thus a homogeneous transplacement of $\kappa(\mathscr{B})$ is of the form

$$\chi_\kappa(\mathbf{X}, t) = \mathbf{x}_0(t) + \mathbf{F}(t)(\mathbf{X} - \mathbf{X}_0), \qquad \det \mathbf{F}(t) \neq 0. \qquad \text{(II.12-1)}$$

In this formula $\mathbf{X}_0$ is a fixed place in $\kappa(\mathscr{B})$; $\mathbf{x}_0$ is a place-valued function of time; and $\mathbf{F}$ is a tensor-valued function of time. By (II.5-1) we see that $\mathbf{F}$ is the local deformation, and that at any one time t it has the same value at all places in $\chi(\mathscr{B}, t)$. This property explains the name "homogeneous transplacement": A transplacement is homogeneous if and only if its local deformation is uniform at each time.

For a given reference placement κ the composition of two homogeneous transplacements is a homogeneous transplacement. For each fixed t the transplacements homogeneous with respect to κ are restrictions of members of the affine group.

If κ_1 and κ_2 are two different reference placements, a motion that gives rise to a transplacement homogeneous with respect to κ_1 generally fails to do the same with respect to κ_2. The class of motions that give rise to transplacements homogeneous with respect to κ_1 coincides with the corresponding class of κ_2 if and only if the differentiable homeomorphism $\kappa_2 \circ \kappa_1^{-1}$ has a constant gradient.

Homogeneous transplacements are most easily visualized as mappings of one vector space into another. Let $\mathbf{p}_\kappa$ denote the field of position vectors in $\kappa(\mathscr{B})$ with respect to the origin $\mathbf{X}_0$, and let $\mathbf{p}$ denote the field of position vectors in $\chi(\mathscr{B}, t)$ with respect to $\mathbf{x}_0(t)$. That is, $\mathbf{p}_\kappa \equiv \mathbf{X} - \mathbf{X}_0$, and $\mathbf{p} \equiv \mathbf{x} - \mathbf{x}_0$. Then (1) may be written in the form

$$\mathbf{p} = \mathbf{F}\mathbf{p}_\kappa, \qquad \text{(II.12-2)}$$

and $\mathbf{F}$ is a function of time only.

Let the two particular position vectors $\mathbf{m}_\kappa$ and $\mathbf{n}_\kappa$ in $\kappa(\mathscr{B})$ be mapped onto $\mathbf{m}$ and $\mathbf{n}$, respectively, by (2). Then

$$\mathbf{m} \cdot \mathbf{n} = \mathbf{m}_\kappa \cdot \mathbf{C}\mathbf{n}_\kappa, \qquad \text{(II.12-3)}$$

$\mathbf{C}$ being the right Cauchy–Green tensor $(\text{II.9-5})_1$. Likewise,

$$\mathbf{m}_\kappa \cdot \mathbf{n}_\kappa = \mathbf{m} \cdot \mathbf{B}^{-1}\mathbf{n}, \qquad \text{(II.12-4)}$$

B being the left Cauchy–Green tensor $(\text{II}.9\text{-}5)_3$. The student will recall that **B** and **C** are symmetric and positive. All vectors parallel to $\mathbf{n}_\kappa$ are increased in length in the same ratio. In particular, if $\mathbf{n}_\kappa$ is a unit vector, generally the **n** corresponding to it through (2) has some length other than 1. This ratio of lengths is called the *stretch* $v_{(\mathbf{n}_\kappa)}$ in the direction of $\mathbf{n}_\kappa$. It may be calculated as follows:

$$v_{(\mathbf{n}_\kappa)} = \sqrt{\mathbf{n}_\kappa \cdot \mathbf{C}\mathbf{n}_\kappa}. \tag{II.12-5}$$

Two orthogonal vectors $\mathbf{m}_\kappa$ and $\mathbf{n}_\kappa$ in $\boldsymbol{\kappa}(\mathscr{B})$ are mapped, generally, into vectors **m** and **n** in $\chi(\mathscr{B}, t)$ that are not orthogonal. This phenomenon is called *shear*, and there are various ways to report it. The angle $\theta_{(\mathbf{n}_\kappa, \mathbf{m}_\kappa)}$ between the images in $\chi(\mathscr{B}, t)$ of two unit vectors $\mathbf{n}_\kappa$ and $\mathbf{m}_\kappa$ in $\boldsymbol{\kappa}(\mathscr{B})$ is one measure of shear. It is determined by the relation

$$\cos\theta_{(\mathbf{n}_\kappa, \mathbf{m}_\kappa)} = \frac{1}{v_{(\mathbf{n}_\kappa)}v_{(\mathbf{m}_\kappa)}}\,\mathbf{n}_\kappa \cdot \mathbf{C}\mathbf{m}_\kappa. \tag{II.12-6}$$

The sphere $|\mathbf{m}_\kappa|^2 = \text{const.}$ in $\boldsymbol{\kappa}(\mathscr{B})$ is mapped into an ellipsoid in $\chi(\mathscr{B}, t)$, and the sphere $|\mathbf{m}|^2 = \text{const.}$ in $\chi(\mathscr{B}, t)$ is the image of an ellipsoid in $\boldsymbol{\kappa}(\mathscr{B})$.

Exercise II.12.1 (CAUCHY). Show that the principal axes of strain, as defined in §II.9, are the principal axes of the ellipsoids just constructed; that the principal stretches are the stretches in the directions of those axes; and that these particular stretches are extremal. Thus resolve a homogeneous transplacement into a translation and a rotation of one set of principal axes into the other, followed or preceded by pure stretches along those axes. Show that the shear of each pair of principal axes is null.

The terms "stretching" and "shearing" in general refer to the rates of change of stretch and shear, respectively, when these latter are defined with respect to the present shape as reference. We may discuss stretching and shearing just as we have discussed stretch and shear, starting from homogeneous transplacements. If we differentiate (2) with **X** held constant, then use (II.11-5), and then use (2) again, we obtain

$$\dot{\mathbf{p}} = \dot{\mathbf{F}}\mathbf{p}_\kappa = \mathbf{G}\mathbf{F}\mathbf{p}_\kappa = \mathbf{G}\mathbf{p}. \tag{II.12-7}$$

Hence by use of (II.11-8) we derive EULER's relation

$$|\mathbf{p}|\,|\mathbf{p}|^{\cdot} = \mathbf{p}\cdot\dot{\mathbf{p}} = \mathbf{p}\cdot\mathbf{D}\mathbf{p}, \tag{II.12-8}$$

D being the stretching tensor; equivalently, if $\mathbf{p} \neq \mathbf{0}$ and if **n** is a unit vector in the direction of **p**, then

$$(\log|\mathbf{p}|)^{\cdot} = \mathbf{n}\cdot\mathbf{D}\mathbf{n}. \tag{II.12-9}$$

Thus the component $\mathbf{n} \cdot \mathbf{Dn}$ of $\mathbf{D}$ is the rate of increase of length, per unit length, of a linear segment in $\boldsymbol{\kappa}(\mathscr{B})$ presently parallel to $\mathbf{n}$ in $\chi(\mathscr{B}, t)$, and this rate is called the *stretching* in the direction of $\mathbf{n}$. The three principal stretchings, which were defined in §II.11, are the extremal stretchings.

Exercise II.12.2 (EULER). For two orthogonal unit vectors $\mathbf{n}_\kappa$ and $\mathbf{m}_\kappa$, by differentiating (6) show that

$$-\dot{\theta}_{(\mathbf{n},\,\mathbf{m})}\big|_{\mathbf{F}=\mathbf{1}} = 2\mathbf{n} \cdot \mathbf{Dm}, \tag{II.12-10}$$

and interpret this result in terms of shearing.

Let $\varphi_{(\mathbf{p},\,\mathbf{m}_\kappa)}$ denote the angle between the position vector $\mathbf{p}$ of $\mathbf{x}$ with respect to $\mathbf{x}_0$ in $\chi(\mathscr{B}, t)$ and the unit vector $\mathbf{m}_\kappa$ in $\boldsymbol{\kappa}(\mathscr{B})$. Then by (2),

$$|\mathbf{p}| \cos \varphi = \mathbf{m}_\kappa \cdot \mathbf{F}\mathbf{p}_\kappa, \tag{II.12-11}$$

in which for simplicity we do not write the subscript $(\mathbf{p}, \mathbf{m}_\kappa)$. Differentiating (11) with respect to t yields

$$\begin{aligned} \cos \varphi\, |\mathbf{p}|^{\cdot} - |\mathbf{p}| \sin \varphi \dot{\varphi} &= \mathbf{m}_\kappa \cdot \dot{\mathbf{F}}\mathbf{p}_\kappa, \\ &= \mathbf{m}_\kappa \cdot \mathbf{G}\mathbf{F}\mathbf{p}_\kappa, \end{aligned} \tag{II.12-12}$$

by (II.11-5). If we now let the value of $\mathbf{p}_\kappa$ be a unit vector orthogonal to $\mathbf{m}_\kappa$, say $\mathbf{n}_\kappa$, and then take the present shape as the reference shape, so that the corresponding value of $\mathbf{p}$ also is $\mathbf{n}$, we find that

$$\dot{\varphi}_{(\mathbf{n},\,\mathbf{m})}\big|_{\mathbf{F}=\mathbf{1}} = -\mathbf{m} \cdot \mathbf{Gn}. \tag{II.12-13}$$

This formula gives the angular rate at which a line segment in $\boldsymbol{\kappa}(\mathscr{B})$ presently parallel to $\mathbf{n}$ turns *away* from the stationary unit vector $\mathbf{m}$ in $\chi(\mathscr{B}, t)$. Likewise, the rate at which a line segment in $\boldsymbol{\kappa}(\mathscr{B})$ presently parallel to $\mathbf{m}$ is turning *toward* the stationary unit vector $\mathbf{n}$ is given by

$$\dot{\varphi}_{(\mathbf{m},\,-\mathbf{n})}\big|_{\mathbf{F}=\mathbf{1}} = +\mathbf{n} \cdot \mathbf{Gm}. \tag{II.12-14}$$

By adding these formulae and using (II.11-8) we obtain

$$\tfrac{1}{2}(\dot{\varphi}_{(\mathbf{n},\,\mathbf{m})} + \dot{\varphi}_{(\mathbf{m},\,-\mathbf{n})})\big|_{\mathbf{F}=\mathbf{1}} = \mathbf{n} \cdot \mathbf{Wm}, \tag{II.12-15}$$

$\mathbf{W}$ being the spin. Thus we have proved a ***fundamental theorem*** of CAUCHY: *The component* $\mathbf{n} \cdot \mathbf{Wm}$ *of* $\mathbf{W}$ *corresponding to the orthogonal unit vectors* $\mathbf{n}$ *and* $\mathbf{m}$ *is the arithmetic mean of the rates of right-handed rotation of a line in* $\boldsymbol{\kappa}(\mathscr{B})$ *presently parallel to* $\mathbf{n}$ *with respect to the direction of* $\mathbf{m}$ *in* $\chi(\mathscr{B}, t)$ *and of a line in* $\boldsymbol{\kappa}(\mathscr{B})$ *presently parallel to* $\mathbf{m}$ *with respect to the direction of* $\mathbf{n}$ *in* $\chi(\mathscr{B}, t)$.

For a general motion the local deformation $\mathbf{F}$ provides a local linear approximation to the transplacement χ_κ. We may say that to within an error that is $\mathbf{o}(\mathbf{X} - \mathbf{X}_0)$ the transplacement χ_κ is approximated at $\mathbf{X}_0$ in $\boldsymbol{\kappa}(\mathscr{B})$, and hence at $\mathbf{x}_0$ in $\chi(\mathscr{B}, t)$, by the homogeneous transplacement (1) which is defined by $\mathbf{F}(\mathbf{X}_0, t)$. Thus the results obtained in this section for homogeneous transplacements may be interpreted in general motions as first-order local approximations to counterparts for the present transplacement of $\boldsymbol{\kappa}(\mathscr{B})$. In loose language, the results valid for all lines in homogeneous transplacement are valid for infinitesimal line segments in any smooth transplacement.

For reference we record here also equations for the velocity field and the acceleration field of the homogeneous transplacement (1):

$$\dot{\mathbf{x}} = \dot{\mathbf{x}}_0 + \dot{\mathbf{F}}\mathbf{F}^{-1}(\mathbf{x} - \mathbf{x}_0),$$
$$\ddot{\mathbf{x}} = \ddot{\mathbf{x}}_0 + \ddot{\mathbf{F}}\mathbf{F}^{-1}(\mathbf{x} - \mathbf{x}_0). \qquad \text{(II.12-16)}$$

That these fields are given by affine functions of place should be obvious without calculation and may be verified also by a glance at (II.11-5) and (II.11-16).

Suppose, conversely, that the velocity field of a motion be affine:

$$\dot{\mathbf{x}} = \mathbf{c}(t) + \mathbf{G}(t)(\mathbf{x} - \mathbf{x}_0), \qquad \text{(II.12-17)}$$

and that $\mathbf{c}$ and $\mathbf{G}$ be continuous functions. Then we can find a place-valued function $\mathbf{x}_0$ and a tensor-valued function $\mathbf{F}$, unique to within initial values $\mathbf{x}_0(0)$ and $\mathbf{F}(0)$, such that

$$\dot{\mathbf{F}} - \mathbf{F}\mathbf{G} = \mathbf{0}, \qquad \dot{\mathbf{x}}_0 - \mathbf{c} = \mathbf{0}. \qquad \text{(II.12-18)}$$

No matter what be $\mathbf{G}$, (II.5-6) shows that $J(t)$ cannot vanish unless $J(0)$ does; hence $\mathbf{F}(t)$ is invertible if $\mathbf{F}(0)$ is. Therefore, the motion of $\mathscr{B}$ corresponding to an affine velocity field is a homogeneous transplacement with respect to the shape $\mathscr{B}$ has at any time.

13. Rates of Change of Integrals over Material Lines, Surfaces, and Regions. Material Vector Lines. The Vorticity Theorems of Helmholtz and Kelvin

A motion carries a given set of body-points into a set of places at the time t. If we are given a set $\mathscr{S}_t$ of places at each time t, and if for a given motion χ it turns out that at each t the set of body-points that occupy $\mathscr{S}_t$ remains the same, the set $\mathscr{S}_t$ is called the locus of a *material set*.

We are often concerned with *material lines, material surfaces,* and *material regions.* For example, a material region is the time sequence of shapes a certain subbody assumes in virtue of the motion it undergoes.

If we define an integral of a spatial field over a material line, surface, or region, its value will generally change in time for two reasons: first, because the field itself changes, and, second, because the domain of integration in $\mathscr{E}$ is changing in consequence of the motion of $\mathscr{B}$.

We have already calculated the time-rate of change of the integral of a function over a material region and recorded the result as (II.6-8). We now address ourselves to the corresponding problem for material lines and surfaces.

We have mentioned that formulae valid strictly for homogeneous transplacements serve as first-order approximations in general. Since only the first-order terms affect the value of an integral, we may derive in this way exact formulae for the time-rate of change of integrals. For example, if $\mathscr{C}$ is a given curve in $\boldsymbol{\kappa}(\mathscr{B})$, the time derivative of a line integral along its shape $\boldsymbol{\chi}(\mathscr{C}, t)$ is obtained by supposing that the material rate of change $\dot{\overline{d\mathbf{x}}}$ of the element of arc $d\mathbf{x}$ is $\mathbf{G}\, d\mathbf{x}$, as (II.12-7) suggests. Thus we infer the following formula, due to KELVIN:

$$\frac{d}{dt}\int_{\mathscr{C}} \mathbf{f} \cdot d\mathbf{x} = \int_{\mathscr{C}} [\dot{\mathbf{f}} \cdot d\mathbf{x} + \mathbf{f} \cdot (\mathbf{G}\, d\mathbf{x})],$$

$$= \int_{\mathscr{C}} (\dot{\mathbf{f}} + \mathbf{G}^{\mathrm{T}}\mathbf{f}) \cdot d\mathbf{x}. \qquad \text{(II.13-1)}$$

The abbreviated notation $\int_{\mathscr{C}}$ denotes integration over the parametric interval of the function $\mathbf{k}$ that defines $\mathscr{C}$ in the reference shape: $\mathbf{X} = \mathbf{k}(l)$. The student should clear the details by solving the following exercise. They will be made obvious anyway by the treatment for the analogous but more complicated problem for surface integrals which we shall give a little further on.

Exercise II.3.1. By transforming line integrals along $\boldsymbol{\chi}(\mathscr{C}, t)$ back into integrals along the stationary curve $\mathscr{C}$ in $\boldsymbol{\kappa}(\mathscr{B})$, construct a formal proof of (1) and so clarify the notation.

Exercise II.13.2. By using (II.12-7), calculate the rate of change of the volume of a material region in a homogeneous transplacement and hence provide another proof of (II.6-8).

Exercise II.13.3. Letting $\int_{\mathscr{C}} \cdots ds$ denote integration with respect to arc length along a material curve $\mathscr{C}$, show that if $\mathbf{t}(s)$ is the unit tangent to the present shape of $\mathscr{C}$ at s, then

$$\frac{d}{dt}\int_{\mathscr{C}} f\, ds = \int_{\mathscr{C}} (\dot{f} + f\mathbf{t}\cdot\mathbf{D}\mathbf{t})\, ds. \tag{II.13-2}$$

Now suppose that a material surface $\mathscr{S}$ has the parametric representation $\mathbf{X} = \mathbf{H}(a, b)$ in $\boldsymbol{\kappa}(\mathscr{B})$. The present shape of this material surface is $\mathbf{x} = \chi_{\boldsymbol{\kappa}}(\mathbf{H}(a, b), t) \equiv \mathbf{h}(a, b, t)$. Let $\partial_a\mathbf{X}$ and $\partial_b\mathbf{X}$ denote the partial derivatives of $\mathbf{H}$; let $\partial_a\mathbf{x}$ and $\partial_b\mathbf{x}$ denote the partial derivatives of $\mathbf{h}$. Then by use of the rule for differentiating composite functions, followed by use of (II.11-5) and properties of the exterior product, we find that

$$\begin{aligned} \partial_a\mathbf{x} &= \mathbf{F}\partial_a\mathbf{X}, \\ (\partial_a\mathbf{x})^{\cdot} &= \dot{\mathbf{F}}\partial_a\mathbf{X} = \dot{\mathbf{F}}\mathbf{F}^{-1}\partial_a\mathbf{x} = \mathbf{G}\partial_a\mathbf{x}, \\ (\partial_a\mathbf{x}\wedge\partial_b\mathbf{x})^{\cdot} &= \mathbf{G}(\partial_a\mathbf{x}\wedge\partial_b\mathbf{x}) + (\partial_a\mathbf{x}\wedge\partial_b\mathbf{x})\mathbf{G}^{\mathrm{T}}. \end{aligned} \tag{II.13-3}$$

Let $\mathbf{S}$ be a skew tensor field. By use of (3) we quickly obtain LAMB's formula

$$\begin{gathered} \frac{d}{dt}\int_{\mathscr{S}} \mathbf{S}\cdot(\partial_a\mathbf{x}\wedge\partial_b\mathbf{x})\, da\, db = \int_{\mathscr{S}} \mathbf{S}^{\mathrm{c}}\cdot(\partial_a\mathbf{x}\wedge\partial_b\mathbf{x})\, da\, db, \\ \mathbf{S}^{\mathrm{c}} \equiv \dot{\mathbf{S}} + \mathbf{S}\mathbf{G} + \mathbf{G}^{\mathrm{T}}\mathbf{S}. \end{gathered} \tag{II.13-4}$$

The integral on the left-hand side is called the *flux* of $\mathbf{S}$ through the present shape $\chi(\mathscr{S}, t)$ of the material surface $\mathscr{S}$. From (4) we read off ***Zorawski's criterion***: *In order for the flux of a skew tensor field* $\mathbf{S}$ *to remain constant in time for each material surface, it is necessary and sufficient that*

$$\mathbf{S}^{\mathrm{c}} = \mathbf{0}. \tag{II.13-5}$$

In this notation the theorem (II.11-34) of D'ALEMBERT, EULER, and BELTRAMI appears as

$$\mathbf{W}^{\mathrm{c}} = \mathbf{W}_2. \tag{II.13-6}$$

If for $\mathbf{S}$ we take $\mathbf{W}$, the integral upon the left-hand side of (4) becomes the *flux of vorticity* through $\chi(\mathscr{S}, t)$. From ZORAWSKI's criterion (5) we then read off a classic vorticity theorem: *In order that the flux of vorticity through each material surface shall remain constant in time, it is necessary and sufficient that*

$$\mathbf{W}_2 = \mathbf{0}. \tag{II.11-31$_{\mathrm{r}}$}$$

The statement that the D'Alembert–Euler condition (II.11-31) is sufficient for constant flux is ***Helmholtz's Third Vorticity Theorem.***

The material derivative $\dot{\mathbf{f}}$, defined by (II.6-3), reflects use of the Euclidean parallel transport. Following the path of a body-point from the place it occupies at the time t to the place it occupies at the time $t + h$, we use the Euclidean parallel transport to translate the value of $\mathbf{f}$ at the latter point back to the former point, subtract from it the value of $\mathbf{f}$ there, divide by h, and pass to the limit to obtain $\dot{\mathbf{f}}$. If we do just the same thing but use the parallelism induced by the motion of the deforming body, we obtain[1]

$$\mathbf{f}^{c} = \mathbf{f}' + (\operatorname{grad} \mathbf{f})\dot{\mathbf{x}} - \mathbf{G}\mathbf{f},$$
$$= \dot{\mathbf{f}} - \mathbf{G}\mathbf{f}. \tag{II.13-7}$$

Thus $\mathbf{f}^{c} = \mathbf{0}$ if and only if $\mathbf{f}$ obeys in all motions the same relation as does the position vector of a body-point in a homogeneous transplacement, namely, $(\text{II}.12\text{-}7)_3$. The quantity $\mathbf{S}^{c}$ defined by $(4)_2$ has a similar interpretation. We may refer to $\mathbf{f}^{c}$ and $\mathbf{S}^{c}$ as the *convected time-fluxes* of $\mathbf{f}$ and $\mathbf{S}$, respectively.

The *vector lines* of a non-vanishing field $\mathbf{f}$ are the curves everywhere tangent to $\mathbf{f}$. Generally these curves move and deform in the course of time. If they do so in such a way as to be occupied always by the same set of body-points, they are material lines. A field of such a kind has *material vector lines*. A material line that once is a vector line of $\mathbf{f}$ is then always a vector line of $\mathbf{f}$.

To determine the non-vanishing fields $\mathbf{f}$ that have material vector lines, we let a curve $\mathscr{C}$ in $\boldsymbol{\kappa}(\mathscr{B})$ be given by the parametric representation $\mathbf{X} = \mathbf{H}(a)$ and proceed as we did above in considering material surfaces. By use of $(\text{II}.13\text{-}3)_4$ we obtain

$$(\mathbf{f} \wedge \partial_a \mathbf{x})^{\cdot} = \dot{\mathbf{f}} \wedge \partial_a \mathbf{x} + \mathbf{f} \wedge \mathbf{G}\partial_a \mathbf{x}. \tag{II.13-8}$$

The material line generated by $\mathscr{C}$ is presently a vector line of $\mathbf{f}$ if and only if there is a non-vanishing scalar field A such that

$$\partial_a \mathbf{x} = A\mathbf{f}. \tag{II.13-9}$$

The material line then remains always a vector line if and only if (9) implies for all t that

$$(\mathbf{f} \wedge \partial_a \mathbf{x})^{\cdot} = \mathbf{0}. \tag{II.13-10}$$

Putting (9) into (8) yields

$$(\mathbf{f} \wedge \partial_a \mathbf{x})^{\cdot} = A(\dot{\mathbf{f}} - \mathbf{G}\mathbf{f}) \wedge \mathbf{f}. \tag{II.13-11}$$

[1] A clear explanation of the idea is included in the paper by BOLDER cited above in §II.6. The standard way to introduce the convected derivative begins from the *Lie derivative* $\mathscr{L}_{\mathbf{v}}$ based upon a vector field $\mathbf{v}$ and then sets $\mathbf{f}^{c} \equiv \mathbf{f}' + \mathscr{L}_{\dot{\mathbf{x}}}\mathbf{f}$. For the Lie derivative a standard old reference is §10 of J. A. SCHOUTEN, *Ricci-Calculus*, Berlin, Springer-Verlag, 1954.

Comparison with (7) yields the ***Helmholtz–Zorawski criterion***: *The non-vanishing field* $\mathbf{f}$ *has material vector lines if and only if*

$$\mathbf{f} \wedge \mathbf{f}^c = \mathbf{0}. \tag{II.13-12}$$

We can express this result equivalently in terms of the unit vector $\mathbf{e}$ in the direction of $\mathbf{f}$, that is, $\mathbf{e} \equiv |\mathbf{f}|^{-1}\mathbf{f}$:

$$\mathbf{f}^c = (\mathbf{e} \cdot \mathbf{f}^c)\mathbf{e}. \tag{II.13-13}$$

In this formula we may, if we like, choose $\mathbf{f}$ to be a field $\mathbf{e}$ of unit magnitude. Then

$$\mathbf{e}^c = (\mathbf{e} \cdot \mathbf{e}^c)\mathbf{e}. \tag{II.13-14}$$

Of course $\mathbf{e} \cdot \dot{\mathbf{e}} = 0$, and from (7) we see that $\mathbf{e} \cdot \mathbf{e}^c = -\mathbf{e} \cdot \mathbf{G}\mathbf{e}$, so in general $\mathbf{e} \cdot \mathbf{e}^c \neq 0$.

By use of (13) we may express the results of §II.12 in more general forms, without recourse either to homogeneous transplacements or to infinitesimals. The student should convince himself of this fact by solving the following exercise.

Exercise II.13.4 (STOKES, GOSIEWSKI, TRUESDELL & TOUPIN, WANG). Let $\mathbf{e}$ be a field of unit vectors having material vector lines. Show that

$$\dot{\mathbf{e}} = [\mathbf{D} + \mathbf{W} - (\mathbf{e} \cdot \mathbf{D}\mathbf{e})\mathbf{1}]\mathbf{e}. \tag{II.13-15}$$

Show that if such an $\mathbf{e}$ presently lies in a principal axis of stretching, it is presently suffering a rigid motion with spin $\mathbf{W}$. Show that if $\mathbf{m}$ and $\mathbf{n}$ are unit vector fields having material vector lines, then

$$\begin{aligned} (\mathbf{m} \cdot \mathbf{n})^{\cdot} &= 2\mathbf{m} \cdot \mathbf{D}\mathbf{n} - (\mathbf{m} \cdot \mathbf{D}\mathbf{m} + \mathbf{n} \cdot \mathbf{D}\mathbf{n})(\mathbf{m} \cdot \mathbf{n}), \\ \mathbf{m} \cdot \dot{\mathbf{n}} - \mathbf{n} \cdot \dot{\mathbf{m}} &= 2\mathbf{m} \cdot \mathbf{W}\mathbf{n} + (\mathbf{m} \cdot \mathbf{D}\mathbf{m} - \mathbf{n} \cdot \mathbf{D}\mathbf{n})(\mathbf{m} \cdot \mathbf{n}). \end{aligned} \tag{II.13-16}$$

Relate these results to those on homogeneous transplacements, given in §II.12.

The *stream lines* of a motion are the vector lines of its velocity field $\dot{\mathbf{x}}$. Generally these lines vary from one time to another; they are not generally the paths of the body-points. It is plain that *the stream lines and the paths of the body-points coincide if and only if both are steady*. In order that a family of lines be steady, it is necessary and sufficient that any tangent field shall suffer change only in magnitude, not in direction. Therefore, in order that the stream lines of a non-vanishing velocity field $\dot{\mathbf{x}}$ be steady, it is necessary and sufficient that

$$\dot{\mathbf{x}} \wedge \dot{\mathbf{x}}' = \mathbf{0}. \tag{II.13-17}$$

This same formula should emerge also as a condition for the stream lines to be material, and it does. Indeed, if we apply (7) to $\dot{\mathbf{x}}$, we find from (II.6-4) that $\dot{\mathbf{x}}^c = \dot{\mathbf{x}}'$, and placing this result in (12) yields (17). Of course (17) is satisfied by any steady flow.

Let $\mathbf{S}$ denote a field of skew tensors. A curve whose tangent at each $\mathbf{x}$ lies in the nullspace of $\mathbf{S}(\mathbf{x})$ is a *vector line* of $\mathbf{S}$. In discussing such vector lines we shall presume that $\dim \mathscr{E} = 3$. Then if $\mathbf{S} \neq \mathbf{0}$, the vector lines of the field $\mathbf{S}$ are the vector lines of the field of axes of $\mathbf{S}$.

Exercise II.13.5. Assuming that $\dim \mathscr{E} = 3$, prove that a material line that is once a vector line of the skew-tensor field $\mathbf{S}$ remains always a vector line of $\mathbf{S}$ if and only if

$$\mathbf{S}\mathbf{S}^c = \mathbf{S}^c\mathbf{S}. \qquad \text{(II.13-18)}$$

Hence in order for the vector lines of $\mathbf{S}$ to be material lines it is sufficient but not necessary that the flux of $\mathbf{S}$ through each material surface shall remain constant in time.

The vector lines of $\mathbf{W}$ are called *vortex lines.* If we take $\mathbf{W}$ for $\mathbf{S}$ in (18) and use (II.11-34), we obtain the condition

$$\mathbf{W}\mathbf{W}_2 = \mathbf{W}_2\mathbf{W}. \qquad \text{(II.13-19)}$$

Since two non-null skew tensors commute if and only if they have the same axis, from (19) we read off a theorem due to POINCARÉ: *In order that a material line which is once a vortex line shall remain always a vortex line, it is necessary and sufficient that the nullspace of* $\mathbf{W}_2$ *shall contain the nullspace of* $\mathbf{W}$, *or vice versa.* If $\mathbf{W}_2 \neq \mathbf{0}$, the condition (18) asserts that *the vector lines of* $\mathbf{W}_2$ *are the vortex lines.* If $\mathbf{W}_2 = \mathbf{0}$, the condition (19) is satisfied, and thus follows the celebrated ***Second Vorticity Theorem*** of HELMHOLTZ: *In a motion that satisfies the D'Alembert–Euler condition* (II.11-31), *a material line that is once a vortex line is always a vortex line.*

A closed line is called a *circuit.* The line integral of the velocity field around a circuit $\mathscr{C}$ is called the *circulation* of that circuit, for the velocity field. It was introduced by KELVIN as a measure of the accumulated tangential velocity of the body-points presently occupying that circuit. If $\mathscr{C} = \partial\mathscr{A}$, KELVIN's transformation of line integrals into surface integrals shows that

$$\int_{\partial\mathscr{A}} \dot{\mathbf{x}} \cdot d\mathbf{x} = \int_{\mathscr{A}} \mathbf{W} \cdot (\partial_a \mathbf{x} \wedge \partial_b \mathbf{x})\, da\, db: \qquad \text{(II.11-21)}_r$$

The circulation of a circuit equals the flux of the spin through any surface of which that circuit is the boundary. The usual condition of orientation is adopted here, and the fields and surfaces are presumed smooth enough to ensure the validity of the transformation.

A surface consisting entirely of vortex lines is a *vortex surface.* From (II.11-21) we see that at a given instant, *a surface is a vortex surface if and only if the circulation of every sufficiently small circuit on it is null.* In particular, *a motion is irrotational if and only if the circulation of every sufficiently small circuit is null.*

A motion such that the circulation of every material circuit is constant in time is called *circulation-preserving*. Because of (II.11-21) we may express HELMHOLTZ's Third Vorticity Theorem and its converse as follows: *The D'Alembert–Euler condition* (II.11-31) *is necessary and sufficient that the motion be circulation-preserving.* KELVIN's proof of this fact amounts to substitution of $\dot{\mathbf{x}}$ for $\mathbf{f}$ in (1) so as to obtain for a material circuit $\mathscr{C}$

$$\frac{d}{dt}\int_{\mathscr{C}} \dot{\mathbf{x}} \cdot d\mathbf{x} = \int_{\mathscr{C}} \left(\ddot{\mathbf{x}} + \operatorname{grad}\frac{1}{2}|\dot{\mathbf{x}}|^2\right) \cdot d\mathbf{x},$$

$$= \int_{\mathscr{C}} \ddot{\mathbf{x}} \cdot d\mathbf{x}, \qquad \text{(II.13-20)}$$

the second step being a consequence of the fact that $\mathscr{C}$ is a circuit. In virtue of a standard theorem on lamellar fields, the integral on the right-hand side vanishes for all $\mathscr{C}$ if and only if grad $\ddot{\mathbf{x}}$ is symmetric. The result then follows by (II.11-29).

There are several ways to see that *every irrotational flow is circulation-preserving.* One way has been indicated in Exercise II.11.7.

Exercise II.13.6 (KELVIN). Directly from the concept of circulation, prove the Helmholtz Theorems and the Lagrange–Cauchy Theorem (Exercise II.11.8) expressed for a material region.

Exercise II.13.7 (APPELL, in principle). Prove that a motion with material vortex lines is circulation-preserving if and only if the vorticity satisfies the differential relation

$$(JW)^{\cdot} = JW\mathbf{n} \cdot \mathbf{D}\mathbf{n}, \qquad \text{(II.11-36)}_r$$

$\mathbf{n}$ being any unit vector in the nullspace of $\mathbf{W}$.

We can now prove an important theorem of APPELL: *A rotational motion with material vortex lines is circulation-preserving if and only if*

$$\int_{\mathscr{C}} \frac{ds}{JW} = \text{const.} \qquad \text{(II.13-21)}$$

for every finite segment $\mathscr{C}$ *of a material vortex line.* Indeed, because of (2)

$$\frac{d}{dt}\int_{\mathscr{C}} \frac{ds}{JW} = \int_{\mathscr{C}} \left[-\frac{(JW)^{\cdot}}{(JW)^2} + \frac{1}{JW}\mathbf{t}\cdot\mathbf{Dt} \right] ds, \qquad \text{(II.13-22)}$$

$\mathbf{t}$ being either of the two continuous fields of unit tangent vectors to the actual shape of $\mathscr{C}$. To conclude the proof, we apply the result of the preceding exercise. The theorem shows that in a circulation-preserving motion the material vortex lines grow longer if W/ρ increases, shorter if W/ρ decreases. The same result may be inferred also from HELMHOLTZ's Second Theorem. If the circulation-preserving motion is isochoric, the result is still simpler: The vortex lines stretch or shrink according as the spin at points upon them increases or decreases.

14. Changes of Frame. Frame-Indifference

The concept of framing has been explained in §I.6, and the transformations induced by a change of frame have been developed in §I.9. The motion (II.1-1) of a body $\mathscr{B}$ is described with respect to a certain framing ϕ; with respect to another framing ϕ^*, it is given by the mapping, say,

$$\mathbf{x}^* = \chi^*(X, t^*). \qquad \text{(II.14-1)}$$

We regard a change of frame (I.9-4) as expressing the relation between the places and times, $(\mathbf{x}, t)$ and $(\mathbf{x}^*, t^*)$, of the same event as it appears to different observers. Thus if (II.1-1) and (1) are to represent the same experiences of a body as apparent to observers in ϕ and ϕ^*, respectively, the motions χ and χ^* must be related by (I.9-11), which we rewrite here:

$$\chi^*(X, t + a) = \mathbf{x}_0^*(t) + \mathbf{Q}(t)(\chi(X, t) - \mathbf{x}_0), \qquad \text{(II.14-2)}$$

$\mathbf{x}_0$ and a being the place and time with respect to ϕ of some assigned event, $\mathbf{x}_0^*$ being a place-valued function, and $\mathbf{Q}$ being a function whose values are orthogonal tensors.

If we choose to describe the motion in terms of a reference placement $\boldsymbol{\kappa}$, the corresponding transplacements $\chi_{\boldsymbol{\kappa}}^*$ and $\chi_{\boldsymbol{\kappa}}$ are related in the same way:

$$\chi_{\boldsymbol{\kappa}}^*(\mathbf{X}, t + a) = \mathbf{x}_0^*(t) + \mathbf{Q}(t)(\chi_{\boldsymbol{\kappa}}(\mathbf{X}, t) - \mathbf{x}_0). \qquad \text{(II.14-3)}$$

In §I.11 we have introduced the concept of frame-indifference. Briefly, a scalar-valued function of place and time is frame-indifferent if it is in fact a function of events, independent of framing. A vector-valued function is frame-indifferent if its value in ϕ^* effects the same translation of the places of events in ϕ^* as its value in ϕ effects upon the places of these same events in $\oint$. A tensor-valued function is frame-indifferent if it transforms each frame-indifferent vector into a frame-indifferent vector. Formally, as we have shown in §I.11, these three conditions for frame-indifference are

$$\begin{aligned} F^* &= F && \text{for scalars,} \\ \mathbf{v}^* &= \mathbf{Qv} && \text{for vectors,} \\ \mathbf{T}^* &= \mathbf{QTQ}^{\mathrm{T}} && \text{for tensors (of second order),} \end{aligned} \tag{II.14-4}$$

the asterisks indicating quantities appropriate to the framing $\oint^*$, and $\mathbf{Q}$ being the orthogonal tensor that occurs in the change of frame (2).

When a quantity is defined by a prescription valid in all frames, conditions such as (II.14-4) may or may not be satisfied. In §I.9 we have calculated the relation (I.9-14) connecting the velocities $\dot{\chi}$ and $\dot{\chi}^*$ as obtained in $\oint$ and $\oint^*$. The result shows that $\dot{\chi}$ and $\dot{\chi}^*$ do not satisfy $(4)_2$, and hence the velocity is not frame-indifferent. Indeed, (I.9-14) shows that the spin $\mathbf{A}$ of $\oint^*$ with respect to $\oint$ gives rise to a velocity in $\oint^*$, which is in fact the velocity corresponding to a rigid motion for which $\oint^*$ is a rest framing (*cf*. §I.10). Likewise, the relation (I.9-21) connecting the accelerations $\ddot{\chi}$ and $\ddot{\chi}^*$ in $\oint$ and $\oint^*$ shows that the acceleration is not frame-indifferent.

Now we shall consider the effect of a change of frame upon quantities for whose definition not only a frame of reference but also a reference placement is employed. We begin with the local deformation. Since the definition (II.5-1) applies both in $\oint$ and in $\oint^*$, we have

$$\mathbf{F} \equiv \nabla\chi_{\kappa}(\mathbf{X}, t), \qquad \mathbf{F}^* \equiv \nabla\chi_{\kappa}^*(\mathbf{X}, t). \tag{II.14-5}$$

Taking the gradient of the relation (3) shows that

$$\mathbf{F}^* = \mathbf{QF}. \tag{II.14-6}$$

Thus the local deformation is not frame-indifferent.

By applying to (6) the polar decomposition (II.9-1), we see that

$$\mathbf{R}^*\mathbf{U}^* = \mathbf{QRU}. \tag{II.14-7}$$

Because $\mathbf{QR}$ is orthogonal and because the polar decomposition of an invertible tensor is unique,

$$\mathbf{R}^* = \mathbf{QR}, \qquad \text{and} \qquad \mathbf{U}^* = \mathbf{U}. \tag{II.14-8}$$

Hence

$$\begin{aligned} \mathbf{V}^* = \mathbf{R}^*\mathbf{U}^*\mathbf{R}^{*\mathrm{T}} &= \mathbf{QRU}(\mathbf{QR})^{\mathrm{T}}, \\ &= \mathbf{QVQ}^{\mathrm{T}}. \end{aligned} \tag{II.14-9}$$

Thus we have shown that $\mathbf{V}$ is frame-indifferent, while $\mathbf{F}$, $\mathbf{R}$, and $\mathbf{U}$ are not. Of course, $\mathbf{C}^* = \mathbf{C}$ and $\mathbf{B}^* = \mathbf{QBQ}^{\mathrm{T}}$, as is immediate by applying $(8)_2$ and $(9)_3$ to the definitions (II.9-5).

If we differentiate (6) with respect to time, we find that

$$\dot{\mathbf{F}}^* = \mathbf{Q}\dot{\mathbf{F}} + \dot{\mathbf{Q}}\mathbf{F}, \tag{II.14-10}$$

but by (II.11-5) $\dot{\mathbf{F}} = \mathbf{GF}$, and $\dot{\mathbf{F}}^* = \mathbf{G}^*\mathbf{F}^*$, so

$$\begin{aligned} \mathbf{G}^*\mathbf{F}^* &= \mathbf{QGF} + \dot{\mathbf{Q}}\mathbf{F}, \\ &= \mathbf{QGQ}^{\mathrm{T}}\mathbf{F}^* + \dot{\mathbf{Q}}\mathbf{Q}^{\mathrm{T}}\mathbf{F}^*. \end{aligned} \tag{II.14-11}$$

Because $\mathbf{F}^*$ is invertible, it may be cancelled from this equation, which by use of the Euler–Cauchy–Stokes Decomposition (II.11-8) becomes

$$\mathbf{D}^* + \mathbf{W}^* = \mathbf{Q}(\mathbf{D} + \mathbf{W})\mathbf{Q}^{\mathrm{T}} + \mathbf{A}, \tag{II.14-12}$$

$\mathbf{A}$ being the spin (I.9-15) of $\mathfrak{f}$ with respect to $\mathfrak{f}^*$:

$$\mathbf{A} \equiv \dot{\mathbf{Q}}\mathbf{Q}^{\mathrm{T}} = -\mathbf{A}^{\mathrm{T}}. \tag{II.14-13}$$

Since a decomposition into symmetric and skew parts is unique,

$$\mathbf{D}^* = \mathbf{QDQ}^{\mathrm{T}}, \qquad \mathbf{W}^* = \mathbf{QWQ}^{\mathrm{T}} + \mathbf{A}. \tag{II.14-14}$$

These formulae embody the ***Theorem of Zaremba and Zorawski***: *The stretching is frame-indifferent, while the spin in $\mathfrak{f}^*$ is the sum of the spin in $\mathfrak{f}$ and the spin of $\mathfrak{f}$ with respect to $\mathfrak{f}^*$.* The assertion is intuitively plain, since a change of frame in effect superimposes a rigid motion, possibly followed by a reflection, neither of which alters the stretchings of elements though the former does rotate the directions in which those stretchings seem to occur. By a result in Exercise I.11.2, the principal stretchings and the principal axes of stretching are likewise frame-indifferent.

If we differentiate (II.9-10) n times with respect to τ and then put $\tau = t$, by appeal to the definition (II.11-17) we conclude that

$$\overset{(n)}{\mathbf{C}} = \mathbf{F}^{\mathrm{T}}\mathbf{A}_n\mathbf{F}, \tag{II.14-15}$$

$\mathbf{A}_n$ being the nth Rivlin–Ericksen tensor. Applying the polar decomposition theorem (II.9-1), then, we conclude that

$$\mathbf{U}^{-1}\overset{(n)}{\mathbf{C}}\mathbf{U}^{-1} = \mathbf{R}^{\mathrm{T}}\mathbf{A}_n\mathbf{R}. \tag{II.14-16}$$

Likewise

$$\mathbf{U}^{*-1}\overset{(n)}{\mathbf{C}}{}^*\mathbf{U}^{*-1} = \mathbf{R}^{*\mathrm{T}}\mathbf{A}_n^*\mathbf{R}^*. \tag{II.14-17}$$

We have shown a little above that $\mathbf{U}^* = \mathbf{U}$ and $\mathbf{C}^* = \mathbf{C}$; hence also $\overset{(n)}{\mathbf{C}}{}^* = \overset{(n)}{\mathbf{C}}$, so the left-hand sides of (16) and (17) are equal. Therefore

$$\mathbf{R}^{*\mathrm{T}}\mathbf{A}_n^*\mathbf{R}^* = \mathbf{R}^{\mathrm{T}}\mathbf{A}_n\mathbf{R}. \tag{II.14-18}$$

By $(8)_1$ we conclude that

$$\mathbf{A}_n^* = \mathbf{Q}\mathbf{A}_n\mathbf{Q}^{\mathsf{T}}. \tag{II.14-19}$$

Thus *the Rivlin–Ericksen tensors are frame-indifferent.* This result generalizes the first assertion in the Zaremba–Zorawski Theorem. The second is equally easy to generalize, but the generalization is not so easy to interpret.

Exercise II.14.1. Prove from (II.8-7) and (6) that $\mathbf{U}_t$ is frame-indifferent and that

$$\mathbf{R}_t^*(\tau) = \mathbf{Q}(\tau)\mathbf{R}_t(\tau)\mathbf{Q}(t)^{\mathsf{T}}. \tag{II.14-20}$$

General References

§§15–25 of NFTM ("The Non-Linear Field Theories of Mechanics," *Handbuch der Physik* $\mathbf{3}_3$, Berlin, Heidelberg, and New York, Springer-Verlag, 1965).

§§13–171 (exhaustive treatment of kinematics in component notation) of CFT ("The Classical Field Theories," *Handbuch der Physik* $\mathbf{3}_1$, Berlin, Göttingen, and Heidelberg, Springer-Verlag, 1960).

Chapter III

The Stress Tensor

Although I envisage here very great generality both in the nature of the fluid and in the forces that act upon each of its particles, I have no fear of those reproaches often levelled with good reason at them who have undertaken to generalize the researches of others. I agree that often an excessive generality obscures rather than enlightens, and that sometimes it leads to calculations so messy as to make it extremely hard to draw any conclusions from them for the simplest cases. When generalizations are subject to this drawback, most certainly we ought abstain from them altogether and limit our studies to particular cases.

But, in the subject I intend to explain, just the opposite happens: The generality that I embrace, far from dazzling our lights, will reveal to us rather the veritable laws of Nature in all their brilliance, and in them we shall find even stronger reason to admire her beauty and her simplicity. It will be an important lesson to learn that some principles till now believed bound to some special case are of greater breadth. Finally, these researches will demand calculations scarcely any more troublesome, and it will be easy to apply them to all special cases we might set up.

EULER
General principles of the state of equilibrium of fluids
Mémoires de l'Académie des Sciences de Berlin **11**, 217–273 (1757)

The geometers who have investigated the equations of equilibrium or motion of thin plates or of surfaces, either elastic or inelastic, have distinguished two kinds of forces, the one produced by dilatation or contraction, the other by the bending of these surfaces .... It has seemed to me that these two kinds of forces could be reduced to a single one, which ought to be called always *tension* or *pressure*, a force which acts upon each element of a section chosen at will, not only in a flexible surface but also in a solid, whether elastic or inelastic, and which is of the same kind as the hydrostatic pressure exerted by a fluid at rest upon the exterior surface of a body, except that the new pressure does not always remain perpendicular to the faces subject to it, nor is it the same in all directions at a given point.

CAUCHY
On the pressure or tension in a solid body
Exercices de Mathématiques, Seconde Année (1827)

In many otherwise good textbooks a standing confusion reigns between three groups of forces: 1. Internal and external forces. 2. Volume and surface forces—a distinction which the mechanics of points is altogether incapable of perceiving. 3. Applied forces and forces of reaction.

HAMEL
On the foundations of mechanics
Mathematische Annalen **66**, 350–397 (1909)

1. Forces and Torques. The Laws of Dynamics. Body Forces and Contact Forces

Forces and torques, like bodies, motions, and masses, are primitive elements of mechanics. They are mathematical quantities introduced *a priori*, represented by symbols, and subjected to mathematical axioms that delimit their properties and render them clear and useful for the description of mechanical phenomena in nature. Axioms for a system of forces in general have been presented in §I.5; torques have been defined as the moments of forces in §I.8; general axioms of dynamics, which relate forces and torques to the motion they effect upon a given body, have been given in §§I.12 and I.13. In the remainder of this book, except in passages where we discuss frame-indifference, we shall suppose that the reference framing ϕ is an inertial one, and we shall base dynamics on ***Euler's Laws of Motion***:

$$\mathbf{f}^a = \dot{\mathbf{m}}, \qquad \mathbf{F}^a = \dot{\mathbf{M}}. \tag{I.13-11$_r$}$$

That is, the rate of increase of the linear momentum of any body equals the applied force $\mathbf{f}^a$ upon it, and the rate of increase of the rotational momentum with respect to $\mathbf{x}_0$ equals the applied torque $\mathbf{F}^a$ upon it, the place $\mathbf{x}_0$ being stationary in the inertial frame.

We begin by restating these laws in more explicit forms, referred to a part $\mathscr{P}$ of $\mathscr{B}$ and to its shape $\chi(\mathscr{P})$ in the inertial framing ϕ. These forms, which follow at once from (I.8-5) and (I.13-10), are

$$\begin{gathered}\int_{\chi(\mathscr{P})} \rho\ddot{\mathbf{x}}\,dV = \mathbf{f}^a(\mathscr{P}),\\ \int_{\chi(\mathscr{P})} (\mathbf{x}-\mathbf{x}_0)\wedge\rho\ddot{\mathbf{x}}\,dV = \mathbf{F}^a(\mathscr{P})_{\mathbf{x}_0};\end{gathered} \tag{III.1-1}$$

we recall that the applied force $\mathbf{f}^a$ and the applied torque $\mathbf{F}^a$ may depend upon the time t, as does the shape $\chi(\mathscr{P})$, though we do not so indicate in the notation. Thus the applied force and torque upon $\mathscr{P}$ are expressed in terms of integrals over the actual shape of $\mathscr{P}$. As always, $\ddot{\mathbf{x}}$ is the acceleration field on $\chi(\mathscr{B})$, and we assume that it is essentially bounded.

In continuum mechanics two different systems of forces are introduced: *body forces* $\mathbf{f}^a_B$, which may be exerted mutually by bodies, whether or not they be in contact, and which are presumed related to the masses of the bodies, and *contact forces* $\mathbf{f}_C$, which are exerted by one body on another through their common surface of contact and are presumed related to that surface, distributed over it, and independent of the masses of the bodies on either side. Specifically, the applied force is assumed to be the sum of two forces of different kinds:

$$\mathbf{f}^{\mathrm{a}} = \mathbf{f}_{\mathrm{B}}^{\mathrm{a}} + \mathbf{f}_{\mathrm{C}}, \tag{III.1-2}$$

where

$$\begin{aligned} \mathbf{f}_{\mathrm{B}}^{\mathrm{a}}(\mathscr{P}) &= \int_{\chi(\mathscr{P})} \rho \mathbf{b}_{\chi(\mathscr{P})}\, dV, \\ \mathbf{f}_{\mathrm{C}}(\mathscr{P}) &= \int_{\partial\chi(\mathscr{P})} \mathbf{t}_{\partial\chi(\mathscr{P})}\, dA, \end{aligned} \tag{III.1-3}$$

$\mathscr{P}$ being any part of the body $\mathscr{B}$. The corresponding torques are given by

$$\begin{aligned} \mathbf{F}_{\mathrm{B}}^{\mathrm{a}}(\mathscr{P})_{\mathbf{x}_0} &= \int_{\chi(\mathscr{P})} (\mathbf{x} - \mathbf{x}_0) \wedge \rho \mathbf{b}_{\chi(\mathscr{P})}\, dV, \\ \mathbf{F}_{\mathrm{C}}(\mathscr{P})_{\mathbf{x}_0} &= \int_{\partial\chi(\mathscr{P})} (\mathbf{x} - \mathbf{x}_0) \wedge \mathbf{t}_{\partial\chi(\mathscr{P})}\, dA, \end{aligned} \tag{III.1-4}$$

and

$$\mathbf{F}_{\mathbf{x}_0}^{\mathrm{a}} = \mathbf{F}_{\mathrm{B}\mathbf{x}_0}^{\mathrm{a}} + \mathbf{F}_{\mathrm{C}\mathbf{x}_0}. \tag{III.1-5}$$

As is clear from $(3)_1$, the applied body force $\mathbf{f}_{\mathrm{B}}^{\mathrm{a}}$ is an absolutely continuous function of volume. For brevity, its density $\mathbf{b}_{\chi(\mathscr{P})}$ with respect to mass will be called henceforth the *body-force field* or even simply the *body force*. Moreover, we shall limit attention in this book to the case when $\mathbf{b}_{\chi(\mathscr{P})}$ itself is an assigned function of place and time and hence independent of $\chi(\mathscr{P})$:

$$\mathbf{b}_{\chi(\mathscr{P})} = \mathbf{b}(\mathbf{x}, t) \qquad \forall\, \chi. \tag{III.1-6}$$

Such fields of body force are called *external*.[1] Commonly the external body force is assumed to be *lamellar*:

$$\mathbf{b} = -\operatorname{grad} \varpi; \tag{III.1-7}$$

[1] Not all external body forces are included in (6). For example, the density of force exerted by a magnetic or electric field is a function of $\dot{\mathbf{x}}$ and of constitutive properties of the body on which it acts. For the purposes of this book (6) is sufficient.

the scalar function[1] ϖ is the *potential* of $\mathbf{b}$. If ϖ is a potential of $\mathbf{b}$, so is $\varpi + h$ if h is a function of time alone. A steady lamellar body force is called *conservative*. A conservative body force has steady potentials, and of course in dealing with such a body force we always choose one of these. If $\mathbf{b}$ is constant in space and time, as is appropriate to heavy bodies near the surface of the earth, it is called the field of *uniform gravity*.[2] For such a $\mathbf{b}$ the potentials are affine functions of the distance $h(\mathbf{x})$ of the place $\mathbf{x}$ from some fixed plane; in the case of uniform gravity a convenient choice is $\varpi = gh$, the constant g being the gravitational acceleration and $h(\mathbf{x})$ being the height of $\mathbf{x}$ above the surface of the earth.

Exercise III.1.1. Two systems of forces applied to the shape of a body are said to be *equipollent* if they give rise to the same resultant force and resultant torque on that shape. Prove that the field of uniform gravity is equipollent to a single force acting at the center of mass of the body, directed parallel to $\mathbf{b}$ and in the same sense ("downward"), and equal in magnitude to the weight of the body.

Body forces are of secondary interest in continuum mechanics, which concerns mainly the effects of contact forces, to which we now address ourselves.

According to $(3)_2$ the resultant contact force $\mathbf{f}_C$ is an absolutely continuous function of the area of the bounding surface $\partial\chi(\mathscr{P})$ on which it acts. The surface density $\mathbf{t}_{\partial\chi(\mathscr{P})}$ is called the *traction field* on $\partial\chi(\mathscr{P})$. If that field be known, the resultant contact force is determined and is independent of whatever may be occurring at places not lying upon $\partial\chi(\mathscr{P})$. In this sense, the traction field is equipollent to the action upon $\mathscr{P}$ of the bodies outside $\mathscr{P}$ and adjacent to it. The assumption that the contact force is of this kind is the ***cut principle*** of EULER and CAUCHY: *Within the shape of a body at any*

[1] In this book we always use the term "function" for a mapping, called in the older literature a "single-valued" function. "Cyclic" or "many-valued" potentials are important in many problems concerning multiply connected regions. Since this book is concerned mainly with local aspects of mechanics, and since "cyclic functions" are locally functions in the ordinary modern sense, we shall not take up the complications that may result from use of body forces with cyclic potentials. The reader already familiar with cyclic potentials can easily state for himself the generalizations to which they give rise in the few theorems in this book where they might be introduced. An example is EULER's theorem in §IV.8.

A clear elementary discussion of cyclic potentials may be found in §§49–54 of H. LAMB, *Hydrodynamics*, 2nd–6th eds., Cambridge, Cambridge University Press, 1895/1932, variously reprinted. A good example of a body force with cyclic potential is discussed in §6 of A. SOMMERFELD's *Mechanics of Deformable Bodies*, New York, Academic Press, 1950.

[2] The force of universal gravitation is a mutual body force, not an external one, and hence is not treated in this book.

given time, conceive a smooth, closed diaphragm; then the action of the part of the body outside that diaphragm and adjacent to it on that inside is equipollent to that of a field of vectors defined on the diaphragm.

Of course, the diaphragm may be chosen as the shape of the boundary of a body, the exterior of which we prefer not to specify, and in this case the cut principle does not furnish an interpretation for the traction $\mathbf{t}_{\partial\chi(\mathscr{B})}$. Rather, tractions upon the boundary of the largest body under consideration are regarded as prescribed by other considerations. For example, so as to represent the application of given forces upon the surface of a given body, without including in the theory such other bodies as may bring those forces to bear, we impose a *boundary condition of traction* by assigning $\mathbf{t}_{\partial\chi(\mathscr{B})}$, or a field closely related to it, on a given boundary surface such as $\partial\chi(\mathscr{B})$ or $\partial\boldsymbol{\kappa}(\mathscr{B})$. Examples of such conditions are given and discussed below, in §§III.8–III.9. In other cases we may leave $\mathbf{t}_{\partial\chi(\mathscr{B})}$ to be determined on such surfaces by imposing a *boundary condition of place*, typically by prescribing on $\partial\chi(\mathscr{B})$ the transplacement $\chi_{\kappa}(\mathbf{X})$ or some quantity derived from it.

We recall that $\mathscr{S}$ has the orientation of the outer normal $\mathbf{n}$ to $\partial\chi(\mathscr{A})$. If $\mathbf{t}_{\mathscr{S}} \cdot \mathbf{n} > 0$, the traction is said to be a *tension*; if $\mathbf{t}_{\mathscr{S}} \cdot \mathbf{n} < 0$, it is a *pressure*.

If we substitute (2)–(5) into EULER's laws (1). we obtain the ***Basic Laws of Motion*** of continuum mechanics, as far as this book is concerned[1]:

$$\int_{\chi(\mathscr{P})} \rho\ddot{\mathbf{x}}\, dV = \int_{\partial\chi(\mathscr{P})} \mathbf{t}_{\partial\chi(\mathscr{P})}\, dA + \int_{\chi(\mathscr{P})} \rho\mathbf{b}\, dV,$$

$$\int_{\chi(\mathscr{P})} (\mathbf{x} - \mathbf{x}_0) \wedge \rho\ddot{\mathbf{x}}\, dV = \int_{\partial\chi(\mathscr{P})} (\mathbf{x} - \mathbf{x}_0) \wedge \mathbf{t}_{\partial\chi(\mathscr{P})}\, dA + \int_{\chi(\mathscr{P})} (\mathbf{x} - \mathbf{x}_0) \wedge \rho\mathbf{b}\, dV, \tag{III.1-8}$$

for all parts $\mathscr{P}$ of all bodies in the universe.

As we have stated in §I.13, all forces are frame-indifferent. Thus, in particular, the contact forces and applied body forces are frame-indifferent. Consequently their densities are frame-indifferent vector fields:

$$\mathbf{b}^* = \mathbf{Q}\mathbf{b}, \qquad \mathbf{t}^*_{\partial\chi^*(\mathscr{P})} = \mathbf{Q}\mathbf{t}_{\partial\chi(\mathscr{P})}, \tag{III.1-9}$$

$\mathbf{Q}$ being the orthogonal tensor occurring in the change of frame (II.14-2).

Of course the forms (8) for the principles of linear and rotational momentum are valid only in an inertial frame. To obtain corresponding forms in a general frame, we need only replace the acceleration field $\ddot{\mathbf{x}}$ by the frame-indifferent vector field $\mathbf{a}_{\mathfrak{f}}$ that reduces to $\ddot{\mathbf{x}}$ when the frame is inertial. That frame-indifferent vector

[1] More general formulations relax the assumption that $\ddot{\mathbf{x}}$ exist everywhere at all times and take account of body couples, couple stresses, multipolar stresses, spin momentum, director stresses, *etc.*, as well as counterparts for diffusion and chemically reacting mixtures.

field we have calculated already and recorded as (II.4-7). With this replacement, the integrals on the left-hand sides of (8) become frame-indifferent, as are all four integrals on the right-hand sides.

The reader who is content to accept these equations, supplemented by axioms endowing $\mathbf{b}$ and $\mathbf{t}_{\partial\chi(\mathscr{P})}$ with some smoothness, may pass straight on to the next section.

The more critical reader will see two objections. First, the resultant contact force $\mathbf{f}_C$ does not define the traction uniquely, since to any $\mathbf{t}_{\partial\chi(\mathscr{P})}$ that satisfies $(3)_2$ we may add $\mathbf{Sn}$ if for $\mathbf{S}$ we take any tensor field such that $\operatorname{div} \mathbf{S} = \mathbf{0}$, and the contact force $\mathbf{f}_C$ will be the same. Second, the resultant body force $\mathbf{f}_B^a$ and resultant contact force $\mathbf{f}_C$ are not clearly related to the general concept of a system of forces, which is a function defined on pairs of separate bodies rather than on single bodies. For such a reader we conclude this section by an analysis which delivers the classical assumptions (3) and (4) as theorems[1] proved from assumptions of continuity phrased in terms of the modern theory of systems of forces, which we outlined in Chapter I. Accordingly, considering a fixed time t and not indicating it in the notation, we shall assume that the system of forces defined on $(\overline{\Omega} \times \overline{\Omega})_0$ is the sum of two such systems:

$$\mathbf{f}(\mathscr{A}, \mathscr{C}) = \mathbf{f}_B(\mathscr{A}, \mathscr{C}) + \mathbf{f}_C(\mathscr{A}, \mathscr{C}), \qquad \text{(III.1-10)}$$

which will be called the *system of body forces* and *system of contact forces*, respectively, and which will be distinguished by different mathematical properties designed to reflect the different interpretations given just after (3). The four axioms for forces laid down in §I.5 are assumed satisfied by both $\mathbf{f}_B$ and $\mathbf{f}_C$ and hence also by their sum, $\mathbf{f}$. In accord with the general theory of mechanics given in §I.12, we assume that the system of forces $\mathbf{f}$ is balanced. In accord with the Axioms of Inertia in §I.13, $\mathbf{f}$ must include the inertial force. Since this force is an absolutely continuous function of volume, it has to furnish a part of $\mathbf{f}_B$ rather than of $\mathbf{f}_C$. Thus what we have called $\mathbf{f}_B^a(\mathscr{P})$ and $\mathbf{f}_C(\mathscr{P})$ in (3) must turn out to be related to $\mathbf{f}_B$ and $\mathbf{f}_C$ as follows:

$$\begin{aligned} \mathbf{f}_B(\mathscr{P}, \mathscr{P}^e) &= \mathbf{f}_B^a(\mathscr{P}) - \int_{\chi(\mathscr{P})} \rho \ddot{\mathbf{x}} \, dV, \\ \mathbf{f}_C(\mathscr{P}, \mathscr{P}^e) &= \mathbf{f}_C(\mathscr{P}). \end{aligned} \qquad \text{(III.1-11)}$$

[1] A number of the arguments here are taken from the paper by M. E. GURTIN & W. O. WILLIAMS, "An axiomatic foundation for continuum thermodynamics," *Archive for Rational Mechanics and Analysis* **26**, 83–117 (1967). In that paper they are phrased in terms of scalar-valued functions having a thermomechanical rather than purely mechanical interpretation, but the mathematics is essentially the same.

We now formulate reasonable conditions on the systems of forces $\mathbf{f}_B$ and $\mathbf{f}_C$ that enable us, if with considerable difficulty, to prove that (11) results with $\mathbf{f}_B^a(\mathscr{P})$ and $\mathbf{f}_C(\mathscr{P})$ having the forms (3). The assumptions we shall make are immediate expressions of the basic ideas of forces exerted upon bodies in virtue of their masses and forces exerted by pairs of bodies in virtue of their areas of contact, namely: *There is a positive number K such that for any separate bodies $\mathscr{A}$ and $\mathscr{C}$*

$$\begin{aligned} |\mathbf{f}_B(\mathscr{A}, \mathscr{C})| &\leqq KM(\mathscr{A}), \\ |\mathbf{f}_C(\mathscr{A}, \mathscr{C})| &\leqq KA(\partial\chi(\mathscr{A}) \cap \partial\chi(\mathscr{C})), \end{aligned} \tag{III.1-12}$$

provided, respectively, that the mass of $\mathscr{A}$ be sufficiently small and that the area of the common boundary of the shapes of $\mathscr{A}$ and $\mathscr{C}$ be sufficiently small. The first requirement is independent of the motion χ; it refers to bodies alone, independently of their shapes. If the second requirement holds for one placement $\chi(\cdot, t)$, it holds for every other, although of course the value of K may depend upon χ and t. All possible body forces acting on $\mathscr{A}$ are ultimately bounded by the mass of $\mathscr{A}$, while all possible contact forces between $\mathscr{A}$ and $\mathscr{C}$ are ultimately bounded by the areas of the boundary they share in any shapes they may assume.

The proof that these assumptions do lead to applied forces of the form (3) is not easy. Beyond (12), we have little to work with: in effect, only that

I. both $\mathbf{f}_B(\cdot, \mathscr{C})$ and $\mathbf{f}_C(\cdot, \mathscr{C})$ are vector-valued measures defined over the parts $\mathscr{P}$ of any body $\mathscr{A}$ that is separate from $\mathscr{C}$, and

II. the sum $\mathbf{f}_B + \mathbf{f}_C$ is balanced.

The first of these follows by the theorem just after the statement of Axiom F4 in §I.5. The second does not imply that either $\mathbf{f}_B$ or $\mathbf{f}_C$ is balanced by itself. Indeed, neither can be in general, since if one were, the resultant force $\mathbf{f}_C(\mathscr{P})$ would vanish on every part $\mathscr{P}$, thereby excluding the possibility that contact forces contribute to the motions of the bodies on which they act.

By NOLL's corollary in §I.5 we can conclude from Statement II that $\mathbf{f}$ is pairwise equilibrated:

$$\mathbf{f}_B(\mathscr{A}, \mathscr{C}) + \mathbf{f}_C(\mathscr{A}, \mathscr{C}) = -\mathbf{f}_B(\mathscr{C}, \mathscr{A}) - \mathbf{f}_C(\mathscr{C}, \mathscr{A}). \tag{III.1-13}$$

A second major property of this sort of system of forces was noted in effect by CAUCHY himself. Namely, while $(12)_1$ is appropriate to body forces in general, when we apply it along with Statement II to continuum mechanics we may invoke (II.2-9) and so conclude that resultant contact force is ultimately bounded by volume: There is a positive number K such that

$$|\mathbf{f}_C(\mathscr{P}, \mathscr{P}^e)| \leqq KV(\chi(\mathscr{P})), \tag{III.1-14}$$

provided $V(\chi(\mathscr{P}))$ be sufficiently small. Hence

$$\lim_{V(\chi(\mathscr{P}))\to 0} \mathbf{f}_C(\mathscr{P}, \mathscr{P}^e) = \mathbf{0}. \tag{III.1-16}$$

Taking up first the system of contact forces $\mathbf{f}_C$, we shall prove that because of the assumption $(12)_2$, $\mathbf{f}_C(\mathscr{A}, \mathscr{C})$ depends upon $\mathscr{A}$ and $\mathscr{C}$ only through the surface of contact of their present shapes.

Lemma (GURTIN & WILLIAMS). *Let $(\mathscr{A}, \mathscr{C})$ and $(\hat{\mathscr{A}}, \hat{\mathscr{C}})$ be pairs of separate bodies; suppose that $\hat{\mathscr{A}} \prec \mathscr{A}$ and $\hat{\mathscr{C}} \prec \mathscr{C}$; suppose further that the shapes of $\mathscr{A}$ and $\mathscr{C}$ share the same boundary as do the shapes of $\hat{\mathscr{A}}$ and $\hat{\mathscr{C}}$:*

$$\partial\chi(\mathscr{A}) \cap \partial\chi(\mathscr{C}) = \partial\chi(\hat{\mathscr{A}}) \cap \partial\chi(\hat{\mathscr{C}}), \tag{III.1-17}$$

and that $\partial\chi(\mathscr{A})$ and $\partial\chi(\hat{\mathscr{A}})$ are like-oriented, as are $\partial\chi(\mathscr{C})$ and $\partial\chi(\hat{\mathscr{C}})$. Then

$$\mathbf{f}_C(\mathscr{A}, \mathscr{C}) = \mathbf{f}_C(\hat{\mathscr{A}}, \hat{\mathscr{C}}). \tag{III.1-18}$$

Proof. We set

$$\mathscr{A}^* \equiv \mathscr{A} \curlywedge \hat{\mathscr{A}}^e, \qquad \mathscr{C}^* \equiv \mathscr{C} \curlywedge \hat{\mathscr{C}}^e, \tag{III.1-19}$$

and note that $(\hat{\mathscr{A}}, \mathscr{C}^*)$ and $(\mathscr{A}^*, \mathscr{C})$ are also pairs of separate bodies. Then a sketch makes it plausible that

$$A(\partial\chi(\hat{\mathscr{A}}) \cap \partial\chi(\mathscr{C}^*)) = 0, \tag{III.1-20}$$

and also that

$$A(\partial\chi(\mathscr{A}^*) \cap \partial\chi(\mathscr{C})) = 0. \tag{III.1-21}$$

In view of $(12)_2$, the contact force between bodies whose surface of contact has area 0 is $\mathbf{0}$. Thus (20) and (21) show that

$$\mathbf{f}_C(\hat{\mathscr{A}}, \mathscr{C}^*) = \mathbf{0}, \qquad \mathbf{f}_C(\mathscr{A}^*, \mathscr{C}) = \mathbf{0}. \tag{III.1-22}$$

We now interrupt the proof of the lemma so as to present an elegant formal proof of the steps (20) and (21), which Mr. WILLIAMS has kindly given me for use in this book. To shorten the formulae, we introduce the temporary notation $\mathsf{A} \equiv \chi(\mathscr{A})$, $\mathsf{C} \equiv \chi(\mathscr{C})$, *etc.*, and we use the fact, which follows easily from (II.1-4), that if $\mathscr{A}$ and $\mathscr{C}$ are separate, their shapes can intersect at boundary points only:

$$\mathsf{A} \cap \mathsf{C} = \partial\mathsf{A} \cap \partial\mathsf{C}. \tag{III.1-23}$$

Since $\hat{\mathscr{A}} \prec \mathscr{A}$, it is enough to prove (21); that is,

$$A(\mathsf{A}^* \cap \mathsf{C}) = 0. \tag{III.1-24}$$

But

$$\mathsf{A}^* \cap \mathsf{C} = \partial\mathsf{A}^* \cap \partial\mathsf{A} \cap \partial\mathsf{C},$$

$$A(\mathsf{A}^* \cap \mathsf{C}) = A(\partial\mathsf{A}^* \cap \partial\mathsf{A} \cap \partial\mathsf{C}), \tag{III.1-25}$$

$$= A(\partial\mathsf{A}^* \cap \partial\hat{\mathsf{A}} \cap \partial\mathsf{C}),$$

so it will suffice to show that

$$A(\partial\mathsf{A}^* \cap \partial\hat{\mathsf{A}} \cap \partial\mathsf{C}) = 0. \tag{III.1-26}$$

For any body $\mathscr{D}$, the set of points on $\partial\mathsf{D}$ at which $\partial\mathsf{D}$ has a tangent plane are typical in the sense that the area of the set of other points is 0. Let $\mathscr{B}_r(\mathbf{x})$ denote the open ball of radius r at $\mathbf{x}$. If $\mathbf{x}$ is a typical point of $\partial\mathsf{D}$, then

$$\lim_{r\to 0} \frac{V(\mathsf{D} \cap \mathscr{B}_r(\mathbf{x}))}{V(\mathscr{B}_r(\mathbf{x}))} = \frac{1}{2}. \tag{III.1-27}$$

Since $V(\partial\mathsf{D}) = 0$, in (27) we may replace D by its interior $\mathring{\mathsf{D}}$ if we wish to.

Now let $\mathbf{x}$ be a point that is typical for $\partial\mathsf{A}^*$, $\partial\hat{\mathsf{A}}$, and $\partial\mathsf{C}$. Since $\mathring{\mathsf{A}}^*$, $\mathring{\hat{\mathsf{A}}}$, and $\mathring{\mathsf{C}}$ are pairwise disjoint,

$$\frac{V(\mathsf{A}^* \cap \mathscr{B}_r(\mathbf{x}))}{V(\mathscr{B}_r(\mathbf{x}))} + \frac{V(\hat{\mathsf{A}} \cap \mathscr{B}_r(\mathbf{x}))}{V(\mathscr{B}_r(\mathbf{x}))} + \frac{V(\mathsf{C} \cap \mathscr{B}_r(\mathbf{x}))}{V(\mathscr{B}_r(\mathbf{x}))} \leqq 1. \tag{III.1-28}$$

Taking the limit as $r \to 0$ leads to a result which contradicts (27). Thus the assumption that $\mathbf{x}$ is typical for $\partial\mathsf{A}^*$, $\partial\hat{\mathsf{A}}$, and $\partial\mathsf{C}$ is false. Therefore, the set of points common to $\partial\mathsf{A}^*$, $\partial\hat{\mathsf{A}}$, and $\partial\mathsf{C}$ contains no point typical for all three. Hence (26) follows.

As we have shown already, (26) implies (21), and so (22) holds.

Taking up the argument again at (22), we see that since

$$\mathscr{A} = \hat{\mathscr{A}} \vee \mathscr{A}^* \quad \text{and} \quad \mathscr{C} = \hat{\mathscr{C}} \vee \mathscr{C}^*, \tag{III.1-29}$$

and since $\mathbf{f}_\mathsf{C}$ is bi-additive on $(\overline{\Omega} \times \overline{\Omega})_0$,

$$\mathbf{f}_\mathsf{C}(\mathscr{A}, \mathscr{C}) = \mathbf{f}_\mathsf{C}(\hat{\mathscr{A}}, \hat{\mathscr{C}} \vee \mathscr{C}^*) + \mathbf{f}_\mathsf{C}(\mathscr{A}^*, \mathscr{C}),$$

$$= \mathbf{f}_\mathsf{C}(\hat{\mathscr{A}}, \hat{\mathscr{C}}) + \mathbf{f}_\mathsf{C}(\hat{\mathscr{A}}, \mathscr{C}^*) + \mathbf{f}_\mathsf{C}(\mathscr{A}^*, \mathscr{C}). \tag{III.1-30}$$

In virtue of (22) this reduces to (18). △

We use the lemma so as to render explicit the nature of the function $\mathbf{f}_\mathsf{C}$. We shall call a subset of the underlying set of a surface a *subsurface* if it is assigned the same orientation as the surface. Letting $\mathscr{S}$ be any subsurface of the boundary $\partial\chi(\mathscr{C})$ of the shape of some body $\mathscr{C}$, we define a new function on the Borel subsets $\mathscr{Q}$ of $\mathscr{S}$. Noting that since $\mathscr{S} \subset \chi(\mathscr{C})$, $\mathscr{Q}$ is a Borel set of $\chi(\mathscr{C}^e)$, we let $\tilde{\mathscr{Q}}$ be the Borel subset of $\mathscr{C}^e$ such that $\mathscr{Q} = \chi(\tilde{\mathscr{Q}})$, and we set

$$\mathbf{f}^\mathsf{S}_\mathsf{C}(\mathscr{Q}, \mathscr{S}) \equiv \mathbf{f}_\mathsf{C}(\tilde{\mathscr{Q}}, \mathscr{C}). \tag{III.1-31}$$

The lemma guarantees that this definition is unambiguous. That is, if $\hat{\mathscr{C}}$ is another body such that $\mathscr{S} \subset \chi(\hat{\mathscr{C}})$, there is a Borel subset $\hat{\mathscr{Q}}$ of $\hat{\mathscr{C}}^e$ such that $\mathscr{Q} = \chi(\hat{\mathscr{Q}})$, and by (18) we conclude that $\mathbf{f}_C(\hat{\mathscr{Q}}, \hat{\mathscr{C}})$ if used in (31) would yield the same function $\mathbf{f}_C^S$. In the same way the lemma may be used to show that if $\tilde{\mathscr{Q}}$ is any Borel set in Ω and is separate from $\mathscr{C}$, then, since $\mathbf{f}_C$ is defined on all Borel sets,

$$\mathbf{f}_C(\tilde{\mathscr{Q}}, \mathscr{C}) = \mathbf{f}_C(\tilde{\mathscr{Q}} \cap \partial\mathscr{C}, \mathscr{C}). \tag{III.1-32}$$

Thus the new function $\mathbf{f}_C^S$, which is defined on surfaces and their subsurfaces by (31), completely determines the old function $\mathbf{f}_C$, defined on $(\bar{\Omega} \times \bar{\Omega})_0$. We have shown, then, that any system of contact forces is defined completely by an appropriate function whose arguments are surfaces and their subsurfaces.

Since, by assumption, $\mathbf{f}_C(\cdot, \mathscr{C})$ is a measure on the subbodies of $\mathscr{C}^e$, the function $\mathbf{f}_C^S(\cdot, \mathscr{S})$ is a measure on the subsurfaces of $\mathscr{S}$. From (31) it follows also that if $\mathscr{S}'$ is a subsurface of $\mathscr{S}$ and if $\mathscr{D}$ is a subsurface of $\mathscr{S}'$, then

$$\mathbf{f}_C^S(\mathscr{D}, \mathscr{S}) = \mathbf{f}_C^S(\mathscr{D}, \mathscr{S}'). \tag{III.1-33}$$

The requirement $(12)_2$ now assumes the form

$$|\mathbf{f}_C^S(\mathscr{D}, \mathscr{S})| \leqq KA(\mathscr{D}), \tag{III.1-34}$$

if $\mathscr{D} \subset \mathscr{S}$ and if $A(\mathscr{D})$ is sufficiently small. By invoking the Radón–Nikodym theorem we see that there is an essentially bounded density $\mathbf{t}_{\mathscr{S}}$ such that

$$\mathbf{f}_C^S(\mathscr{D}, \mathscr{S}) = \int_{\mathscr{D}} \mathbf{t}_{\mathscr{S}}\, dA. \tag{III.1-35}$$

Moreover, it follows from (33) that if the surfaces $\mathscr{S}$ and $\mathscr{S}'$ have the same orientation and if $\mathscr{S}' \subset \mathscr{S}$, then

$$\mathbf{t}_{\mathscr{S}'} = \mathbf{t}_{\mathscr{S}} \tag{III.1-36}$$

at almost every point of $\mathscr{S}'$. Now going back to the definition (31) of $\mathbf{f}_C^S$, we may interpret (35) as asserting the following

Traction Theorem (GURTIN & WILLIAMS). *If a system of forces $\mathbf{f}_C$ on $(\bar{\Omega} \times \bar{\Omega})_0$ satisfies $(12)_2$, there is an essentially bounded density $\mathbf{t}_{\mathscr{S}}$ such that*

$$\mathbf{f}_C(\mathscr{A}, \mathscr{C}) = \int_{\mathscr{S}} \mathbf{t}_{\mathscr{S}}\, dA, \qquad \mathscr{S} \equiv \partial\chi(\mathscr{A}) \cap \partial\chi(\mathscr{C}); \tag{III.1-37}$$

moreover, $\mathbf{t}_{\mathscr{S}'} = \mathbf{t}_{\mathscr{S}}$ if $\mathscr{S}'$ is a subsurface of $\mathscr{S}$.

Putting $\mathscr{C} = \mathscr{A}^e$ in this theorem yields $(3)_2$.

Thus far we have not used the assumption that the system of forces is balanced. Through its corollary (13), this assumption enables us to prove the following

Theorem of Action and Reaction (NOLL). *Both the system of body forces and the system of contact forces are pairwise equilibrated:*

$$\left.\begin{aligned} \mathbf{f}_B(\mathscr{A},\mathscr{C}) &= -\mathbf{f}_B(\mathscr{C},\mathscr{A}) \\ \mathbf{f}_C(\mathscr{A},\mathscr{C}) &= -\mathbf{f}_C(\mathscr{C},\mathscr{A}) \end{aligned}\right\} \quad \forall(\mathscr{A},\mathscr{C}) \in (\overline{\Omega}\times\overline{\Omega})_0. \qquad \text{(III.1-38)}$$

Proof. Again we let $\mathscr{S}$ denote the surface of contact between the separate bodies $\mathscr{A}$ and $\mathscr{C}$. We know from geometry that we may choose sequences of parts $\mathscr{A}_n$ and $\mathscr{C}_n$ of $\mathscr{A}$ and $\mathscr{C}$, respectively, such that in the limit as $n\to\infty$ the volumes of their shapes vanish, yet they retain $\mathscr{S}$ as their surface of contact. Formally,

$$\mathscr{A}_n \prec \mathscr{A}, \qquad \mathscr{C}_n \prec \mathscr{C}, \qquad \mathscr{S} = \partial\chi(\mathscr{A}_n) \cap \partial\chi(\mathscr{C}_n), \qquad \text{(III.1-39)}$$

and, since in continuum mechanics mass is an absolutely continuous function of volume (§II.2),

$$M(\mathscr{A}_n) \to 0, \qquad M(\mathscr{C}_n) \to 0. \qquad \text{(III.1-40)}$$

The lemma of GURTIN & WILLIAMS shows that

$$\begin{aligned} \mathbf{f}_C(\mathscr{A}_n,\mathscr{C}_n) &= \mathbf{f}_C(\mathscr{A},\mathscr{C}), \\ \mathbf{f}_C(\mathscr{C}_n,\mathscr{A}_n) &= \mathbf{f}_C(\mathscr{C},\mathscr{A}). \end{aligned} \qquad \text{(III.1-41)}$$

Of course $\mathscr{A}_n$ and $\mathscr{C}_n$, being parts of separate bodies, are separate, so we may substitute $\mathscr{A}_n$ for $\mathscr{A}$ and $\mathscr{C}_n$ for $\mathscr{C}$ in (13) and by use of (41) obtain

$$\mathbf{f}_B(\mathscr{A}_n,\mathscr{C}_n) + \mathbf{f}_C(\mathscr{A},\mathscr{C}) = -\mathbf{f}_B(\mathscr{C}_n,\mathscr{A}_n) - \mathbf{f}_C(\mathscr{C},\mathscr{A}). \qquad \text{(III.1-42)}$$

As $n\to\infty$, both body forces $\mathbf{f}_B(\mathscr{A}_n,\mathscr{C}_n)$ and $\mathbf{f}_B(\mathscr{C}_n,\mathscr{A}_n)$ vanish because of $(12)_1$, so $(38)_2$ follows. By (13) we obtain also $(38)_1$. △

Corollary. *If* $-\mathscr{S}$ *denotes the surface having the same underlying set as* $\mathscr{S}$ *but opposite orientation, then*

$$\mathbf{t}_{-\mathscr{S}} = -\mathbf{t}_{\mathscr{S}}. \qquad \text{(III.1-43)}$$

Exercise III.1.2. Using the Lebesgue Differentiation Theorem[1], prove that (43) follows from $(38)_2$ and (37).

[1] *E.g.,* Theorem 41.3 in the book by M. E. MUNROE, *Introduction to Measure and Integration,* Reading, Massachusetts, Addison-Wesley, 1953.

We may now address ourselves to the system of body forces. By $(38)_1$ and the theorem of NOLL and GURTIN & WILLIAMS in §I.5 we see that the resultant body force is additive on separate bodies. That is, if

$$\mathbf{f}_B(\mathscr{B}) \equiv \mathbf{f}_B(\mathscr{B}, \mathscr{B}^e), \tag{III.1-44}$$

then

$$\mathbf{f}_B(\mathscr{B} \curlyvee \mathscr{C}) = \mathbf{f}_B(\mathscr{B}) + \mathbf{f}_B(\mathscr{C}) \qquad \forall (\mathscr{B}, \mathscr{C}) \in (\overline{\Omega} \times \overline{\Omega})_0 . \tag{III.1-45}$$

According to a theorem of GURTIN & WILLIAMS[1], such a function if also ultimately bounded by volume is the restriction to $\chi(\mathscr{B})$ of a Borel measure which is ultimately bounded by volume. We may use the same symbol $\mathbf{f}_B$ to denote that measure. By the Radón–Nikodym Theorem, it follows that $\mathbf{f}_B$ has an essentially bounded density $\mathbf{b}_{tot}$:

$$\mathbf{f}_B(\mathscr{B}) = \int_{\chi(\mathscr{B})} \rho \mathbf{b}_{tot} \, dV. \tag{III.1-46}$$

According to the Axioms of Inertia (§I.13), the inertial forces have the density $-\mathbf{a}_{\oint}$ with respect to mass, $\mathbf{a}_{\oint}$ being that frame-indifferent vector field which in an inertial frame $\oint$ reduces to $\ddot{\mathbf{x}}$, so if we set

$$\mathbf{b} \equiv \mathbf{b}_{tot} + \mathbf{a}_{\oint}, \tag{III.1-47}$$

we obtain $(11)_1$ and hence also $(8)_1$. A specific expression for $\mathbf{a}_{\oint}$ has been recorded as (II.4-7). The density $\mathbf{b}$ may include effects of mutual body forces[2]. but in this book, as has been said already, no body forces other than external ones will be considered. Since $\mathbf{b}_{tot}$ is essentially bounded and since we have assumed that $\ddot{\mathbf{x}}$ in an inertial frame is essentially bounded, so is $\mathbf{b}$.

2. Reactions upon Containers and Submerged Obstacles

With little more than the concept of a system of contact forces and the theorem of action and reaction in the form (III.1-38) we can sometimes evaluate the force and the torque that a deforming body exerts upon a container or an object submerged in it. Analyses of this kind go back to the earliest days of mechanics; they remain of great utility to engineers, because they require very little detailed knowledge of either the body or its motion;

[1] *Op. cit.*, Appendix.

[2] A correspondingly general and precise theory of mutual body forces, such as those resulting from Newtonian gravity, may be read off from the scalar counterparts for heating established by W. O. WILLIAMS, "On internal interactions and the concept of thermal isolation," *Archive for Rational Mechanics and Analysis* **34**, 245–258 (1969), and "Thermodynamics of rigid continua," *ibid.* **36**, 270–284 (1970).

and for just the same reason they sometimes provide essential steps in the precise, mathematical treatment of qualitative problems. Here we shall consider only three examples, the simplest. We present them upon a general framework due, more or less, to v. MISES, CISOTTI, and BOGGIO.

First we substitute (II.6-12), (III.1-2), and (III.1-5) into Euler's Laws (I.13-11), thus obtaining expressions for the resultant contact force $\mathbf{f}_C$ and resultant contact torque $\mathbf{F}_C$ on the present shape $\mathscr{S}$ of the part $\mathscr{P}$ of a body:

$$\begin{aligned} -\mathbf{f}_C &= -\int_{\mathscr{S}} (\rho\dot{\mathbf{x}})' \, dV - \int_{\partial\mathscr{S}} \rho\dot{\mathbf{x}} \otimes \dot{\mathbf{x}}\mathbf{n} \, dA + \mathbf{f}_B^a, \\ -\mathbf{F}_C &= -\int_{\mathscr{S}} \mathbf{p} \wedge (\rho\dot{\mathbf{x}})' \, dV - \int_{\partial\mathscr{S}} \mathbf{p} \wedge \rho\dot{\mathbf{x}} \otimes \dot{\mathbf{x}}\mathbf{n} \, dA + \mathbf{F}_B^a. \end{aligned} \tag{III.2-1}$$

Here $\mathbf{F}_C$ is taken with respect to a fixed place $\mathbf{x}_0$ in an inertial frame, and $\mathbf{p} \equiv \mathbf{x} - \mathbf{x}_0$ as in §§I.13 and II.12. Because of (III.1-38), the left-hand sides of (1) are the contact force and torque, respectively, exerted by $\mathscr{P}$ upon its exterior $\mathscr{P}^e$. It is these quantities that we wish to evaluate in special cases. Together they are called the *reaction* of $\mathscr{P}$ upon its exterior.

In general terms (1) asserts that if the resultant applied force and torque upon $\mathscr{P}$ are known, *then the reaction of $\mathscr{P}$ upon its exterior is determined by the fields of density and velocity over the present shape of $\mathscr{P}$*. Kinematical data thus determine the reaction.

The essential field $\rho\dot{\mathbf{x}}$, which has the dimensions of momentum per unit volume, is called the *mass flow*. In the three examples we shall give now we shall assume that the mass flow is steady: $(\rho\dot{\mathbf{x}})' = \mathbf{0}$.

Example 1. *Flow in a Stationary Container.* Let $\mathscr{P}$ be confined by a bounded, stationary container. On the walls $\partial\mathscr{S}$ of the container the condition (II.6-17) is satisfied, so the integral over $\partial\mathscr{S}$ vanishes. A little reflection enables us to derive the same result even if $\mathscr{P}$ does not fill the container entirely. We have shown that *a body in motion with steady mass flow within a bounded, stationary container exerts upon that container just the reaction it would exert, were it at rest.*

Example 2. *Flow in a Pipe.* Suppose that a body is flowing through a stationary pipe of arbitrary form. We consider the part $\mathscr{P}$ contained in the portion of the pipe cut off by two surfaces, the inlet $\mathscr{S}_i$ and the outlet $\mathscr{S}_o$. Upon the walls of the pipe (II.6-17) is satisfied, so, if $(\rho\dot{\mathbf{x}})' = \mathbf{0}$, (1) becomes

$$\begin{aligned} -\mathbf{f}_C &= \int_{\mathscr{S}_i} \rho\dot{\mathbf{x}} \otimes \dot{\mathbf{x}}\mathbf{n} \, dA - \int_{\mathscr{S}_o} \rho\dot{\mathbf{x}} \otimes \dot{\mathbf{x}}\mathbf{n} \, dA + \mathbf{f}_B^a, \\ -\mathbf{F}_C &= \int_{\mathscr{S}_i} \mathbf{p} \wedge \rho\dot{\mathbf{x}} \otimes \dot{\mathbf{x}}\mathbf{n} \, dA - \int_{\mathscr{S}_o} \mathbf{p} \wedge \rho\dot{\mathbf{x}} \otimes \dot{\mathbf{x}}\mathbf{n} \, dA + \mathbf{F}_B^a. \end{aligned} \tag{III.2-2}$$

In writing the integral over $\mathscr{S}_i$ we have taken the normal $\mathbf{n}$ as directed inward, so as to emphasise that a difference is being calculated.

We have not yet called upon the Traction Theorem. Doing so, we use (III.1-37) and (III.1-38)$_2$ to express the contact force and torque $\mathbf{f}_p$ and $\mathbf{F}_p$ on the pipe alone:

$$\mathbf{f}_C = -\mathbf{f}_p - \int_{\mathscr{S}_i} \mathbf{t}_{\mathscr{S}_i}\, dA + \int_{\mathscr{S}_o} \mathbf{t}_{\mathscr{S}_o} dA,$$
$$\mathbf{F}_C = -\mathbf{F}_p - \int_{\mathscr{S}_i} \mathbf{p} \wedge \mathbf{t}_{\mathscr{S}_i}\, dA + \int_{\mathscr{S}_o} \mathbf{p} \wedge \mathbf{t}_{\mathscr{S}_o}\, dA. \tag{III.2-3}$$

Combining this result with (2), we find that

$$\mathbf{f}_p = \int_{\mathscr{S}_i} (\rho\dot{\mathbf{x}} \otimes \dot{\mathbf{x}}\mathbf{n} - \mathbf{t}_{\mathscr{S}_i})\, dA - \int_{\mathscr{S}_o} (\rho\dot{\mathbf{x}} \otimes \dot{\mathbf{x}}\mathbf{n} - \mathbf{t}_{\mathscr{S}_o})\, dA + \mathbf{f}_B^a,$$
$$\mathbf{F}_p = \int_{\mathscr{S}_i} \mathbf{p} \wedge (\rho\dot{\mathbf{x}} \otimes \dot{\mathbf{x}}\mathbf{n} - \mathbf{t}_{\mathscr{S}_i})\, dA - \int_{\mathscr{S}_o} \mathbf{p} \wedge (\rho\dot{\mathbf{x}} \otimes \dot{\mathbf{x}}\mathbf{n} - \mathbf{t}_{\mathscr{S}_o})\, dA + \mathbf{F}_B^a. \tag{III.2-4}$$

From these formulae we see that *measurement of* ρ, $\dot{\mathbf{x}}$, *and* $\mathbf{t}_{\mathscr{S}}$ *at the inlet and the outlet suffice to determine the reaction exerted on a stationary pipe by a body moving through it with steady mass flow*, the applied force and torque on the body being known.

Various simplifying assumptions reduce the general expressions (4) to examples of great use in hydraulics. Special cases of (4)$_1$ are called "Bernoulli's theorem", "the flow energy theorem", "the impulse theorem", *etc*. For the truth of the result it is not necessary that the body fill the pipe or that the fields ρ and $\dot{\mathbf{x}}$ be smooth within it.

Example 3. *Reaction upon a Submerged Object.* In this example we shall suppose that $\rho' = 0$, $\dot{\mathbf{x}}' = \mathbf{0}$, $\mathbf{f}_B^a = \mathbf{0}$, and $\mathbf{F}_B^a = \mathbf{0}$. We consider a body filling all of space except for a stationary, rigid, bounded object. The shape of the object need not be specified in the present context, for upon it the boundary condition (II.6-17) is satisfied, so the integrals of integrands proportional to $\dot{\mathbf{x}}$ over the boundary of the obstacle vanish. We consider the region $\mathscr{R}_c$ between the obstacle and a closed *control surface* $\mathscr{S}_c$ so large as to contain the obstacle entirely. We denote by $\mathscr{P}$ the part of the body whose shape is the region $\mathscr{R}_c$. We may take for $\mathscr{S}_c$ the surface of a sphere if we wish to. From (1) we conclude that

$$-\mathbf{f}_C = -\int_{\mathscr{S}_c} \rho\dot{\mathbf{x}} \otimes \dot{\mathbf{x}}\mathbf{n}\, dA,$$
$$-\mathbf{F}_C = -\int_{\mathscr{S}_c} \mathbf{p} \wedge (\rho\dot{\mathbf{x}} \otimes \dot{\mathbf{x}}\mathbf{n})\, dA. \tag{III.2-5}$$

Exercise III.2.1. Prove that if $\mathbf{v}$ is any constant vector field,

$$\int_{\mathscr{S}_c} \rho\dot{\mathbf{x}} \otimes \dot{\mathbf{x}}\mathbf{n}\, dA = \int_{\mathscr{S}_c} \rho(\dot{\mathbf{x}} - \mathbf{v}) \otimes (\dot{\mathbf{x}} - \mathbf{v})\mathbf{n}\, dA + \left[\int_{\mathscr{S}_c} \rho(\dot{\mathbf{x}} - \mathbf{v}) \otimes \mathbf{n}\, dA\right]\mathbf{v},$$

$$\int_{\mathscr{S}_c} \mathbf{p} \wedge (\rho\dot{\mathbf{x}} \otimes \dot{\mathbf{x}}\mathbf{n})\, dA = \int_{\mathscr{S}_c} \mathbf{p} \wedge [\rho(\dot{\mathbf{x}} - \mathbf{v}) \otimes (\dot{\mathbf{x}} - \mathbf{v})\mathbf{n}]\, dA \qquad \text{(III.2-6)}$$

$$+ \int_{\mathscr{S}_c} \mathbf{p} \wedge [\rho(\dot{\mathbf{x}} - \mathbf{v}) \otimes \mathbf{v}\mathbf{n}]\, dA + \mathbf{m} \wedge \mathbf{v},$$

$\mathbf{m}$ being the momentum of $\mathscr{P}$.

We now call upon the Traction Theorem. If we write $\mathbf{f}_{\text{obs}}$ and $\mathbf{F}_{\text{obs}}$ for the force and torque exerted by $\mathscr{P}$ upon the obstacle, then (III.1-37) and $(\text{III.1-38})_2$ show that

$$\mathbf{f}_C = -\mathbf{f}_{\text{obs}} + \int_{\mathscr{S}_c} \mathbf{t}_{\mathscr{S}_c}\, dA,$$

$$\mathbf{F}_C = -\mathbf{F}_{\text{obs}} + \int_{\mathscr{S}_c} \mathbf{p} \wedge \mathbf{t}_{\mathscr{S}_c}\, dA. \qquad \text{(III.2-7)}$$

Putting (5), (6), and (7) together, we obtain finally

$$\mathbf{f}_{\text{obs}} = -\int_{\mathscr{S}_c} \rho(\dot{\mathbf{x}} - \mathbf{v}) \otimes (\dot{\mathbf{x}} - \mathbf{v})\,\mathbf{n}\, dA - \left[\int_{\mathscr{S}_c} \rho(\dot{\mathbf{x}} - \mathbf{v}) \otimes \mathbf{n}\, dA\right]\mathbf{v}$$

$$+ \int_{\mathscr{S}_c} (\mathbf{t}_{\mathscr{S}_c} + P\mathbf{n})\, dA, \qquad \text{(III.2-8)}$$

$$\mathbf{F}_{\text{obs}} = -\int_{\mathscr{S}_c} \mathbf{p} \wedge \rho(\dot{\mathbf{x}} - \mathbf{v}) \otimes (\dot{\mathbf{x}} - \mathbf{v})\mathbf{n}\, dA$$

$$- \int_{\mathscr{S}_c} \mathbf{p} \wedge \rho(\dot{\mathbf{x}} - \mathbf{v}) \otimes \mathbf{v}\mathbf{n}\, dA - \mathbf{m} \wedge \mathbf{v} + \int_{\mathscr{S}_c} \mathbf{p} \wedge (\mathbf{t}_{\mathscr{S}_c} + P\mathbf{n})\, dA;$$

here we have added to each right-hand side the resultant force and resultant torque of a constant scalar pressure P, both of these resultants being null.

These formulae serve to evaluate the reaction exerted by the motion of $\mathscr{P}$ upon a stationary, rigid obstacle immersed in it. The obstacle itself seems not to enter the final results. All we need know is the steady kinematical fields ρ and $\dot{\mathbf{x}}$ and the traction field $\mathbf{t}_{\mathscr{S}_c}$ upon the control surface $\mathscr{S}_c$. The choice of $\mathscr{S}_c$ is ours. Generally it is advantageous to choose it very large; we expect then that the effect of the obstacle upon the fields of ρ and $\dot{\mathbf{x}}$ be lessened. To evaluate the integrals we may adjust as we like the form of $\mathscr{S}_c$ and the values of the arbitrary constants $\mathbf{v}$ and P.

The most celebrated example is provided by the steady, uniform flow of a body past an obstacle. Then $\dot{\mathbf{x}} \to \mathbf{v}$, say, at ∞. If also the traction field at great distances from the obstacle is approximately a uniform hydrostatic pressure P, then $\mathbf{t}_{\mathscr{S}_e} \to -P\mathbf{n}$ at ∞. So as to model this condition we consider the body $\mathscr{P}_r$ that presently occupies the space between the obstacle and the surface $\mathscr{S}_r$ of the sphere of radius r, centered upon some fixed point. To $\mathscr{P}_r$ for any large enough r we may apply (8). If the integrals in (8) converge as $r \to \infty$, we obtain definite expressions for $\mathbf{f}_{\text{obs}}$ and $\mathbf{F}_{\text{obs}}$. For example, if as $r \to \infty$

$$\dot{\mathbf{x}} - \mathbf{v} = \mathbf{o}(r^{-2}), \qquad \rho = O(1), \qquad \mathbf{t}_{\mathscr{S}_r} + P\mathbf{n} = \mathbf{o}(r^{-2}), \qquad \text{(III.2-9)}$$

then all the integrals in $(8)_1$ converge to null as $r \to \infty$, so

$$\mathbf{f}_{\text{obs}} = \mathbf{0}. \qquad \text{(III.2-10)}$$

That is, *the conditions* (9) *are sufficient that the infinite body with steady density in steady flow past the obstacle exert no resultant force on the obstacle.*

The conditions (9) are of the essence for the result. Without some conditions of this kind, no such conclusion follows. They assert that the disturbance due to the presence of the obstacle falls off rapidly at great distances from it; indeed, they specify the rate at which it falls off. In §VI.6 we shall show that they are satisfied for a certain kind of body. For other bodies they are not. In general, $\mathbf{f}_{\text{obs}} \neq \mathbf{0}$.

EULER, treating a very special case, was the first to obtain (10). His reasoning provides a primitive example of that which we have given in general terms. D'ALEMBERT, much later, rediscovered or appropriated the result; his unduly special reasoning applies only to obstacles of great symmetry. The fact itself he announced as a paradox. The name has stuck: "The D'Alembert paradox". Both the name and the fact have given rise to perennial confusion.

If as $r \to \infty$

$$\dot{\mathbf{x}} - \mathbf{v} = \mathbf{o}(r^{-3}), \qquad \rho = O(1), \qquad \mathbf{t}_{\partial\mathscr{S}_r} + P\mathbf{n} = \mathbf{o}(r^{-3}), \qquad \text{(III.2-11)}$$

then of course (10) holds, and also the integrals in $(8)_2$ converge. The infinite body occupying the region $\mathscr{S}_\infty$ outside the obstacle has finite *relative momentum* $\mathbf{m}_\infty$, given by

$$\mathbf{m}_\infty \equiv \int_{\mathscr{S}_\infty} \rho(\dot{\mathbf{x}} - \mathbf{v})\, dV, \qquad \text{(III.2-12)}$$

and from $(8)_2$ we obtain

$$\mathbf{F}_{\text{obs}} = \mathbf{v} \wedge \mathbf{m}_\infty. \qquad \text{(III.2-13)}$$

Here, not in the proof of (10), belong appeals to symmetry, for they suffice to show that $\mathbf{F}_{\text{obs}} = \mathbf{0}$ for some obstacles, though in general $\mathbf{F}_{\text{obs}} \neq \mathbf{0}$.

3. The Traction Field. Cauchy's Postulate and the Hamel–Noll Theorem

In §III.1 we have expressed the principles of balance of linear and rotational momentum in terms of the traction field $\mathbf{t}_{\partial\chi(\mathscr{P})}$ on the boundary $\partial\chi(\mathscr{P})$ of the shape of each part $\mathscr{P}$ of $\mathscr{B}$:

$$\int_{\chi(\mathscr{P})} \rho\ddot{\mathbf{x}}\, dV = \int_{\partial\chi(\mathscr{P})} \mathbf{t}_{\partial\chi(\mathscr{P})}\, dA + \int_{\chi(\mathscr{P})} \rho\mathbf{b}\, dV,$$

$$\int_{\chi(\mathscr{P})} (\mathbf{x} - \mathbf{x}_0) \wedge \rho\ddot{\mathbf{x}}\, dV = \int_{\partial\chi(\mathscr{P})} (\mathbf{x} - \mathbf{x}_0) \wedge \mathbf{t}_{\partial\chi(\mathscr{P})}\, dA \qquad \text{(III.1-8)}_r$$

$$+ \int_{\chi(\mathscr{P})} (\mathbf{x} - \mathbf{x}_0) \wedge \rho\mathbf{b}\, dV;$$

here we continue to consider a particular time t but do not indicate it in the notation. In order to reduce these integral equations to equivalent field equations, we must express the traction field $\mathbf{t}_{\mathscr{S}}$, which is defined only upon $\mathscr{S}$, in terms of fields defined in an open set containing $\mathscr{S}$ in its interior. To an extension of this kind we now address ourselves.

A place $\mathbf{x}$ on $\partial\chi(\mathscr{P})$ obviously lies also upon the boundaries $\partial\chi(\mathscr{Q})$ of infinitely many parts $\mathscr{Q}$ of $\mathscr{P}$. The traction $\mathbf{t}$ for these various surfaces having the point $\mathbf{x}$ in common depends, in general, upon the surface $\mathscr{S}$. In §III.1 we have shown that if $\mathscr{S}'$ is a subsurface of $\mathscr{S}$, then $\mathbf{t}_{\mathscr{S}'} = \mathbf{t}_{\mathscr{S}}$, but we have not established any relation between $\mathbf{t}_{\mathscr{S}'}$ and $\mathbf{t}_{\mathscr{S}}$ for more general pairs of surfaces $\mathscr{S}$ and $\mathscr{S}'$, for example in the case when $\mathscr{S}'$ and $\mathscr{S}$ have in common only the one place $\mathbf{x}$ we are considering. In the classical continuum mechanics of CAUCHY and his successors it was always assumed that *the traction on all like-oriented surfaces with a common tangent plane at* $\mathbf{x}$ *was the same*. That is, $\mathbf{t}_{\mathscr{S}}$ at $\mathbf{x}$ was assumed to depend upon $\mathscr{S}$ only through the oriented normal $\mathbf{n}$ of $\mathscr{S}$ at $\mathbf{x}$:

$$\mathbf{t}_{\mathscr{S}} = \mathbf{t}(\mathbf{x}, \mathbf{n}). \qquad \text{(III.3-1)}$$

This is ***Cauchy's Postulate.*** $\mathscr{S}$ is oriented so that its normal $\mathbf{n}$ points out of $\chi(\mathscr{B})$ if $\mathscr{S}$ is a part of $\partial\chi(\mathscr{B})$. Thus $\mathbf{t}(\mathbf{x}, -\mathbf{n})$ is the traction at $\mathbf{x}$ on all surfaces $\mathscr{S}$ tangent to $\partial\chi(\mathscr{B})$ and forming parts of the boundaries of bodies in the exterior $\chi(\mathscr{B}^e)$ of $\chi(\mathscr{B})$. In this sense $\mathbf{t}(\mathbf{x}, \mathbf{n})$ is the traction exerted on $\mathscr{B}$ at $\mathbf{x}$ by the contiguous bodies outside $\mathscr{B}$, while $\mathbf{t}(\mathbf{x}, -\mathbf{n})$ is the traction exerted there by $\mathscr{B}$ on the contiguous bodies outside it.

As a trivial corollary of (III.1-43) follows ***Cauchy's Fundamental Lemma***:

$$\mathbf{t}(\mathbf{x}, -\mathbf{n}) = -\mathbf{t}(\mathbf{x}, \mathbf{n}). \qquad \text{(III.3-2)}$$

For those readers who have not stopped to follow the theorems given in §III.1 we include here a sketch of CAUCHY's own argument to prove (2) as a consequence of his assumption (1) and the balance of linear momentum, of course without use of (III.1-43).

Proof. In view of (1), it suffices to consider the case when $\mathscr{S}$ is an oriented disk of sufficiently small radius, centered at $\mathbf{x}$; then $-\mathscr{S}$ is the oppositely oriented disk. Assuming that the universe Ω is rich enough in sets that every right-circular cylinder of sufficiently small base and altitude is the shape of some body, for $\chi(\mathscr{B})$ we take a circular cylinder which is normal to $\mathscr{S}$ and is bisected transversely by $\mathscr{S}$. If ε denotes the altitude of this cylinder, then $V(\chi(\mathscr{B})) = \varepsilon A(\mathscr{S})$. We assume that $\mathbf{b} - \ddot{\mathbf{x}}$ is essentially bounded. We apply the balance of linear momentum as expressed by (III.1-8) to $\chi(\mathscr{B})$ and then take the limit as $\varepsilon \to 0$, the disk $\mathscr{S}$ being kept fixed, so of course its area remains constant. The limit of the difference of the volume integrals is $\mathbf{0}$, so

$$\lim_{\varepsilon \to 0} \int_{\partial\chi(\mathscr{B})} \mathbf{t}(\mathbf{x}, \mathbf{n})\, dA = \mathbf{0}. \tag{III.3-3}$$

(Of course this result is a special case of (III.1-16), but the present proof is intended for the reader who skipped the part of §III.1 that follows (III.1-9).) In the passage to the limit $\mathbf{n}$ does not vary, but the set of $\mathbf{x}$ over which the integral is taken shrinks down to $\mathscr{S}$, twice over. If we assume that $\mathbf{t}(\cdot, \mathbf{n})$ is an essentially bounded function of $\mathbf{x}$, the limit of the integral over the mantle of the cylinder is $\mathbf{0}$, since the area of that mantle tends to 0. Thus only the limits of the two integrals over $\mathscr{S}$ remain to be considered. If we assume further that $\mathbf{t}(\cdot, \mathbf{n})$ is a continuous function of $\mathbf{x}$, then the limits of these integrals equal the integrals of the limit functions in the two cases:

$$\int_{\mathscr{S}} [\mathbf{t}(\mathbf{x}, \mathbf{n}) + \mathbf{t}(\mathbf{x}, -\mathbf{n})]\, dA = \mathbf{0}. \tag{III.3-4}$$

Since $\mathscr{S}$ is any sufficiently small disk normal to $\mathbf{n}$ at $\mathbf{x}$, (2) follows. △

The reader who is content to lay down CAUCHY's Postulate (1) and to make the assumptions of smoothness concerning $\mathbf{b} - \ddot{\mathbf{x}}$ and $\mathbf{t}$ that we have stated in the course of proving CAUCHY's Fundamental Lemma should now pass on to the next section. On the other hand, the reader who has followed the development in §III.1 will have noted that one of the assumptions of smoothness made just now is unnecessarily strong, while another has already been proved to hold in the mathematical theory based on (III.1-12). In fact, as we shall see now, CAUCHY's Postulate can be proved true as a consequence of the principle of linear momentum and very mild assumptions of smoothness.

Theorem (HAMEL, NOLL). *Suppose that the contact force* $\mathbf{f}_C(\mathscr{A}, \mathscr{C})$ *exerted upon any part* $\mathscr{A}$ *of* $\mathscr{B}$ *by the separate body* $\mathscr{C}$ *be determined by a traction field* $\mathbf{t}_{\mathscr{S}}$ *through* (III.1-37). *Then Cauchy's Postulate* (1) *holds almost everywhere on every surface* $\mathscr{S}$ *interior to* $\chi(\mathscr{B})$.

The reader should recall that (III.1-37) has been proved to hold as a consequence of (III.1-12). We assume here also another consequence of (III.1-12), namely, that the body force has a bounded density $\mathbf{b}$. We assume also that $\ddot{\mathbf{x}}$ is bounded.

Proof of the Hamel–Noll theorem. By (III.1-37) and the Lebesgue differentiation theorem we know that at almost all points of the surface $\mathscr{S}$

$$\mathbf{t}_{\mathscr{S}}(\mathbf{x}) = \lim_{m\to\infty} \frac{\int_{\mathscr{U}_m} \mathbf{t}_{\mathscr{S}}\, dA}{A(\mathscr{U}_m)} \tag{III.3-5}$$

if $\mathscr{U}_m$ is a suitably selected sequence of sets on $\mathscr{S}$ shrinking down to $\mathbf{x}$. We are to show that if $\mathscr{T}$ and $\mathscr{S}$ have a common oriented normal $\mathbf{n}$ at $\mathbf{x}$, and if both $\mathbf{t}_{\mathscr{T}}(\mathbf{x})$ and $\mathbf{t}_{\mathscr{S}}(\mathbf{x})$ exist, then

$$\mathbf{t}_{\mathscr{T}}(\mathbf{x}) = \mathbf{t}_{\mathscr{S}}(\mathbf{x}). \tag{III.3-6}$$

The common value of the two functions $\mathbf{t}_{\mathscr{T}}$ and $\mathbf{t}_{\mathscr{S}}$ at $\mathbf{x}$ is then a function of $\mathbf{n}$ only and may be denoted by $\mathbf{t}(\mathbf{x}, \mathbf{n})$; if so, the HAMEL–NOLL theorem will have been proved.

If $\mathscr{S}$ and $\mathscr{T}$ coincide near $\mathbf{x}$, the result is trivial. Otherwise, at the regular point $\mathbf{x}$ common to $\mathscr{S}$ and $\mathscr{T}$ we describe a circular cylinder of small radius Δr about the common normal $\mathbf{n}$ and denote the parts of its interior lying between $\mathscr{S}$ and $\mathscr{T}$ by $\Delta\mathscr{D}$. We denote the cylindrical part of $\partial\Delta\mathscr{D}$ by $\Delta\mathscr{A}^*$ and the parts of $\partial\Delta\mathscr{D}$ common to $\mathscr{S}$ and $\mathscr{T}$, respectively, by $\Delta\mathscr{A}$ and $\Delta\mathscr{A}'$.

First suppose that $\mathscr{T}$ is the tangent plane to $\mathscr{S}$ at $\mathbf{x}$, that $\mathbf{x}$ is an elliptic point for $\mathscr{S}$, and that (5) holds at $\mathbf{x}$ for $\mathscr{S}$. Then not only does $\mathscr{S}$ lie entirely on one side of $\mathscr{T}$ near $\mathbf{x}$, but also we may construct a paraboloid of revolution $\mathscr{S}^*$ with vertex $\mathbf{x}$, with $\mathscr{T}$ as its tangent plane, and such as to include between itself and $\mathscr{T}$ all of $\mathscr{S}$, for sufficiently small Δr. Specifically, if $z = f(x, y)$ is the rectangular Cartesian equation of $\mathscr{S}$ near $\mathbf{x}$, the co-ordinates x and y being in $\mathscr{T}$ and z being distance along the normal to $\mathscr{T}$, then $\partial_x^2 f \geqq 0$, and $\partial_y^2 f \geqq 0$ at $\mathbf{x}$, and if we set

$$K \equiv \max(\partial_x^2 f, \partial_x \partial_y f, \partial_y^2 f) \qquad \text{when} \quad x^2 + y^2 \leqq \Delta r^2, \tag{III.3-7}$$

a paraboloid of the kind desired is given by

$$z = 2K(x^2 + y^2). \tag{III.3-8}$$

The area of the cylindrical part $\Delta\mathscr{A}^*$ of $\partial\Delta\mathscr{D}$ is not greater than that of the part of the cylinder between the plane and the paraboloid. Thus

$$\begin{aligned} A(\Delta\mathscr{A}^*) &\leqq (2\pi\Delta r)\cdot 2K(\Delta r)^2, \\ &= o(\Delta r^2) \end{aligned} \tag{III.3-9}$$

as $\Delta r \to 0$. Likewise the volume of $\Delta\mathscr{D}$ is bounded by that of the region between $\mathscr{S}^*$, the cylinder, and the plane. Thus

$$\begin{aligned} V(\Delta\mathscr{D}) &\leqq \pi K \Delta r^4, \\ &= o(\Delta r^3). \end{aligned} \tag{III.3-10}$$

Of course

$$A(\Delta\mathscr{A}') = \pi\Delta r^2. \tag{III.3-11}$$

Exercise III.3.1. Prove that

$$A(\Delta\mathscr{A}) = \pi\Delta r^2 + o(\Delta r^2). \tag{III.3-12}$$

Hence

$$\begin{aligned} A(\partial\Delta\mathscr{D}) &= 2\pi\Delta r^2 + o(\Delta r^2) \quad \text{as} \quad \Delta r \to 0, \\ V(\Delta\mathscr{D}) &= o(A(\partial\Delta\mathscr{D})) \quad \text{as} \quad \Delta r \to 0. \end{aligned} \tag{III.3-13}$$

We assume that the universe Ω is rich enough in sets that $\Delta\mathscr{D}$, no matter what be Δr, is the shape of some body $\mathscr{P}_{\Delta r}$. Because of $(13)_2$ and (III.1-14)

$$\lim_{\Delta r \to 0} \frac{f_C(\mathscr{P}_{\Delta r}, \mathscr{P}^e_{\Delta r})}{A(\partial\Delta\mathscr{D})} = 0. \tag{III.3-13A}$$

Orienting the tangent plane $\mathscr{T}$ so that its normal points into $\mathscr{S}$ at $\mathbf{x}$, we conclude that for almost all $\mathbf{x}$ on $\mathscr{S}$

$$\lim_{\Delta r \to 0} \frac{1}{A(\partial\Delta\mathscr{D})} \left(\int_{\Delta\mathscr{A}} \mathbf{t}_{\mathscr{T}}\, dA + \int_{\Delta\mathscr{A}'} \mathbf{t}_{\mathscr{T}}\, dA + \int_{\Delta\mathscr{A}^*} \mathbf{t}_{\Delta\mathscr{A}^*}\, dA \right) = \mathbf{0}. \tag{III.3-14}$$

In view of (9) and $(13)_1$ the limit of the third term vanishes, so by (11) and (12) we see that

$$\lim_{\Delta r \to 0} \frac{1}{A(\Delta\mathscr{A})} \int_{\Delta\mathscr{A}} \mathbf{t}_{\mathscr{S}}\, dA = \lim_{\Delta r \to 0} \frac{1}{A(\Delta\mathscr{A}')} \int_{\Delta\mathscr{A}'} \mathbf{t}_{\mathscr{T}}\, dA, \tag{III.3-15}$$

provided either limit exist. Now the value of the limit on the left-hand side is $\mathbf{t}_{\mathscr{S}}(\mathbf{x})$ by assumption, so the limit on the right-hand side is proved to exist by the argument given. The limit on the right-hand side is independent of the choice of $\mathscr{S}$. Hence $\mathbf{t}_{\mathscr{S}}(\mathbf{x})$ is the same for all surfaces $\mathscr{S}$ that are elliptic at $\mathbf{x}$, provided $\mathbf{x}$ be a place where (5) holds for $\mathscr{S}$.

Exercise III.3.2. Complete the proof of the HAMEL–NOLL theorem by considering the case when $\mathbf{x}$ is a saddle point for $\mathscr{S}$ or for $\mathscr{T}$ or for both. △

4. Cauchy's Fundamental Theorem: Existence of the Stress Tensor

CAUCHY's Postulate (III.3-1) and its consequence, CAUCHY's Fundamental Lemma (III.3-2), enable us to determine the way the traction field $\mathbf{t}$ depends upon $\mathbf{n}$. Indeed, it is a linear function of $\mathbf{n}$, as shown by

Cauchy's Fundamental Theorem. *If* $\mathbf{t}(\cdot, \mathbf{n})$ *is a continuous function of* $\mathbf{x}$, *there is a tensor field* $\mathbf{T}(\cdot)$ *such that*

$$\mathbf{t}(\mathbf{x}, \mathbf{n}) = \mathbf{T}(\mathbf{x})\mathbf{n}. \tag{III.4-1}$$

CAUCHY himself interpreted this theorem as stating that the tractions on any three linearly independent planes at a point determine the traction on all surfaces at that point. Indeed, let $\{\mathbf{e}_k\}$ be a basis in $\mathscr{V}$, so that $\mathbf{n} = n^k\mathbf{e}_k$. Then (1) asserts that

$$\mathbf{t} = \mathbf{T}(n^k\mathbf{e}_k) = n^k(\mathbf{T}\mathbf{e}_k) = n^k\mathbf{t}_k, \tag{III.4-2}$$

$\mathbf{t}_k$ being the traction on a plane whose outer unit normal is $\mathbf{e}_k$. The value $\mathbf{T}(\mathbf{x})$ of the tensor field $\mathbf{T}$ at $\mathbf{x}$ is called the *stress tensor*, and CAUCHY's Fundamental Theorem asserts *the existence of the stress-tensor field.* The letter $\mathbf{T}$ should recall "tension", since $\mathbf{n} \cdot \mathbf{T}(\mathbf{x})\mathbf{n} > 0$ if and only if the traction $\mathbf{t}(\mathbf{x})$ is a tension. Sometimes, accordingly, $-\mathbf{T}$ is called the *pressure tensor.*

CAUCHY himself proved his theorem by applying (III.1-14) to a tetrahedron, three of whose four faces were mutually perpendicular. In this way he concluded that if $\{\mathbf{e}_k\}$ was an orthonormal basis at $\mathbf{x}$, then

$$\mathbf{T} = \sum_{k=1}^{3} \mathbf{t}(\mathbf{x}, \mathbf{e}_k) \otimes \mathbf{e}_k, \tag{III.4-3}$$

a statement equivalent to (2). CAUCHY's proof, which suggests a method for discovering the theorem, has been reproduced again and again in the textbooks. Here we shall give a proof due to NOLL[1], which is similar to

[1] We follow the presentation by M. E. GURTIN in §15 of "The linear theory of elasticity," FLÜGGE's *Handbuch der Physik* **VIa/2,** ed. C. TRUESDELL, Berlin and New York, Springer-Verlag, 1972. In the same section GURTIN gives a rigorous and efficient version of CAUCHY's original proof.

CAUCHY's in resting essentially upon (III.1-14) but differs in using a construction employing any two linearly independent vectors rather than an orthogonal triad.

Proof. The function $\mathbf{t}(\mathbf{x}, \cdot)$ is defined by (III.3-1) only for arguments that are unit vectors. We may extend it as follows to all of $\mathscr{V}$:

$$\mathbf{t}(\mathbf{x}, \mathbf{v}) \equiv \begin{cases} |\mathbf{v}|\, \mathbf{t}\left(\mathbf{x}, \dfrac{\mathbf{v}}{|\mathbf{v}|}\right) & \text{if } \mathbf{v} \neq \mathbf{0} \\ \mathbf{0} & \text{if } \mathbf{v} = \mathbf{0}. \end{cases} \tag{III.4-4}$$

If $A > 0$ and $\mathbf{v} \neq \mathbf{0}$, then by $(4)_1$

$$\mathbf{t}(\mathbf{x}, A\mathbf{v}) = |A\mathbf{v}|\, \mathbf{t}\left(\mathbf{x}, \frac{A\mathbf{v}}{|A\mathbf{v}|}\right) = A\mathbf{t}(\mathbf{x}, \mathbf{v}), \tag{III.4-5}$$

and the same result follows trivially if $\mathbf{v} = \mathbf{0}$ or $A = 0$. If $A < 0$, then (5) and CAUCHY's Fundamental Lemma (III.3-2) show that

$$\mathbf{t}(\mathbf{x}, A\mathbf{v}) = \mathbf{t}(\mathbf{x}, -|A|\, \mathbf{v}) = |A|\, \mathbf{t}(\mathbf{x}, -\mathbf{v}) = A\mathbf{t}(\mathbf{x}, \mathbf{v}). \tag{III.4-6}$$

Thus $\mathbf{t}(\mathbf{x}, \cdot)$ is a homogeneous function of vectors.

We wish now to show that $\mathbf{t}(\mathbf{x}, \cdot)$ is additive:

$$\mathbf{t}(\mathbf{x}, \mathbf{v}_1 + \mathbf{v}_2) = \mathbf{t}(\mathbf{x}, \mathbf{v}_1) + \mathbf{t}(\mathbf{x}, \mathbf{v}_2). \tag{III.4-7}$$

If $\mathbf{v}_1$ and $\mathbf{v}_2$ are linearly dependent, (7) follows at once from the homogeneity of $\mathbf{t}(\mathbf{x}, \cdot)$. We suppose, then, that $\mathbf{v}_1$ and $\mathbf{v}_2$ are linearly independent. At a given place $\mathbf{x}_0$ the planes $\mathbb{P}_1$ and $\mathbb{P}_2$ normal to $\mathbf{v}_1$ and $\mathbf{v}_2$, respectively, are distinct. We set

$$\mathbf{v}_3 \equiv -(\mathbf{v}_1 + \mathbf{v}_2) \tag{III.4-8}$$

and consider the wedge $\mathscr{A}$ that is bounded by these two planes, the plane $\mathbb{P}_3$ normal to $\mathbf{v}_3$ at the place $\mathbf{x}_0 + \varepsilon\mathbf{v}_3$, the planes $\mathbb{P}_4$ and $\mathbb{P}_5$ distant δ from $\mathbf{x}_0$ and parallel to the plane of $\mathbf{v}_1$, $\mathbf{v}_2$, and $\mathbf{v}_3$. We suppose both ε and δ small enough that $\mathscr{A}$ be the configuration of some part of $\mathscr{B}$, and we denote by $\partial_i\mathscr{A}$ the portion of the plane $\mathbb{P}_i$ that makes a part of the boundary of $\mathscr{A}$. We shall hold δ fixed and let ε approach 0. $\partial_5\mathscr{A}$ is a triangle in the plane of $\mathbf{v}_1$, $\mathbf{v}_2$, and $\mathbf{v}_3$. If the lengths of its sides normal to these vectors are, respectively, l_1, l_2, and l_3, then consideration of similar triangles shows that $l_1/l_3 = |\mathbf{v}_1|/|\mathbf{v}_3|$ and $l_2/l_3 = |\mathbf{v}_2|/|\mathbf{v}_3|$. Also $l_3 = O(\varepsilon)$. Hence if we write A_i for the area of $\partial_i\mathscr{A}$, we see that

$$A_1 = \frac{|\mathbf{v}_1|}{|\mathbf{v}_3|} A_3, \qquad A_2 = \frac{|\mathbf{v}_2|}{|\mathbf{v}_3|} A_3,$$
$$A_3 = O(\varepsilon) \qquad \text{as} \quad \varepsilon \to 0, \tag{III.4-9}$$
$$V(\mathscr{A}) = \tfrac{1}{2}\varepsilon\, |\mathbf{v}_3|\, A_3 = 2\delta A_4 = 2\delta A_5.$$

We set

$$\mathbf{c} \equiv \frac{|\mathbf{v}_3|}{A_3} \int_{\partial \mathscr{A}} \mathbf{t}(\mathbf{x}, \mathbf{n})\, dA. \tag{III.4-10}$$

From (9) and the assumption that $\mathbf{t}(\cdot, \mathbf{n})$ is continuous we see that

$$\mathbf{c} = \sum_{i=1}^{3} \frac{|\mathbf{v}_i|}{A_i} \int_{\partial_i \mathscr{A}} \mathbf{t}\left(\mathbf{x}, \frac{\mathbf{v}_i}{|\mathbf{v}_i|}\right) dA + O(\varepsilon) \qquad \text{as} \quad \varepsilon \to 0, \tag{III.4-11}$$

where we have used CAUCHY's Fundamental Lemma (III.3-2) so as to incorporate into the remainder the integrals over $\partial_4 \mathscr{A}$ and $\partial_5 \mathscr{A}$. Since $\mathbf{t}$ is a homogeneous function of its second argument and a continuous function of its first argument,

$$\mathbf{c} \to \sum_{k=1}^{3} \mathbf{t}(\mathbf{x}_0, \mathbf{v}_k) \qquad \text{as} \quad \varepsilon \to 0. \tag{III.4-12}$$

On the other hand, by $(9)_4$ and (III.1-14), which is a consequence of the balance of momentum, we see that $\mathbf{c} \to \mathbf{0}$ as $\varepsilon \to 0$. Therefore, since the sum in (12) is independent of ε, it must vanish:

$$\sum_{k=1}^{3} \mathbf{t}(\mathbf{x}_0, \mathbf{v}_k) = \mathbf{0}. \tag{III.4-13}$$

By (8), $\mathbf{t}(\mathbf{x}_0, \cdot)$ is additive. Because every homogeneous additive function is a tensor, we have shown that $\mathbf{t}(\mathbf{x}, \mathbf{v}) = \mathbf{T}(\mathbf{x})\mathbf{v}$, and restriction of this result to unit vectors yields CAUCHY's Fundamental Theorem (1). △

CAUCHY's Fundamental Theorem must not be confused with any standard theorem in measure theory. The proof rests essentially upon (III.1-14), which reflects the balance of linear momentum. Of course the theorem could be phrased more abstractly, without verbal reference to mechanics, but the following exercise shows that if we do not impose (III.1-14), then $\mathbf{t}$ need not be a linear function of $\mathbf{n}$. While the example provided there serves to illustrate the point of mathematics being made, it does not seem to arise in any context in mechanics. However, in some important theories of continuum mechanics there are, indeed, traction fields that are not delivered by a stress tensor,[1] but we shall not take up such theories in this book.

[1] *Cf.* R. A. TOUPIN, "Elastic materials with couple stresses," *Archive for Rational Mechanics and Analysis* **11**, 387–414 (1962). TOUPIN calls the tensor whose contravariant components he denotes by t^{ij} the stress tensor, but according to his Equation $(7.19)_2$ the traction vector on a boundary surface is given only in part by $t^{ij}n_j$.

Exercise III.4.1 (GURTIN). Letting $\mathbf{k}$ be a non-vanishing vector, show that if $\mathbf{t}(\mathbf{x}, \mathbf{n}) = (\mathbf{k} \cdot \mathbf{n})\mathbf{n}$, then (III.1-16) holds, but (III.1-14) and CAUCHY's Fundamental Lemma do not.

In the proof of CAUCHY's Fundamental Theorem the assumption that $\mathbf{t}(\cdot, \mathbf{n})$ be *continuous* is crucial. Then there is a sequence of sets $\mathscr{A}_m$ in the plane through $\mathbf{x}$ and normal to $\mathbf{n}$ such that

$$\mathbf{t}(\mathbf{x}, \mathbf{n}) = \lim_{m \to \infty} \frac{\int_{\mathscr{A}_m} \mathbf{t}_{\mathscr{A}_m}\, dA}{A(\mathscr{A}_m)}, \tag{III.4-14}$$

$\mathbf{t}_{\mathscr{A}_m}$ being the traction as defined by (III.1-37). This fact suggests that we define as follows the *average traction* $\mathbf{t}_r(\mathbf{x}, \mathbf{n})$ over a closed disk $\mathscr{D}_r$ of positive radius r, centered at $\mathbf{x}$ and normal to $\mathbf{n}$:

$$\mathbf{t}_r(\mathbf{x}, \mathbf{n}) \equiv \frac{\int_{\mathscr{D}_r} \mathbf{t}_{\mathscr{D}_r}\, dA}{A(\mathscr{D}_r)}. \tag{III.4-15}$$

We say that $\mathbf{f}_C$ has *uniform average traction* in the actual shape $\chi(\mathscr{B})$ if for each fixed $\mathbf{n}$ the one-parameter family of functions $\{\mathbf{t}_r(\mathbf{x}, \mathbf{n})\}$ is, as $r \to 0$, uniformly Cauchy-convergent in the set of $\mathbf{x}$ belonging to any compact subset $\chi(\mathscr{B})$. Using $(\text{III.1-8})_1$ and (III.2-2), one can then show that the map $\mathbf{x} \mapsto \mathbf{t}_r(\mathbf{x}, \mathbf{n})$ is continuous.

Theorem (GURTIN & MARTINS[1]). *The following two statements are equivalent:*

(i) $\mathbf{f}_C$ *has uniform average traction.*
(ii) $\mathbf{t}_r(\mathbf{x}, \mathbf{n}) \to$ *a limit for every* $\mathbf{x}$ *and every* $\mathbf{n}$; *if*

$$\mathbf{t}(\mathbf{x}, \mathbf{n}) \equiv \lim_{r \to 0} \mathbf{t}_r(\mathbf{x}, \mathbf{n}), \tag{III.4-16}$$

then $\mathbf{t}(\cdot, \mathbf{n})$ *is a continuous function of* $\mathbf{x}$.

Thus the continuity of $\mathbf{t}(\cdot, \mathbf{n})$ is characterized directly in terms of the nature of the system of contact forces $\mathbf{f}_C$.

Neither condition (i) nor condition (ii) is implied by our basic assumptions concerning $\mathbf{f}_C$, set forth in §III.1. We state without proof the following *generalization* of CAUCHY's Theorem, which, altogether dispensing with the assumption of continuity, refers directly to the existence of the average traction.

Theorem (GURTIN & MARTINS). *At all places where the limit exists, let* $\mathbf{t}$ *be defined by* (16). *Then on* $\chi(\mathscr{B})$ *there is a tensor field* $\mathbf{T}$ *such that on a set in* $\chi(\mathscr{B})$ *that differs from* $\chi(\mathscr{B})$ *by a set of volume* 0

$$\mathbf{t}(\mathbf{x}, \mathbf{n}) = \mathbf{T}(\mathbf{x})\mathbf{n} \tag{III.4-1$_r$}$$

for every unit vector $\mathbf{n}$.

[1] M. E. GURTIN & L. C. MARTINS, "Cauchy's theorem in classical physics," *Archive for Rational Mechanics and Analysis*, **60**, 305–324 (1976).

If $\mathbf{t}(\cdot, \mathbf{n})$ as so defined is a continuous function of $\mathbf{x}$, then $\mathbf{T}$ is defined everywhere and is the stress tensor delivered by CAUCHY's Fundamental Theorem.

The normal component $\mathbf{n} \cdot \mathbf{t}(\mathbf{x}, \mathbf{n})$ of the traction corresponding to $\mathbf{n}$ at $\mathbf{x}$ is called the *normal traction* at $\mathbf{x}$ on the surfaces that have the outer unit normal $\mathbf{n}$ at $\mathbf{x}$, while the tangential component $\mathbf{t} - (\mathbf{n} \cdot \mathbf{t})\mathbf{n}$ is called the *shear traction* on those surfaces. From CAUCHY's Fundamental Theorem (1) we see that the normal traction is the normal component of $\mathbf{T}$:

$$\mathbf{n} \cdot \mathbf{t} = \mathbf{n} \cdot \mathbf{T}\mathbf{n}. \tag{III.4-17}$$

If $\mathbf{n}$, $\mathbf{e}$, $\mathbf{f}$ is an orthonormal basis at $\mathbf{x}$, then

$$\mathbf{t} - (\mathbf{n} \cdot \mathbf{t})\mathbf{n} = (\mathbf{e} \cdot \mathbf{T}\mathbf{n})\mathbf{e} + (\mathbf{f} \cdot \mathbf{T}\mathbf{n})\mathbf{f}. \tag{III.4-18}$$

The components $\mathbf{e} \cdot \mathbf{T}\mathbf{n}$ and $\mathbf{f} \cdot \mathbf{T}\mathbf{n}$ of the shear traction, accordingly, are called the *shear stresses* in the directions of $\mathbf{e}$ and $\mathbf{f}$, respectively, on a surface normal to $\mathbf{n}$ at $\mathbf{x}$.

Exercise III.4.2. Prove ***Cauchy's Reciprocal Theorem:***

$$\mathbf{n} \cdot \mathbf{t}(\cdot, \mathbf{m}) = \mathbf{m} \cdot \mathbf{t}(\cdot, \mathbf{n}) \quad \forall \mathbf{m}, \mathbf{n} \quad \Leftrightarrow \quad \mathbf{T} = \mathbf{T}^{\mathsf{T}}. \tag{III.4-19}$$

In view of CAUCHY's Fundamental Theorem (1) we may express the principles of balance of linear and rotational momentum $(\text{III.1-8})_{1,2}$ as follows in terms of the stress tensor $\mathbf{T}$:

$$\int_{\chi(\mathscr{P}, t)} \rho \ddot{\mathbf{x}}\, dV = \int_{\partial\chi(\mathscr{P}, t)} \mathbf{T}\mathbf{n}\, dA + \int_{\chi(\mathscr{P}, t)} \rho \mathbf{b}\, dV,$$

$$\int_{\chi(\mathscr{P}, t)} (\mathbf{x} - \mathbf{x}_0) \wedge \rho \ddot{\mathbf{x}}\, dV = \int_{\partial\chi(\mathscr{P}, t)} (\mathbf{x} - \mathbf{x}_0) \wedge (\mathbf{T}\mathbf{n})\, dA + \int_{\chi(\mathscr{P}, t)} (\mathbf{x} - \mathbf{x}_0) \wedge \rho \mathbf{b}\, dV, \tag{III.4-20}$$

where now we have restored the time t in the notation. These are the forms in which the two principles of momentum are commonly stated in continuum mechanics.

Since all forces are frame-indifferent, under change of frame from $\mathfrak{f}$ to $\mathfrak{f}^*$

$$\mathbf{t}^* = \mathbf{Q}\mathbf{t}, \qquad \mathbf{b}^* = \mathbf{Q}\mathbf{b}, \tag{III.4-21}$$

$\mathbf{Q}(t)$ being the orientation of $\mathfrak{f}^*$ relative to $\mathfrak{f}$ at the time t. As we have seen in Exercise I.11-3, the unit normal $\mathbf{n}$ to a surface in $\mathscr{E}$ is frame-indifferent:

$$\mathbf{n}^* = \mathbf{Q}\mathbf{n}. \tag{III.4-22}$$

CAUCHY's Fundamental Theorem (1) applies in $\mathfrak{f}^*$ as well as in $\mathfrak{f}$. Hence $\mathbf{T}$ transforms frame-indifferent vectors into frame-indifferent vectors. From the analysis leading to (I.9-10) we conclude that *the stress tensor is frame-indifferent*:

$$\mathbf{T}^* = \mathbf{Q}\mathbf{T}\mathbf{Q}^{\mathrm{T}}. \qquad \text{(III.4-23)}$$

5. The General Balance

The integral equation $(\text{III.4-20})_1$, like other fundamental equations of classical physics, has the form of an *equation of balance*:

$$\frac{d}{dt}\int_{\mathscr{P}} \Psi \, dM = \int_{\chi(\mathscr{P},\, t)} \rho\dot{\Psi} \, dV = \int_{\partial\chi(\mathscr{P},\, t)} \mathfrak{T}\mathbf{n} \, dA + \int_{\chi(\mathscr{P},\, t)} \rho\mathfrak{s} \, dV. \qquad \text{(III.5-1)}$$

Here Ψ and $\mathfrak{s}$ are tensor fields of the same order, defined over $\mathscr{P}$ and $\chi(\mathscr{P}, t)$, respectively, and $\mathfrak{T}$ is a tensor field of order greater by 1 than that of Ψ and $\mathfrak{s}$. One trivial case has been encountered still earlier, for the principle of conservation of mass (§I.4) may be expressed in this form by taking $\Psi = 1$, $\mathfrak{T} = 0$, $\mathfrak{s} = 0$. More generally, we interpret the equation of balance (1) as asserting that the rate of increase of the total Ψ in a part $\mathscr{P}$ of a body may be expressed as the sum of two effects: inflow through the boundary of the shape of $\mathscr{P}$ and growth at places within that shape. If a result of the form (1) holds, $\mathfrak{T}$ is called an *efflux* of Ψ and $\mathfrak{s}$ is called a *supply* of Ψ.

Of course, neither $\mathfrak{T}$ nor $\mathfrak{s}$ is determined uniquely by (1). For example, any divergenceless tensor of the same order as $\mathfrak{T}$ may be added to $\mathfrak{T}$ without effecting any alteration of the other two terms in (1).

Equations of balance are commonly applied in two special cases: in regions where the fields occurring in them are smooth, and at certain kinds of discontinuities. We consider the former case here.

Specifically, we assume that $\mathfrak{T}$ is continuously differentiable in $\chi(\mathscr{P}, t)$ and continuous on $\partial\chi(\mathscr{P}, t)$. Then Green's transformation may be applied to regions with smooth boundaries:

$$\int_{\partial\chi(\mathscr{P},\, t)} \mathfrak{T}\mathbf{n} \, dA = \int_{\chi(\mathscr{P},\, t)} \operatorname{div} \mathfrak{T} \, dV. \qquad \text{(III.5-2)}$$

Thus the general balance (1) becomes

$$\int_{\chi(\mathscr{P},\, t)} (\rho\dot{\Psi} - \operatorname{div} \mathfrak{T} - \rho\mathfrak{s}) \, dV = 0. \qquad \text{(III.5-3)}$$

Now a principle of balance is asserted to hold *for all bodies* and hence for all shapes of all parts $\mathscr{P}$ of a given body. In particular, then, (3) follows for all parts $\mathscr{P}$ whose shapes at the time t are sufficiently small spheres about the place $\mathbf{x}$ in the interior of $\chi(\mathscr{B}, t)$. If the integral of a continuous function f over every small sphere about $\mathbf{x}$ vanishes, the function itself vanishes at $\mathbf{x}$. Conversely, of course, if f vanishes at all points, so does its integral over every region contained in its domain. Thus we obtain the following

Theorem. *If the equation of balance* (1) *holds for all parts* $\mathscr{P}$ *of* $\mathscr{B}$, *then at all interior points of regions of* $\chi(\mathscr{B}, t)$ *where* $\rho\dot{\Psi} - \operatorname{div} \text{♃} - \rho\text{♂}$ *is continuous*

$$\rho\dot{\Psi} = \operatorname{div} \text{♃} + \rho\text{♂}. \tag{III.5-4}$$

Conversely, if (4) *holds at all interior points of a region, and if* ♃ *is continuous on the boundary of that region, the general balance* (1) *holds at the time t for the body occupying that region.*

The differential equation (4) is called the ***General Field Equation***. In the spatial description the material derivative $\dot{\Psi}$ is to be calculated by (II.6-3) or one of its generalizations.

6. Cauchy's Laws of Motion

We now obtain local forms of the principles of balance of linear and rotational momentum. First, by choosing $\Psi = \dot{\mathbf{x}}$, ♃ $= \mathbf{T}$, and ♂ $= \mathbf{b}$ we reduce (III.5-1) to $(\text{III.4-20})_1$, so by (III.5-4) we obtain ***Cauchy's First Law of Motion***:

$$\rho\ddot{\mathbf{x}} = \operatorname{div} \mathbf{T} + \rho\mathbf{b}, \tag{III.6-1}$$

as a necessary and sufficient condition that linear momentum be balanced for all subbodies in the interior of a region where $\rho\ddot{\mathbf{x}}$, $\rho\mathbf{b}$, $\mathbf{T}$, and $\operatorname{div} \mathbf{T}$ are continuous.

The treatment of $(\text{III.4-20})_2$ is not quite so immediate.

Exercise III.6.1. Prove that for any tensor field $\mathbf{S}$ continuously differentiable on the regular region $\mathscr{R}$

$$\int_{\partial\mathscr{R}} (\mathbf{x} - \mathbf{x}_0) \wedge \mathbf{S}\mathbf{n}\, dA = \int_{\mathscr{R}} [(\mathbf{x} - \mathbf{x}_0) \wedge \operatorname{div} \mathbf{S} + \mathbf{S}^T - \mathbf{S}]\, dV. \tag{III.6-2}$$

If we substitute (2) into $(\text{III}.4\text{-}20)_2$ and suppose that CAUCHY's First Law (1) holds, as a necessary and sufficient condition for the balance of rotational momentum we obtain

$$\int_{\chi(\mathscr{P},\,t)} (\mathbf{T}^{\mathrm{T}} - \mathbf{T})\, dV = \mathbf{0}. \tag{III.6-3}$$

Since $\mathbf{T}$ is continuous,

$$\mathbf{T}^{\mathrm{T}} = \mathbf{T}. \tag{III.6-4}$$

This is ***Cauchy's Second Law of Motion***: *Under the hypotheses leading to the First Law, and on the assumption that linear momentum is balanced, the balance of rotational momentum is equivalent to symmetry of the stress tensor.*

CAUCHY's Laws assert that $\mathbf{T}^{\mathrm{T}} - \mathbf{T}$ and $\rho\ddot{\mathbf{x}} - \operatorname{div}\mathbf{T} - \rho\mathbf{b}$ vanish in an inertial frame. Since these quantities are frame-indifferent in the Galilean class of any given frame, they vanish in one inertial frame if and only if they vanish in all inertial frames. *Cf.* §I.13.

Since $\mathbf{T}$ is frame-indifferent, so is $\mathbf{T}^{\mathrm{T}} - \mathbf{T}$. Thus CAUCHY's Second Law is a frame-indifferent statement. That is, it holds for one framing if and only if it holds for all framings.

CAUCHY's First Law as stated is not frame-indifferent, but of course it can be modified so as to become so. Indeed, both $\operatorname{div}\mathbf{T}$ and $\rho\mathbf{b}$ are frame-indifferent, reflecting the fact that all forces and masses are frame-indifferent. The acceleration $\ddot{\mathbf{x}}$, however, is not frame-indifferent, as we have discussed at length in §§I.9 and I.11. In accord with the frame-indifferent statement of the Axioms of Inertia in §I.13, CAUCHY's First Law in a general framing $\mathfrak{f}^*$ assumes the form

$$\rho\mathbf{a}_{\mathfrak{f}} = \operatorname{div}\mathbf{T}^* + \rho\mathbf{b}^*; \tag{III.6-5}$$

here

$$\mathbf{a}_{\mathfrak{f}} = \ddot{\mathbf{x}}^* - \ddot{\mathbf{x}}_0^* - 2\mathbf{A}(\dot{\mathbf{x}}^* - \dot{\mathbf{x}}_0^*) - (\dot{\mathbf{A}} - \mathbf{A}^2)(\mathbf{x}^* - \mathbf{x}_0^*), \qquad (\text{II}.4\text{-}7)_{\mathrm{r}}$$

$$\mathbf{T}^* = \mathbf{Q}\mathbf{T}\mathbf{Q}^{\mathrm{T}}, \qquad (\text{III}.4\text{-}23)_{\mathrm{r}}$$

$$\mathbf{b}^* = \mathbf{Q}\mathbf{b}, \qquad (\text{III}.4\text{-}21)_{2\mathrm{r}}$$

$\mathbf{Q}$, $\mathbf{x}_0$, and $\mathbf{x}_0^*$ being the quantities defining the change of frame (II.14-2), $\mathbf{A} \equiv \dot{\mathbf{Q}}\mathbf{Q}^{\mathrm{T}}$, and $\dot{\mathbf{x}}^*$ and $\ddot{\mathbf{x}}^*$ being the fields of velocity and acceleration relative to $\mathfrak{f}^*$.

Some authors prefer to transfer the terms following the minus signs in (II.4-7) to the right-hand side of CAUCHY's First Law and call them "forces" or "apparent forces". According to their point of view, CAUCHY's First Law in the form (1) is valid in all frames, but the body force $\mathbf{b}$ must be augmented for forces "due to the motion" of the observer's frame with respect to an inertial frame. Since just the same equations result, this point of view is legitimate, but it is scarcely felicitous, since it obscures the basic nature of frames and of systems of forces.

In this book, except in passages where requirements of frame-indifference are developed, we shall always presume that the frame used is inertial.

CAUCHY's two Laws of Motion have been proved by arguments that apply only to interior points. On the boundary of a body not in contact with any other body, those arguments have no force. We shall assume that the stress field $\mathbf{T}(\mathbf{x}, t)$ is continuous on the closure $\overline{\chi(\mathscr{B}, t)}$ of the shape of $\mathscr{B}$ at the time t. Thus the second law (4) holds on the boundary. At boundary points we shall assume that the first law (1) holds in the sense of an interior limit.

In some recent theories of continuum mechanics stress tensors that are not symmetric appear. In these theories either there are torques that are not moments of forces, or the density of rotational momentum is not simply the moment of the density of linear momentum, or both. We shall not have need of these more general ideas in this course, for which the classical laws of CAUCHY will suffice as local statements of the principles of linear and rotational momentum.

Since the stress tensor is symmetric, it has a spectral decomposition of the form

$$\mathbf{T} = \sum_{k=1}^{3} t_k \mathbf{e}_k \otimes \mathbf{e}_k, \tag{III.6-6}$$

$\{\mathbf{e}_k\}$ being a suitably selected orthonormal basis at $(\mathbf{x}, t)$. The numbers t_k are called the *principal stresses*, and the directions of the $\mathbf{e}_k$ are called the *principal axes of stress*. If the three principal stresses are distinct, the principal axes of stress are unique; otherwise, $\mathbf{T}$ has infinitely many triads of principal axes. In general, there are no surfaces[1] everywhere normal to the fields $\mathbf{e}_k$, but of course there are infinitely many surfaces normal to each $\mathbf{e}_k$ at any one given place $\mathbf{x}$ at the time t. On such a surface at the place $\mathbf{x}$ and the time t, the principal stress t_k is the normal traction, and all shear stresses vanish.

[1] A vector field $\mathbf{v}$ possesses a normal congruence of surfaces if and only if there are scalar functions A and B such that $\mathbf{v} = A\,\nabla B$. Equivalently, $\mathbf{v}$ must be normal to the null-space of $\nabla\mathbf{v} - (\nabla\mathbf{v})^{\mathsf{T}}$. Fields enjoying this property are called "complex-lamellar". *Cf.* CFT, § App. 33. In works on differential geometry may be found necessary and sufficient conditions that an orthogonal basis field be locally the natural basis of a curvilinear co-ordinate system, but we shall not need them in this book. For any purpose of the theory anholonomic components do just as well as components with respect to a co-ordinate system. An example of the use of anholonomic components to solve a special problem is given in Exercise V.4.1 in Volume 2.

Exercise III.6.2. The stress is said to be *hydrostatic* at $\mathbf{x}$ if, for all $\mathbf{n}$

$$|\mathbf{t}(\mathbf{x}, \mathbf{n})| \text{ is independent of } \mathbf{n}, \quad \text{and} \quad \mathbf{n} \cdot \mathbf{t}(\mathbf{x}, \mathbf{n}) \text{ is of one sign,} \tag{III.6-7}$$

and

$$\mathbf{n} \cdot \mathbf{t}(\mathbf{x}, \mathbf{n}) \text{ is independent of } \mathbf{n}, \tag{III.6-8}$$

and

$$\mathbf{t}(\mathbf{x}, \mathbf{n}) \text{ is parallel to } \mathbf{n}. \tag{III.6-9}$$

Show that, in view of CAUCHY's Second Law, any one of these statements implies the others and also

$$\mathbf{T} = -p(\mathbf{x})\mathbf{1}. \tag{III.6-10}$$

What relations can be established between these results if $\mathbf{T}$ is not symmetric? A particular time t, not indicated in the notation, is understood throughout the argument.

The working W of a system of forces has been defined by (I.8-7). In an inertial frame W is expressed in terms of the kinetic energy K and the power P by (I.14-1). The power (I.14-2) in continuum mechanics is the rate of working of the contact force and the body force:

$$P = \int_{\partial\chi(\mathscr{P}, t)} \dot{\mathbf{x}} \cdot \mathbf{t} \, dA + \int_{\chi(\mathscr{P}, t)} \rho \dot{\mathbf{x}} \cdot \mathbf{b} \, dV. \tag{III.6-11}$$

Exercise III.6.3 (PIOLA's power theorem). Prove that if $\mathbf{t}$ and $\mathbf{b}$ are regarded as given fields in (11), then in order that $P = 0$ in every instantaneously rigid motion of $\mathscr{P}$, it is necessary and sufficient that the linear momentum and rotational momentum of $\mathscr{P}$ both be constant.

Exercise III.6.4 (STOKES's power formula). Prove that

$$W = \int_{\chi(\mathscr{P}, t)} w \, dV, \tag{III.6-12}$$

where w, which is called the *stress power*, is given by

$$w = \mathbf{T} \cdot \mathbf{G} = \mathbf{T} \cdot \mathbf{D}. \tag{III.6-13}$$

Hence show that $w = 0$ in a rigid motion, and also in an isochoric motion subject to hydrostatic stress. Prove also that w is a frame-indifferent scalar.

Exercise III.6.5 (Balance of mechanical energy). Set

$$P_C = \int_{\partial\chi(\mathscr{P}, t)} \dot{\mathbf{x}} \cdot \mathbf{t} \, dA; \tag{III.6-14}$$

assume that the field of body force is lamellar, and define the corresponding *potential energy* $V(\mathscr{P}, t)$ as follows:

$$V \equiv \int_{\chi(\mathscr{P}, t)} \rho \varpi \, dV. \tag{III.6-15}$$

Prove that

$$P = P_C - \dot{V} + \int_{\chi(\mathscr{P}, t)} \rho \varpi' \, dV, \tag{III.6-16}$$

and hence that

$$W = P_C - (\dot{K} + \dot{V}) + \int_{\chi(\mathscr{P}, t)} \rho \varpi' \, dV. \tag{III.6-17}$$

Recalling the definition of "mechanically perfect" in §I.14, conclude that in a mechanically perfect motion of a body subject to conservative body force and to boundary tractions normal to the velocities at the points where they act,

$$K + V = \text{const.}, \tag{III.6-18}$$

provided, of course, that we select a steady ϖ.

The result stated in the foregoing exercise is a theorem of conservation of purely mechanical energy. It provides motivation for our having called "conservative" a body force that is steady as well as lamellar. Its hypotheses hold in some cases governed by some classical theories of continua, but not very generally.

7. Mean Values and Lower Bounds for the Stress Field

SIGNORINI remarked that since

$$\operatorname{div}(\Psi \mathbf{T}) = \mathbf{T} \operatorname{grad} \Psi + \Psi \operatorname{div} \mathbf{T}, \tag{III.7-1}$$

CAUCHY's First Law (III.6-1) yields

$$\mathbf{T} \operatorname{grad} \Psi = \operatorname{div}(\Psi \mathbf{T}) + \rho \Psi (\mathbf{b} - \ddot{\mathbf{x}}), \tag{III.7-2}$$

so if we integrate this identity over the present shape $\mathscr{S}$ of a body and then use the divergence theorem, we obtain

$$\int_{\mathscr{S}} \mathbf{T} \operatorname{grad} \Psi \, dV = \int_{\partial \mathscr{S}} \Psi \mathbf{T} \mathbf{n} \, dA + \int_{\mathscr{S}} \rho \Psi (\mathbf{b} - \ddot{\mathbf{x}}) \, dV. \tag{III.7-3}$$

The left-hand side is proportional to a certain weighted mean of the stress field over the shape $\mathscr{S}$. It is determined by the value of $\mathbf{T}$ upon the boundary $\partial\mathscr{S}$ and by the corresponding mean of $\rho\Psi(\mathbf{b}-\ddot{\mathbf{x}})$. The result seems to be of interest mainly in the case of a body at rest, so $\ddot{\mathbf{x}}=\mathbf{0}$. Then it expresses the *mean values of the stress field in terms of the load alone*: $\mathbf{Tn}$ upon the boundary, $\rho\mathbf{b}$ in the interior.

An example due to CHREE and FINGER makes the point clear. If we take for Ψ the position vector $\mathbf{p}$, and if

$$\mathbf{L} \equiv \frac{1}{V(\mathscr{S})}\left[\int_{\partial\mathscr{S}} \mathbf{p}\otimes\mathbf{Tn}\,dA + \int_{\mathscr{S}} \rho\mathbf{p}\otimes\mathbf{b}\,dV\right], \tag{III.7-4}$$

then the mean value $\overline{\mathbf{T}}$ of the stress field in $\mathscr{S}$ is given by

$$\overline{\mathbf{T}} = \mathbf{L}. \tag{III.7-5}$$

The skew part of this equation merely reaffirms Cauchy's Second Law, but the symmetric part has some interesting applications.

First let us suppose that $\mathscr{S}$ is the region between a surface subject to hydrostatic pressure p_o and entirely within it a surface subject to hydrostatic pressure p_i. We suppose also that $\mathbf{b}=\mathbf{0}$, and we write $V(\mathscr{C})$ for the volume of the cavity. Then $\mathbf{L}$ is easy to evaluate, and (5) yields

$$-\overline{\mathbf{T}} = \left[p_o + \frac{V(\mathscr{C})}{V(\mathscr{S})}(p_o - p_i)\right]\mathbf{1}. \tag{III.7-6}$$

Thus *hydrostatic loading gives rise to a stress field that is hydrostatic in mean.* If $p_o = p_i$, the mean stress is the applied pressure. If $p_o > p_i$, the mean pressure always exceeds p_o.

Next we consider a body in a shape $\mathscr{S}$ subject to surface tractions alone, all of which are parallel to a certain vector $\mathbf{e}$. A loading of this kind is called *uniaxial.* If $\mathbf{f}$ is any vector normal to $\mathbf{e}$, then $\mathbf{Tf}=\mathbf{0}$ on $\partial\mathscr{S}$, so from (4) we obtain

$$\mathbf{Lf} = \frac{1}{V(\mathscr{S})}\int_{\partial\mathscr{S}}(\mathbf{Tn}\cdot\mathbf{f})\mathbf{p}\,dA = \frac{1}{V(\mathscr{S})}\int_{\partial\mathscr{S}}(\mathbf{n}\cdot\mathbf{Tf})\mathbf{p}\,dA = \mathbf{0}. \tag{III.7-7}$$

Thus (5) yields

$$\overline{\mathbf{T}}\mathbf{f} = \mathbf{0}. \tag{III.7-8}$$

Thus *the stress field corresponding to uniaxial surface load is uniaxial in mean.*

Exercise III.7.1. Let $\mathscr{A}$ be a portion of a plane normal to $\mathbf{e}$, and upon $\mathscr{A}$ let $\mathbf{Te}$ be a constant multiple of $\mathbf{e}$. Show that if $F\mathbf{e}$ denotes the resultant contact load upon $\mathscr{A}$, and if the centroid of $\mathscr{A}$ is at $\mathbf{p}_o(\mathscr{A})$, then $(\int_{\mathscr{A}}\mathbf{p}\otimes\mathbf{Te}\,dA)\mathbf{e} = F\mathbf{p}_o(\mathscr{A})$. Hence show that if the shape $\mathscr{S}$ of a body at rest has two plane, parallel faces normal to **e**, upon each of which a uniform tensile load is applied, and if the body is otherwise free, then the mean tensile stress is given by

$$\mathbf{e} \cdot \overline{\mathbf{T}}\mathbf{e} = \frac{Fd}{V(\mathscr{S})}, \tag{III.7-9}$$

F being the resultant tensile force applied to either face, and d being the distance between the plane faces.

Numerous other results of this kind were obtained by SIGNORINI and his school. They studied also *moments of stress* $\overline{\mathbf{p} \otimes \mathbf{p} \otimes \cdots \otimes \mathbf{p} \otimes \mathbf{T}}$ and showed that many components of those moments can be determined from the moments of the load.

Exercise III.7.2 (SIGNORINI). Let L be the third-order tensor whose components L_{kmq} are defined as follows in terms of the components p_s of the position vector:

$$\begin{aligned} L_{kpq} &= \tfrac{1}{2}(M_{qkp} + M_{pkq} - M_{pqk}), \\ M_{rst} &\equiv \frac{1}{V(\mathscr{S})}\left[\int_{\partial\mathscr{S}} p_r p_s T_{tq} n_q \, dA + \int_{\mathscr{S}} \rho p_r p_s (b_t - \ddot{x}_t) \, dV\right]. \end{aligned} \tag{III.7-10}$$

Prove that

$$\overline{\mathbf{p} \otimes \mathbf{T}} = \mathsf{L}. \tag{III.7-11}$$

SIGNORINI showed how to obtain lower bounds for the mean stresses in terms of other, more accessible means such as **L** and L. His method was extended by GRIOLI. The results are most easily expressed if we regard **T** as a 6-dimensional vector field, which we shall do for the remainder of this section.

Let the functions $F_{\mathfrak{a}}$, $\mathfrak{a} = 1, 2, \ldots, m$, be orthonormal in mean over the present shape $\mathscr{S}$ of a body:

$$\overline{F_{\mathfrak{a}} F_{\mathfrak{b}}} = \delta_{\mathfrak{a}\mathfrak{b}}. \tag{III.7-12}$$

Let **K** be any symmetric, non-negative tensor over the space of 6-dimensional vectors, and let $\mathbf{C}_{\mathfrak{a}}$, $\mathfrak{a} = 1, 2, \ldots, m$, be vectors in that space. Then

$$0 \leqq \mathbf{K} \cdot (\mathbf{T} - \sum_{\mathfrak{a}=1}^{m} F_{\mathfrak{a}} \mathbf{C}_{\mathfrak{a}}) \otimes (\mathbf{T} - \sum_{\mathfrak{b}=1}^{m} F_{\mathfrak{b}} \mathbf{C}_{\mathfrak{b}}). \tag{III.7-13}$$

Calculating the mean value of this expression, we obtain

$$0 \leqq \mathbf{K} \cdot \overline{\mathbf{T} \otimes \mathbf{T}} + \mathbf{K} \cdot \sum_{\mathfrak{a}=1}^{m} \mathbf{C}_{\mathfrak{a}} \otimes (\mathbf{C}_{\mathfrak{a}} - 2\overline{F_{\mathfrak{a}} \mathbf{T}}). \tag{III.7-14}$$

The vectors $\mathbf{C}_{\mathfrak{a}}$ have been arbitrary so far. We now choose them as follows:

$$\mathbf{C}_{\mathfrak{a}} = \overline{F_{\mathfrak{a}} \mathbf{T}}. \tag{III.7-15}$$

Then (14) reduces to

$$\mathbf{K} \cdot \overline{\mathbf{T} \otimes \mathbf{T}} \geqq \mathbf{K} \cdot \sum_{a=1}^{m} \overline{F_a \mathbf{T}} \otimes \overline{F_a \mathbf{T}}. \qquad \text{(III.7-16)}$$

The non-negative tensor $\mathbf{K}$ and the orthonormal functions F_a remain ours to choose. Thus (16) provides infinitely many lower bounds for the components of $\overline{\mathbf{T} \otimes \mathbf{T}}$, bounds which depend upon the shape of the body and the loads applied to it. One general result is worth noting before we descend to particular applications: If $\mathbf{K}$ is positive rather than merely non-negative, (13) and (15) show that equality is achieved in (16) if and only if

$$\mathbf{T} = \sum_{a=1}^{m} F_a \overline{F_a \mathbf{T}}. \qquad \text{(III.7-17)}$$

Therefore, among all stress fields that have in common the m means $\overline{F_a \mathbf{T}}$ for a given set of orthonormal functions F_a, such fields as satisfy (17) give the minimum value to $\mathbf{K} \cdot \overline{\mathbf{T} \otimes \mathbf{T}}$ for every choice of the positive tensor $\mathbf{K}$.

For example, we may take for $\mathbf{K}$ the tensor whose components with respect to a particular basis are all 0 except for K_{kk}, which has the value 1. Since $T_k^2 \leqq \max \overline{T_k^2}$, from (16) we see that

$$\max T_k^2 \geqq \overline{T_k^2} \geqq \sum_{a=1}^{m} (\overline{F_a T_k})^2, \qquad k = 1, 2, \ldots, 6. \qquad \text{(III.7-18)}$$

Thus lower bounds for the magnitude of every component of $\mathbf{T}$ with respect to a constant basis field have been obtained.

The bounds we have demonstrated are expressed in terms of the means $\overline{F_a \mathbf{T}}$ and thus might seem more difficult to calculate than such direct means as $\overline{\mathbf{T} \otimes \mathbf{T}}$. On the contrary, the results obtained earlier in this chapter show that for suitable choices of the functions F_a the means $\overline{F_a \mathbf{T}}$ can be evaluated in terms of the applied loads. We have obtained two results of this kind, namely, (5) and (11). We shall see now that they may be used so as to evaluate the right-hand side of (16) in terms of the shape of the body and the loads acting upon it.

To do so, we are guided by the properties of the center of mass and the Euler tensor of a body (§§I.8, I.10). These assure us that by choice of a system of rectangular Cartesian co-ordinates we can satisfy the relations $\int_{\mathscr{S}} z_k \, dV = 0$ and $\int_{\mathscr{S}} z_p z_q \, dV = 0$ if $p \neq q$. We could describe these co-ordinates as having their origin at the centroid of $\mathscr{S}$ and their axes parallel to principal axes of inertia of a body of uniform density having the shape $\mathscr{S}$. To express the result, it is convenient to write A_k for the reciprocal of the square root of the kth axial momentoid of inertia:

$$A_k \equiv 1/\sqrt{\overline{z_k^2}}, \qquad k = 1, 2, 3. \qquad \text{(III.7-19)}$$

Then the following 4 functions F_α are orthonormal in mean over $\mathscr{S}$:

$$F_0 \equiv 1, \qquad F_k \equiv A_k z_k, \qquad k = 1, 2, 3. \tag{III.7-20}$$

The relation (5) may be expressed as $\overline{F_0 \mathbf{T}} = \mathbf{L}$. In the present notation, which regards $\mathbf{T}$ as a 6-dimensional vector field, the third-order tensor L defined in terms of the applied loads by (10) becomes a triple of vectors $\mathbf{L}_1$, $\mathbf{L}_2$, $\mathbf{L}_3$, and we may express (11) in the form $\overline{z_k \mathbf{T}} = \mathbf{L}_k$, $k = 1, 2, 3$. Thus, finally, if

$$\mathbf{N}_\alpha \equiv \begin{cases} \mathbf{L} & \text{if} \quad \alpha = 0, \\ A_k \mathbf{L}_k & \text{if} \quad \alpha = 1, 2, 3, \end{cases} \tag{III.7-21}$$

by using in (16) the particular set of orthonormal functions (20) we obtain an elegant inequality discovered by SIGNORINI:

$$\mathbf{K} \cdot \overline{\mathbf{T} \otimes \mathbf{T}} \geqq \sum_{\alpha=0}^{3} \mathbf{N}_\alpha \cdot \mathbf{K} \mathbf{N}_\alpha . \tag{III.7-22}$$

The 4 vectors $\mathbf{N}_\alpha$ on the right-hand side are determined by (5) and (11) from the shape of the body and the loads acting upon it.

The result (22) may be rendered more specific by considering special loadings upon special shapes. Perhaps the most interesting application, however, is the most trivial. Various older theories of plasticity lay down an axiom that for an appropriate choice of $\mathbf{K}$ there is a constant C such that

$$\mathbf{T} \cdot \mathbf{K}\mathbf{T} \leqq C. \tag{III.7-23}$$

SIGNORINI's inequality (22) show that this inequality cannot hold unless the loads and the shape are such that

$$\sum_{\alpha=0}^{4} \mathbf{N}_\alpha \cdot \mathbf{K} \mathbf{N}_\alpha \leqq C. \tag{III.7-24}$$

Thus (23) cannot serve as a general assumption in any theory designed to represent the behavior of bodies of arbitrary shape subject to arbitrary loads.

8. Load. Boundary Condition of Traction

The applied force $\mathbf{f}^{\mathrm{a}}$ and applied torque $\mathbf{F}^{\mathrm{a}}$ acting upon the shape $\chi(\mathscr{P}, t)$ appear on the right-hand sides of (III.4-20). When ρ is regarded as given, both of them are determined by the field $\mathbf{b}$ upon $\chi(\mathscr{P}, t)$ and the field $\mathbf{Tn}$ upon $\partial\chi(\mathscr{P}, t)$. These two fields constitute the *load* on $\mathscr{P}$ in its shape

$\chi(\mathscr{P}, t)$. In many particular problems of continuum mechanics the load on some given shape is prescribed. The condition

$$\mathbf{Tn} = \mathbf{t} \qquad \text{upon} \quad \partial\chi(\mathscr{B}, t), \tag{III.8-1}$$

$\mathbf{t}$ being a given function of $\mathbf{x}$ and t, is the boundary condition of *prescribed traction.* Such a condition supplements the kinematical boundary conditions mentioned in §II.6.

For example, if p is a prescribed scalar field, and if

$$\mathbf{Tn} = -p\mathbf{n} \qquad \text{upon} \quad \partial\chi(\mathscr{B}, t), \tag{III.8-2}$$

the body is subject to the *pressure* p upon its boundary. Of course this condition does not require the stress field throughout $\chi(\mathscr{B}, t)$ to be hydrostatic (Exercise III.6.2).

When the field p in (2) has a constant value on $\partial\chi(\mathscr{D}, t)$, the body is subject to *uniform pressure.* This boundary condition often is regarded as a model for the contact load exerted by a quiet body of gas upon a body submerged in it. A field of pressure proportional to the distance from a fixed plane provides a common model for the contact load exerted by a quiet body of *heavy liquid* of uniform density upon a body partly or wholly submerged in it. The fixed plane represents the horizontal upper surface of the body of liquid.

Exercise III.8.1 (ARCHIMEDES, STEVIN, EULER). Let a body whose shape is a bounded, regular region be submerged partly or wholly in a heavy liquid of uniform density. The centroid of the part of the shape below the horizontal upper surface of the liquid is called the *center of buoyancy.* Assuming that the part of the boundary above the upper surface is free of traction, prove that *the resultant contact load on the body is equipollent to a force applied at the center of buoyancy directed upwards, of magnitude equal to the weight of the fluid displaced* by the body. If the body force applied to the body is the same gravitational field as that which acts on the liquid, then, calling the line connecting the center of buoyancy to the center of mass the *hydrostatic axis*, prove that the body is isolated, as defined by (I.13-11A), if and only if

1. the hydrostatic axis is vertical, and
2. the weight of the displaced fluid equals the weight of the body.

An important special kind of pressure is that exerted by *surface tension.* In contrast to the other special cases just presented, this one reflects the nature of the body as well as its shape and the nature of its surroundings:

$$\begin{aligned} p &= 2\sigma k, \\ k &= \text{mean curvature of } \partial\chi(\mathscr{B}, t), \\ \sigma &= \text{const.} \end{aligned} \tag{III.8-3}$$

The constant σ, called the *coefficient* of surface tension, is adjustable so as to model, more or less, the nature of the parts of $\mathscr{B}$ adjacent to the inside of $\partial\chi(\mathscr{B}, t)$ and the parts of the surroundings adjacent to the outside of $\partial\chi(\mathscr{B}, t)$. It is the first example of a *constitutive modulus* in this book. Others, referring to the material of which $\mathscr{B}$ is composed, will appear in later chapters.

Exercise III.8.2. Prove that the contact load of surface tension upon a shape whose boundary has a continuous unit normal field is null. Thus *a body loaded by surface tension alone is isolated.*

A body $\mathscr{B}$ subjected to null loads, that is,

$$\mathbf{b} = \mathbf{0} \quad \text{in } \chi(\mathscr{B}, t), \qquad \mathbf{Tn} = \mathbf{0} \quad \text{upon } \partial\chi(\mathscr{B}, t), \tag{III.8-4}$$

is *free*. Of course a free body is isolated, but an isolated body need not be free.

9. Motion of a Free Body

The theory presented thus far is so general as to impose little restriction upon the motions of a body. Theories of particular materials, developed in the succeeding chapter and applied throughout the remainder of this book, impose systematic restrictions upon bodies by requiring that they consist of particular materials. In addition to these *constitutive* restrictions, or in some cases instead of them, kinematic assumptions are sometimes imposed directly, and these may severely limit the kind of motion possible. For example, rigid motions have been discussed in §I.10, circulation-preserving motions in §II.13, and some effects of kinematic boundary conditions have been demonstrated in §II.11. The dynamic boundary conditions discussed in the preceding section also have their effects, as we shall see now in what appears to be the simplest case, namely, the motion of a free body.

In the mechanics of systems of mass-points the motion of a free body is trivial. In contrast, a free rigid body may rotate about its center of mass in a most complicated way. When we come to deformable bodies, the problem of free motion becomes indeterminate. Nevertheless, something definite can be learned about it. For example, by putting the position vector $\mathbf{p}$ for Ψ in (III.7-3) we see that

$$\overline{\mathbf{T}} = -\overline{\rho\mathbf{p} \otimes \ddot{\mathbf{x}}}, \tag{III.9-1}$$

by which the mean stress at each time restricts, or is restricted by, the acceleration field. In particular, if we denote by p the arithmetic mean of the normal tractions, $p \equiv -\frac{1}{3}\,\mathrm{tr}\,\mathbf{T}$, then from (1) it follows that

$$\bar{p} = \overline{\tfrac{1}{3}\rho\mathbf{p}\cdot\ddot{\mathbf{x}}}. \tag{III.9-2}$$

DAY, acknowledging influence of SUNDMAN's classic work on the three-body problem of analytical dynamics, has exploited (2) so as to prove a theorem relating the diameter d of the shape of a free body to $\bar{p}$. The body is supposed free when $t \geqq 0$. By its diameter $d(t)$ is meant the supremum of the distances between the places occupied by its body-points at the time t. The position vector $\mathbf{p}(t)$ will henceforth be taken with respect to the center of mass of the free body in its shape at the time t.

Theorem. *Let* $\mathbf{M}$ *be the rotational momentum of* $\mathscr{B}$ *with respect to its center of mass. If* $\mathbf{M} \neq \mathbf{0}$, *then either* $d(t) \to \infty$ *or there is a positive time* t^* *such that* $\bar{p}(t^*) < 0$.

In particular, in the interior of a free body that remains within a bounded region of space for all time, a region in which at least one of the principal stresses is a tension must develop unless the rotational momentum of the body is null.

Lemma 1. *If*

$$P(t) \equiv \int_{\chi(\mathscr{B},\,t)} \rho\,|\mathbf{p}|^2\,dV, \tag{III.9-3}$$

then

$$P \leqq \tfrac{1}{2}Md^2. \tag{III.9-4}$$

Exercise III.9.1. Prove the lemma.

Lemma 2. *If* K *is the kinetic energy of* $\mathscr{B}$,

$$4PK \geqq \tfrac{1}{2}\dot{P}^2 + |\mathbf{M}|^2. \tag{III.9-5}$$

Proof. By the Cauchy–Schwarz inequality

$$|\mathbf{M}|^2 = \left|\int \rho\mathbf{p}\wedge\dot{\mathbf{x}}\,dV\right|^2 \leqq \left(\int \rho\,|\mathbf{p}|^2\,dV\right)\left(\int \frac{\rho\,|\mathbf{p}\wedge\dot{\mathbf{x}}|^2}{|\mathbf{p}|^2}\,dV\right),$$

$$= P\int \frac{\rho\,|\mathbf{p}\wedge\dot{\mathbf{x}}|^2}{|\mathbf{p}|^2}\,dV, \tag{III.9-6}$$

all integrations being carried out over $\chi(\mathscr{B}, t)$. Now for any real number λ and any vectors $\mathbf{a}$ and $\mathbf{b}$, $\mathbf{b}$ not being $\mathbf{0}$,

$$|\lambda \mathbf{a} - \mathbf{b}|^2 \geqq \frac{|\mathbf{a} \wedge \mathbf{b}|^2}{2|\mathbf{b}|^2}, \tag{III.9-7}$$

so by (6)

$$\begin{aligned} \lambda^2 P + 2K - \lambda \dot{P} &= \int \rho(\lambda^2 |\mathbf{p}|^2 + |\dot{\mathbf{x}}|^2 - 2\lambda \mathbf{p} \cdot \dot{\mathbf{x}})\, dV \\ &\geqq \tfrac{1}{2}|\mathbf{M}|^2/P. \end{aligned} \tag{III.9-8}$$

To prove (5), we make the choice $\lambda = \frac{1}{2}\dot{P}/P$. △

Proof of the theorem. We take the material derivative of (3), recall that $\dot{\mathbf{p}} = \dot{\mathbf{x}}$, and use (2) to obtain

$$\ddot{P} = 4K + 6\bar{p}V, \tag{III.9-9}$$

V being the volume of the present shape of $\mathscr{B}$. If $\bar{p}(t) \geqq 0$ when $t \geqq 0$, then from (9) it follows that $\ddot{P} \geqq 4K$. Thus (5) yields

$$2P\ddot{P} \geqq \dot{P}^2 + 2|\mathbf{M}|^2. \tag{III.9-10}$$

Hence

$$(P^2)^{\cdot\cdot} = 2P\ddot{P} + 2\dot{P}^2 \geqq 2|\mathbf{M}|^2 > 0. \tag{III.9-11}$$

Integrating this inequality twice shows that $P(t) \to \infty$. A glance at (4) suffices to prove that $d(t) \to \infty$. △

General References

§§199–238 of CFT. The treatment of stress is exhaustive for 1960 but now partly outmoded. Various difficulties result from an attempt to confine attention to resultant forces without bringing into the open the systems of forces giving rise to those resultants.

W. NOLL, "The foundations of classical mechanics in the light of recent advances in continuum mechanics," in *The Axiomatic Method, with Special Reference to Geometry and Physics* (1957), Amsterdam, North-Holland Publ., 1959, pp. 266–281. Reprinted in NOLL's *The Foundations of Mechanics and Thermodynamics*, New York, Heidelberg, and Berlin, Springer-Verlag, 1974.

Chapter IV

Constitutive Relations

If geometry is to serve as a model for the treatment of axioms of physics, we shall try first to cover with a few axioms as large a class of physical phenomena as possible, and then by adjoining further axioms, one after another, to arrive at the more special theories .... Also the mathematician will have to take account not only of those theories that come near to reality but also, as in geometry, of all logically possible theories, and he must always be careful to obtain a complete survey of the consequences implied by the system of axioms laid down.

Further, it is the task of the mathematician, complementing the physicist's way of looking at things, in each instance to examine exactly whether the further axioms be compatible with the foregoing ones. The physicist regards himself often as being compelled by the results of his experiments to make new hypotheses from time to time *during* the development of his theory; ... he calls only upon those experiments or a certain physical intuition, a practice which in the rigorously logical erection of a theory is not admissible.

HILBERT
in regard to his Sixth Problem,
"Mathematical Treatment of the
Axioms of Physics" (1900)
Mathematical problems,
Archiv für Mathematik und Physik
(3)**1**, 44–63, 213-237 (1901)

1. Dynamic Processes

A motion χ assigns to a body $\mathscr{B}$ a shape $\chi(\mathscr{B}, t)$ at the time t. At a point $\mathbf{x}$ of that shape, the stress tensor $\mathbf{T}(\mathbf{x}, t)$ determines the traction on every surface that is then the boundary of the shape of some part $\mathscr{P}$ of $\mathscr{B}$. In this sense the stress field $\mathbf{T}$ is assigned to the body in its motion. The ordered pair $(\chi, \mathbf{T})$ is called a *dynamic process* for $\mathscr{B}$ if χ and $\mathbf{T}$ are related in such a way as to satisfy the principles of balance of linear and rotational momentum.

At interior points of regions where χ and $\mathbf{T}$ are sufficiently smooth, the principles of linear and rotational momentum are expressed by CAUCHY's Laws of Motion. The second law (III.6-4) requires that the stress be symmetric. The first law (III.6-1) relates the stress field to the acceleration $\ddot{\mathbf{x}}$ in an inertial frame, provided the body force field $\mathbf{b}$ be known. We regard $\mathbf{b}$, which represents the action on $\mathscr{B}$ of unspecified bodies exterior to $\mathscr{B}$, as assignable. While in practice only a few special body forces like that of gravity are available in laboratories or daily life—indeed, typically in specific problems of continuum mechanics we consider only the case in which $\mathbf{b} = \mathbf{0}$—in principle we have no way of delimiting the class of all *possible* fields of body force. Therefore, in arguments concerning the totality of all possible motions of a body, we shall necessarily think of $\mathbf{b}$ as being unrestricted. Whatever be χ and $\mathbf{T}$, a field $\mathbf{b}$ such as to satisfy the principle of balance of linear momentum is determined by (III.6-1), or, if the frame of reference is not inertial, by (III.6-5). Thus CAUCHY's First Law imposes no restriction at all upon χ and $\mathbf{T}$.

Accordingly, any ordered pair $\{\chi, \mathbf{T}\}$ consisting in a smooth motion of $\mathscr{B}$ and a symmetric tensor field defined on $\chi(\mathscr{B}, t)$ for each t is a *dynamic process*.

A dynamic process is defined in terms of a framing $\mathfrak{f}$. Thus, properly, we should refer to $\{\chi, \mathbf{T}\}$ as being a *dynamic process in* $\mathfrak{f}$. Suppose now we consider another framing $\mathfrak{f}^*$. We have reason to regard the motion χ^* obtained from χ by the transformation (I.9-11) as being the very same motion as apparent in $\mathfrak{f}^*$:

$$\begin{aligned} \mathbf{x}^* = \chi^*(X, t^*) = \mathbf{x}_0^*(t) + \mathbf{Q}(t)(\chi(X, t) - \mathbf{x}_0), \\ t^* = t + a, \end{aligned} \qquad \text{(I.9-11)}_r$$

a, $\mathbf{x}_0$, $\mathbf{x}_0^*(t)$, and $\mathbf{Q}(t)$ being prescribed. Likewise, as we have seen in §III.5, the stress $\mathbf{T}$ is a frame-indifferent quantity:

$$\mathbf{T}^*(\mathbf{x}^*, t^*) = \mathbf{Q}(t)\mathbf{T}(\mathbf{x}, t)\mathbf{Q}(t)^{\mathsf{T}}; \qquad \text{(IV.1-1)}$$

here $\mathbf{x}^*$ and t^* are determined from $\mathbf{x}$ and t by (I.9-11). Finally, if the dynamic process $(\chi, \mathbf{T})$ determines a body force $\mathbf{b}$ in $\mathfrak{f}$, then $\mathbf{b}^*$ as given by

$$\mathbf{b}^*(\mathbf{x}^*, t^*) = \mathbf{Q}(t)\mathbf{b}(\mathbf{x}, t) \qquad \text{(IV.1-2)}$$

serves to balance $\{\chi^*, \mathbf{T}^*\}$ in $\mathfrak{f}^*$, and, of course, CAUCHY's First Law is understood to hold in the frame-indifferent form (III.6-5). Thus not only is $\{\chi^*, \mathbf{T}^*\}$ a dynamic process defined in terms of $\mathfrak{f}^*$, but also the body force $\mathbf{b}^*$ corresponding to it is the same, in the sense of frame-indifference, as the body force required to equilibrate $\{\chi, \mathbf{T}\}$ in $\mathfrak{f}$. Thus the definition of a dynamic process is frame-indifferent, and the process $\{\chi^*, \mathbf{T}^*\}$ in $\mathfrak{f}^*$ may be regarded as describing the same phenomena of nature as does $\{\chi, \mathbf{T}\}$ in $\mathfrak{f}$. We shall say formally that $\{\chi^*, \mathbf{T}^*\}$ is the process in $\mathfrak{f}^*$ that is *equivalent* to the process $\{\chi, \mathbf{T}\}$ in $\mathfrak{f}$ if the two are related through (I.9-11) and (1).

2. Constitutive Relations. Noll's Axioms

The principles and definitions so far presented express properties *common* to all bodies and motions. The *diversity* of natural bodies, which arises from the differences among the materials that make up those bodies, we represent in the theory by *constitutive relations*. In mechanics, a constitutive relation is a restriction upon the forces or the motions or both. In popular terms, forces applied to a body "cause" it to undergo a motion, and the motion "caused" differs according to the nature of the body.

In this regard some constitutive relations are trivial, in the sense that a constant function is a trivial special case of a function. External body forces are of this kind. The assumption that the body force is external, since it restricts the class of body forces to those unaffected by the motions of such bodies as may occupy the part of space in which they act, is a constitutive relation, but it is not of the kind subjected to study in continuum mechanics, in which, in typical problems, we simply assume that $\mathbf{b} = \text{const.}$ or even $\mathbf{0}$ but go on to analyse in detail the different responses to these trivial body forces shown by bodies in which there are different kinds of contact forces.

Indeed, the only forces of much interest in continuum mechanics are contact forces. As we have seen in §III.4, these are determined from the stress tensor field $\mathbf{T}$. Just as different figures are defined in geometry as idealizations of certain important natural objects, in continuum mechanics *ideal materials* are defined by particular relations between the stress tensor and the motion of the body. Some special materials, like some special figures, are important in themselves, but it is more efficient to study infinite classes of geometric objects and infinite classes of materials, distinguished by properties of symmetry and invariance. *A general theory of constitutive relations* lays down overriding restrictions that all constitutive relations must satisfy in order to represent mathematically the sort of behavior observed in materials in nature. In the class of all such constitutive relations a rational scheme of classification is then introduced, and theorems characterizing or describing the members of this class are then proved.

The approach is precisely that of Euclidean geometry, in which, after statement of the axioms satisfied by all geometric objects, theorems characterizing and relating classes of figures are proved. However, since mechanics is a discipline vastly more subtle and sensitive than geometry, the parallel stops here and does not extend to the theorems themselves or even to the methods of constructing proofs.

Continuum mechanics, like any other branch of mathematics, has its own characteristic concepts and methods. These were created in large part by JAMES BERNOULLI, EULER, CAUCHY, GREEN, STOKES, KELVIN, MAXWELL, and HUGONIOT, but only in recent years have they been subjected to general and collective scrutiny and forged into a unified doctrine.

The further development of continuum mechanics in this book will fall within the axioms laid down by NOLL in 1958. These, which we now state, while they are by no means the most general considered today, are nevertheless far more general than necessary for our purpose in this introductory book, but they are so clear and easy to grasp that more special statement here would only blunt them.

Axiom N1. Principle of Determinism. *The stress at the place occupied by the body-point X at the time t is determined by the history χ^t of the motion of $\mathscr{B}$ up to the time t:*

$$\mathbf{T}(\chi(X, t), t) = \mathfrak{F}(\chi^t; X, t). \tag{IV.2-1}$$

Here $\mathfrak{F}$ denotes a mapping of histories χ^t, body-points X, and times t onto symmetric tensors. The domain of the first argument χ^t is the set of possible motions of $\mathscr{B}$ (and not merely their restrictions to the body-point X). The range of $\mathfrak{F}$ is some subset of the set of symmetric tensors. The mapping $\mathfrak{F}$ is the *constitutive mapping* of the body-point X; and the body-point X itself is now said to be a *material point* of $\mathscr{B}$, which is composed of the *material* defined by $\mathfrak{F}$; the relation (1) is the *constitutive relation* of the material defined by $\mathfrak{F}$. The mapping $\mathfrak{F}$ is neither more nor less than a rule which, for each body-point and at each time, assigns to the history up to the time t of each conceivable motion of $\mathscr{B}$ a unique stress tensor $\mathbf{T}(\chi(X, t), t)$ at the place $\mathbf{x}$ occupied by X at the time t. As X ranges over $\mathscr{B}$, the value of $\mathfrak{F}$ at the time t delivers the stress field $\mathbf{T}(\mathbf{x}, t)$ acting upon $\chi(\mathscr{B}, t)$. In rough terms, *the past and present placements given by the motion of $\mathscr{B}$ to the body-points it comprises determine the stress field over its present shape $\chi(\mathscr{B}, t)$.*

The concept of material here defined represents the common observation that many natural bodies exhibit *memory* of their past experiences, sometimes continuing to respond to the effects of change of form long after the change itself took place. For this reason $\mathfrak{F}$ is often called a *memory functional.* Of course, those special $\mathfrak{F}$ that depend on χ only through its present value, which model materials without memory, or through the present values of its time derivatives, which model materials with short-range memory, are not excluded.

Only framings preserving the sense of time are allowed in mechanics, as has been stated in §I.6. In view of this fact and the definition (II.10-1) of the history χ^t, the constitutive relation (1) respects the sense of time. While past and present motion determine present stress, it by no means follows that future and present motion determine present stresses. In the materials of nature the past of a specimen cannot generally be reconstructed from its present and future conditions, and irreversibility of this kind is allowed for by the mathematical theory from the start. Indeed, irreversibility is the rule, not the exception, in continuum mechanics, and the study of various precise senses of that word is one of the main aims of the theory. In this continuum mechanics departs strongly from the tradition of analytical dynamics, in which, in typical cases, such as that presented above in §I.14, past and future are interchangeable.

It is possible that (1) be invertible in the sense that the motion χ of a body is determined, conversely, from the history $\mathbf{T}^t$ of the stress field defined over it. However, such cannot be the case in general, since in Eulerian hydrodynamics, defined by the special constitutive relation (IV.4-4) below, a knowledge of the pressure field for all times and at all places determines nothing more about the placement $\chi(\cdot, t)$ than its mass density ρ. Thus an inverted relation giving χ in terms of $\mathbf{T}^t$ *cannot possibly be general.*

Axiom N2. Principle of Local Action. In the principle of determinism the motions of body-points Z that lie far away from X are allowed to affect the stress at X. The notion of contact force makes it natural to exclude action at a distance as a material property. Accordingly, we assume a second constitutive axiom: *The motion of body-points at a finite distance from X in some shape of $\mathscr{B}$ may be disregarded in calculating the stress at X.* (Of course, by the smoothness assumed for χ, body-points once a finite distance apart are always a finite distance apart.) Formally, if χ and $\bar{\chi}$ are motions such that for some neighborhood $\mathscr{N}(X)$

$$\bar{\chi}^t(Z, s) = \chi^t(Z, s) \quad \forall s \geqq 0, \quad \forall Z \in \mathscr{N}(X), \tag{IV.2-2}$$

then

$$\mathfrak{F}(\bar{\chi}^t; X, t) = \mathfrak{F}(\chi^t; X, t). \tag{IV.2-3}$$

Axiom N3. Principle of Material Frame-Indifference. We have said that we shall regard two equivalent dynamic processes as being really the same phenomenon, viewed against two frames of reference as background. We regard material properties as being likewise indifferent to the choice of frame. Since constitutive equations are designed to express idealized material properties, we require they shall be frame-indifferent. That is, *if the constitutive relation* (1) *is satisfied by the dynamic process* $\{\chi, \mathbf{T}\}$, *it is satisfied by every equivalent process* $\{\chi^*, \mathbf{T}^*\}$. Formally, the constitutive mapping $\mathfrak{F}$ in (1) must satisfy the identity

$$\mathbf{T}(\chi^*(X, t^*), t^*)^* = \mathfrak{F}(\chi^{*t^*}; X, t^*) \qquad \text{(IV.2-4)}$$

for all $\mathbf{T}^*$, χ^*, and t^* that may be obtained from $\mathbf{T}$, χ, and t by transformations of the forms (IV.1-1) and (I.9-11).

While some steps may be taken to delimit the class of all constitutive mappings that satisfy Axioms N1–N3, in this book we shall consider only *simple materials.* To this special class, which is still general enough to include all the older theories of continua and most of the more recent ones, we now address ourselves.

3. Simple Materials

The constitutive axioms N1 and N2 state that the history of the motion of an arbitrarily small neighborhood of a material point determines the stress at the present place of that point. The first approximation to the deformation χ_κ at $\mathbf{X}$ is provided, at each t, by the local deformation $\mathbf{F}_\kappa(\mathbf{X}, t)$, the properties of which we have discussed in §II.5 and thereafter. Thus the history of $\mathbf{F}_\kappa$, which we shall denote by $\mathbf{F}_\kappa^t(\mathbf{X})$ or simply $\mathbf{F}_\kappa^t$, provides a first approximation near $\mathbf{X}$ to the history χ_κ^t of the referential description χ_κ of $\mathscr{B}$. If a knowledge of this first approximation suffices to determine the stress at $\mathbf{X}$, the corresponding material point X is called *simple*. Formally, (IV.2-1) becomes in this special case

$$\mathbf{T}(\mathbf{x}, t) = \mathbf{T}(\chi_\kappa(\mathbf{X}, t), t) = \mathfrak{G}_\kappa(\mathbf{F}_\kappa^t(\mathbf{X}), \mathbf{X}). \qquad \text{(IV.3-1)}$$

Clearly the principles of determinism and local action, Axioms N1 and N2, are satisfied. We shall consider Axiom N3 presently.

The mapping $\mathfrak{G}_\kappa$ is called the *response with respect to* $\boldsymbol{\kappa}$. If it is such as to satisfy Axiom N3, the Principle of Material Frame-Indifference, it defines a particular *simple material*; otherwise, it does not. The domain of its first argument, for fixed $\mathbf{X}$, is a suitable class of histories of invertible tensors. Its range is some subset of the set of all symmetric tensor fields over the present shape $\chi(\mathscr{B}, t)$ of $\mathscr{B}$ in $\mathscr{E}$. In other words, at a given time t and at a fixed place $\mathbf{X}$ in the reference shape it maps the history of an invertible tensor function of time onto a symmetric tensor at the place $\mathbf{x}$ presently occupied by X.

As in §II.7, let $\boldsymbol{\lambda}$ map the reference shape $\boldsymbol{\kappa}_1(\mathscr{B})$ onto another one, $\boldsymbol{\kappa}_2(\mathscr{B})$, and let $\mathbf{P} \equiv \nabla\boldsymbol{\lambda}$. Thus $\mathbf{P}$ is a given function of $\mathbf{X}$. Substituting (II.7-5) into (1) yields

$$\mathfrak{G}_{\kappa_1}(\mathbf{F}^t_{\kappa_1}) = \mathfrak{G}_{\kappa_1}(\mathbf{F}^t_{\kappa_2}\mathbf{P}), \tag{IV.3-2}$$

in a notation which omits the place $\mathbf{X}$ in $\boldsymbol{\kappa}_1(\mathscr{B})$ that the material point X occupies. Thus if for any invertible history $\mathbf{F}^t$ we set

$$\mathfrak{G}_{\kappa_2}(\mathbf{F}^t) \equiv \mathfrak{G}_{\kappa_1}(\mathbf{F}^t\mathbf{P}), \tag{IV.3-3}$$

the constitutive relation (1) assumes the form

$$\mathbf{T}(\mathbf{x}, t) = \mathfrak{G}_{\kappa_2}(\mathbf{F}^t(\mathbf{X}), \mathbf{X}), \tag{IV.3-4}$$

provided now $\mathbf{F}$ be interpreted as the gradient of χ_{κ_2} at $\mathbf{X}$, and $\mathbf{x} = \chi_{\kappa_2}(\mathbf{X}, t)$. Thus $\mathbf{T}$ is determined just as well by the history of the local deformation of $\boldsymbol{\kappa}_2(\mathscr{B})$ as by the history of the local deformation of $\boldsymbol{\kappa}_1(\mathscr{B})$. Although the response $\mathfrak{G}_{\kappa_2}$ is not generally the same mapping as is the response $\mathfrak{G}_{\kappa_1}$, the existence of such a mapping is a fact independent of the choice of reference placement. Therefore, the definition of a simple material, while it mentions a reference placement, does not depend upon that placement and hence could be expressed without any use of reference placements.

Homogeneous transplacements were defined and analysed in §II.12. A homogeneous deformation history $\mathbf{F}^t$ can be constructed from any invertible tensor function $\mathbf{F}$. By exhausting the class of homogeneous deformation histories, we exhaust the domain of the response $\mathfrak{G}_{\kappa}(\cdot, \mathbf{X})$. Thus *the constitutive mapping of a simple material point is determined for all deformation histories by its restriction to the histories of homogeneous transplacements.*

In laboratories of experimental mechanics great weight is laid upon homogeneous transplacements, and the results of more complicated transplacements are commonly explained in terms of them. In this sense, though doubtless unconsciously, experimenters tend to presume that a material found in the laboratory may be modelled sufficiently well in the mathematical theory by some simple material. In this book we shall sometimes use the term "experiment" in an ideal sense. We shall imagine an experiment in which a subbody containing X is subjected to a particular transplacement history χ_{κ}^{t} with respect to the reference placement $\boldsymbol{\kappa}$, and we shall suppose that the resulting stress $\mathbf{T}$ is then measured. As χ_{κ}^{t} ranges over various transplacement histories, various values of $\mathbf{T}$ result. We shall describe the constitutive relation itself as expressing the outcome of these experiments. In this sense we may say that *the material point X of the body $\mathscr{B}$ is simple if the outcomes of all experiments at X are determined by the outcomes of all experiments on homogeneous transplacements of parts of $\mathscr{B}$ near X*. In §IV.9, so as to delimit the ideal experimental program suggested by this fact, we shall determine the homogeneous transplacements that can be produced by the action of uniform body forces.

The definition of a simple material and its interpretations are independent of the choice of reference placement $\boldsymbol{\kappa}$. The response $\mathfrak{G}_{\kappa}$ with respect to $\boldsymbol{\kappa}$ is not, nor are the experiments just mentioned. A motion of $\mathscr{B}$ that is a homogeneous transplacement of $\boldsymbol{\kappa}_1(\mathscr{B})$ is not a homogeneous transplacement of $\boldsymbol{\kappa}_2(\mathscr{B})$ unless $\boldsymbol{\kappa}_2 \circ \boldsymbol{\kappa}_1^{-1}$ is an affine mapping. The responses $\mathfrak{G}_{\kappa_1}$ and $\mathfrak{G}_{\kappa_2}$ are in general different mappings, each being determined uniquely from the other by (3).

Henceforth in this book we shall consider only simple materials. When, as will usually be the case, a reference placement $\boldsymbol{\kappa}$ is selected once and for all, we shall write the constitutive relation (1) of a simple material in the abbreviated form

$$\mathbf{T}(t) = \mathfrak{G}(\mathbf{F}^t). \qquad \text{(IV.3-5)}$$

The theory of simple materials includes all the common purely mechanical theories of continua studied in works on mechanics, engineering, physics, applied mathematics, *etc*. In Volume 3 we shall generalize this definition so as to take account of the effects of heating and change of temperature. While modern studies of continuum physics include microstructure, electromagnetism, chemical reactions, diffusion, and relativistic phenomena, we shall not consider them in this book.

4. Some Classical Special Cases. Specimens of the Effect of the Axiom of Frame-Indifference

In this section we shall define some of the special materials, the theory of which has furnished the main subject of study in continuum mechanics in former times, and we shall use them to illustrate the power of the Principle of Material Frame-Indifference in reducing the apparent generality of a class of hypothetical constitutive relations. The reader who is already familiar with classical theories or who desires only a consecutive, systematic development of continuum mechanics should skip this section and pass to the next.

An *elastic material* is defined by the special case when the mapping $\mathfrak{G}$ in (IV.3-5) reduces to a function $\mathfrak{g}$ of the present local deformation $\mathbf{F}(\mathbf{X}, t)$, irrespective of the values $\mathbf{F}^t(\mathbf{X}, s)$ of the history $\mathbf{F}^t$ in the past, *i.e.* when $s > 0$:

$$\mathbf{T} = \mathfrak{g}(\mathbf{F}, \mathbf{X}), \tag{IV.4-1}$$

$\mathfrak{g}(\cdot, \mathbf{X})$ being a function which maps invertible tensors $\mathbf{F}$ onto symmetric tensors $\mathbf{T}$. Not all such functions define elastic materials, however, since the Principle of Material Frame-Indifference, stated in §II.2 as Axiom N3, is not satisfied unless $\mathfrak{g}$ is of a special kind, as delimited in the following

Theorem (CELLERIER, RICHTER). *Let the polar decomposition of the local deformation* $\mathbf{F}$ *be* $\mathbf{F} = \mathbf{RU}$. *Then the constitutive relation of an elastic material is of the form*

$$\mathbf{T} = \mathbf{R}\mathfrak{g}(\mathbf{U}, \mathbf{X})\mathbf{R}^{\mathrm{T}}. \tag{IV.4-2}$$

in which $\mathfrak{g}(\cdot, \mathbf{X})$ *maps positive symmetric tensors onto symmetric tensors. Conversely, any such* $\mathfrak{g}$ *serves by means of* (2) *to define a particular elastic material.*

Proof. We invoke Axiom N3 only in a weakened form. Indeed, since (1) involves $\mathbf{F}^t$ only through $\mathbf{F}$, which is $\mathbf{F}^t(0)$, we need specify nothing about the orthogonal tensor history $\mathbf{Q}^t$ mentioned in Axiom N3 except its present value $\mathbf{Q}^t(0)$, which we shall denote by $\mathbf{Q}$. Under a change of frame $\mathbf{F}$ obeys the transformation rule (II.14-6). Thus, according to (IV.2-4),

$$\mathfrak{g}(\mathbf{QRU}) = \mathbf{Q}\mathfrak{g}(\mathbf{RU})\mathbf{Q}^{\mathrm{T}}, \tag{IV.4-3}$$

$\mathbf{X}$ being omitted from the notation since it is held fixed in this proof. The identity (3) must hold for all orthogonal $\mathbf{Q}$, all orthogonal $\mathbf{R}$, and all positive and symmetric $\mathbf{U}$. In particular, (3) must hold if we choose $\mathbf{Q} = \mathbf{R}^{\mathrm{T}}$. Hence (2) follows as a necessary condition. That it is also sufficient, is trivial to verify. △

In the special case when $\mathfrak{g}(\mathbf{U}) = g(\det \mathbf{U})\mathbf{1}$, (2) reduces in view of (II.9-8) to

$$\mathbf{T} = -p(\rho)\mathbf{1}; \tag{IV.4-4}$$

p is the *pressure field*, which, as the notation indicates, is determined by the density field. In this case the material is called an *elastic fluid* or *Eulerian fluid*. Much of classical hydrodynamics rests upon this constitutive relation.

The velocity field of a fluid is called a *flow*. The term "flow" also has a popular or physical meaning, and so as to reconcile common language with kinematics, sometimes the capacity of a fluid to flow is attributed to its failure to sustain shear stress when at rest in any configuration whatever. We shall see in §IV.16 that this property, while common to all simple fluids, does not suffice to define them. The Eulerian fluid satisfies it *a fortiori*, since it never sustains shear stress, whether it be at rest or in motion.

A class of fluids more general than the Eulerian ones and not included among elastic materials as special cases may be defined by the functional relation

$$\mathbf{T} = \mathfrak{h}(\mathbf{G}, \rho, \dot{\mathbf{x}}, \mathbf{x}, t), \tag{IV.4-5}$$

the first argument, $\mathbf{G}$, being the velocity gradient (II.11-7). We shall see now that the Principle of Material Frame-Indifference forces the last three arguments to drop out and imposes further restrictions upon the function $\mathfrak{h}$. Indeed, Axiom N3 requires that for an arbitrary orthogonal tensor function of time $\mathbf{Q}$, an arbitrary place-valued function of time $\mathbf{x}_0^*$, an arbitrary place $\mathbf{x}_0$, and an arbitrary constant a, the function $\mathfrak{h}$ shall satisfy for all arguments $\mathbf{G}$, ρ, $\dot{\mathbf{x}}$, $\mathbf{x}$, t the identity

$$\begin{aligned}
&\mathfrak{h}(\mathbf{G}, \rho, \dot{\mathbf{x}}, \mathbf{x}, t)\\
&\quad = \mathbf{Q}^{\mathsf{T}}\mathfrak{h}(\mathbf{D}^* + \mathbf{W}^*, \rho^*, \dot{\mathbf{x}}^*, \mathbf{x}^*, t^*)\mathbf{Q},\\
&\quad = \mathbf{Q}^{\mathsf{T}}\mathfrak{h}(\mathbf{Q}\mathbf{D}\mathbf{Q}^{\mathsf{T}} + \mathbf{Q}\mathbf{W}\mathbf{Q}^{\mathsf{T}} + \mathbf{A}, \rho, \mathbf{Q}\dot{\mathbf{x}}\\
&\qquad + \dot{\mathbf{x}}_0^* + \mathbf{A}(\mathbf{x}^* - \mathbf{x}_0^*), \mathbf{x}_0^* + \mathbf{Q}(\mathbf{x} - \mathbf{x}_0), t + a)\mathbf{Q},
\end{aligned} \tag{IV.4-6}$$

to explicate which we have used (II.11-8), (II.14-14), (I.9-11), and (I.9-14). Let us consider particular, fixed arguments $\mathbf{G}$, ρ, $\dot{\mathbf{x}}$, $\mathbf{x}$, t and choose a function $\mathbf{Q}$ such that $\mathbf{Q}(t) = \mathbf{1}$, $\mathbf{A} = \dot{\mathbf{Q}}(t) = -\mathbf{W}$; a function $\mathbf{x}_0^*$ such that $\dot{\mathbf{x}}_0^*(t) = -\dot{\mathbf{x}} - \mathbf{A}(\mathbf{x}^* - \mathbf{x}_0^*(t))$, $\mathbf{x}_0^*(t) = \mathbf{x}_0 - (\mathbf{x} - \mathbf{x}_0)$; and a constant a such that $a = -t$. Then (6) yields the necessary condition, for each argument of $\mathfrak{h}$, that

$$\mathfrak{h}(\mathbf{G}, \rho, \dot{\mathbf{x}}, \mathbf{x}, t) = \mathfrak{h}(\mathbf{D}, \rho, \mathbf{0}, \mathbf{x}_0, 0), \tag{IV.4-7}$$

where $\mathbf{x}_0$ is any fixed place. Thus $\mathfrak{h}$ reduces to a function of $\mathbf{D}$ and ρ alone:

$$\mathfrak{h}(\mathbf{G}, \rho, \dot{\mathbf{x}}, \mathbf{x}, t) = \mathfrak{h}(\mathbf{D}, \rho). \tag{IV.4-8}$$

Roughly, we may describe the formal reasoning just given as showing that since the spin and the velocity may be transformed away by a suitable change of frame, and since any place and time may similarly be transformed into any other, these four arguments cannot enter a frame-indifferent constitutive relation of the class asserted in (5). But this is not all. If we substitute (8) back into (6), we obtain the relation

$$\mathfrak{h}(\mathbf{Q}\mathbf{D}\mathbf{Q}^{\mathrm{T}}, \rho) = \mathbf{Q}\mathfrak{h}(\mathbf{D}, \rho)\mathbf{Q}^{\mathrm{T}}. \quad \text{(IV.4-9)}$$

This identity must be satisfied for all symmetric tensors $\mathbf{D}$ and all orthogonal tensors $\mathbf{Q}$. Conversely, if it is satisfied, so also is (6). Thus we have the following

Theorem (NOLL). *In order that the relation* (5) *satisfy the Principle of Material Frame-Indifference, it is necessary and sufficient that* $\mathfrak{h}$ *reduce to a function of* $\mathbf{D}$ *and* ρ *alone and also satisfy* (9) *as an identity in* $\mathbf{Q}$ *and* $\mathbf{D}$.

A function $\mathfrak{h}$ that maps tensors onto tensors and satisfies the functional equation (9) is called *isotropic*. In a sense which we shall make precise in §IV.14, NOLL's theorem asserts that *all materials whose constitutive relations are included in* (5) *are isotropic materials.*

A material having a constitutive relation in the class defined by (5) when $\mathfrak{h}$ is restricted to be an affine function of its first argument is called a *linearly viscous fluid.* By NOLL's theorem, such a fluid must have a constitutive relation of the form

$$\mathbf{T} = \mathfrak{h}(\mathbf{D}, \rho), \quad \text{(IV.4-10)}$$

in which $\mathfrak{h}$ is an affine isotropic mapping of symmetric tensors onto symmetric tensors.

We shall now determine the most general function of that kind. For later use we shall at first leave aside the condition that $\mathfrak{h}$ be affine. The dimension of the vector space considered plays no part in the results or the analysis, so long as it be finite.

Lemma (RIVLIN & ERICKSEN, SERRIN, NOLL). *Let* $\mathfrak{h}$ *map symmetric tensors, skew tensors, or orthogonal tensors onto tensors. If for all* $\mathbf{A}$ *of one of these kinds and for all orthogonal* $\mathbf{Q}$

$$\mathfrak{h}(\mathbf{Q}\mathbf{A}\mathbf{Q}^{\mathrm{T}}) = \mathbf{Q}\mathfrak{h}(\mathbf{A})\mathbf{Q}^{\mathrm{T}}, \quad \text{(IV.4-11)}$$

then every proper vector of $\mathbf{A}$ *is a proper vector of* $\mathfrak{h}(\mathbf{A})$.

Proof. Let $\mathbf{e}$ be a proper vector of $\mathbf{A}$, and let $\mathbf{R}_{\mathbf{e}}$ be the reflection across the plane normal to $\mathbf{e}$:

$$\mathbf{R}_{\mathbf{e}} = \mathbf{1} - 2\mathbf{e} \otimes \mathbf{e}. \quad \text{(IV.4-12)}$$

Then for any tensor $\mathbf{A}$

$$\mathbf{R_e A R_e^T} = \mathbf{A} - 2\mathbf{e} \otimes \mathbf{A}^T\mathbf{e} - 2\mathbf{Ae} \otimes \mathbf{e} + 4(\mathbf{e} \cdot \mathbf{Ae})(\mathbf{e} \otimes \mathbf{e}). \quad \text{(IV.4-13)}$$

If $\mathbf{A}^T = \mathbf{A}$ and $\mathbf{Ae} = a\mathbf{e}$, it follows that

$$\mathbf{R_e A R_e^T} = \mathbf{A}. \quad \text{(IV.4-14)}$$

If $\mathbf{A}^T = -\mathbf{A}$, then $\mathbf{Ae} = \mathbf{0}$, so (14) follows again. If $\mathbf{A}^T = \mathbf{A}^{-1}$, then $\mathbf{Ae} = \mathbf{A}^T\mathbf{e} = \pm\mathbf{e}$, so (14) follows yet again. Since $\mathbf{R_e}$ is orthogonal, (11) requires that

$$\begin{aligned} \mathbf{R_e}\mathfrak{h}(\mathbf{A})\mathbf{R_e^T} &= \mathfrak{h}(\mathbf{R_e A R_e^T}), \\ &= \mathfrak{h}(\mathbf{A}), \end{aligned} \quad \text{(IV.4-15)}$$

the second step being a consequence of (14). Thus $\mathbf{R_e}$ commutes with $\mathfrak{h}(\mathbf{A})$, and so

$$\mathbf{R_e}\mathfrak{h}(\mathbf{A})\mathbf{e} = \mathfrak{h}(\mathbf{A})\mathbf{R_e e} = -\mathfrak{h}(\mathbf{A})\mathbf{e}. \quad \text{(IV.4-16)}$$

That is, $\mathbf{R_e}$ transforms $\mathfrak{h}(\mathbf{A})\mathbf{e}$ into its negative. Hence $\mathfrak{h}(\mathbf{A})\mathbf{e}$ is parallel to $\mathbf{e}$. △

Theorem (CAUCHY). *In order that a function $\mathfrak{h}$ mapping symmetric tensors onto symmetric tensors be both isotropic and affine, it is necessary and sufficient that it be of the form*

$$\mathfrak{h}(\mathbf{A}) = (\alpha + \beta \operatorname{tr} \mathbf{A})\mathbf{1} + \gamma\mathbf{A}, \quad \text{(IV.4-17)}$$

in which α, β, and γ are constants.

Proof (GURTIN). The projection $\mathbf{P_e}$ has as proper vectors $\mathbf{e}$ itself and all vectors normal to $\mathbf{e}$. By the lemma, these are proper vectors of $\mathfrak{h}(\mathbf{P_e})$. The spectral representation of $\mathfrak{h}(\mathbf{P_e})$ is therefore

$$\mathfrak{h}(\mathbf{P_e}) = \beta(\mathbf{e})\mathbf{1} + \gamma(\mathbf{e})\mathbf{P_e}, \quad \text{(IV.4-18)}$$

$\beta(\mathbf{e})$ being the proper number that corresponds to the vectors normal to $\mathbf{e}$, and $\beta(\mathbf{e}) + \gamma(\mathbf{e})$ being the proper number that corresponds to $\mathbf{e}$. It is sufficient to restrict the argument of β and γ to unit vectors. If $\mathbf{f}$ is any unit vector, there is an orthogonal tensor $\mathbf{Q}$ such that $\mathbf{Qe} = \mathbf{f}$. Then $\mathbf{P_f} = \mathbf{QP_eQ}^T$. Using this fact and (11) and then appealing to (18) twice, we show that

$$\begin{aligned} \mathbf{0} &= \mathbf{Q}\mathfrak{h}(\mathbf{P_e})\mathbf{Q}^T - \mathfrak{h}(\mathbf{QP_eQ}^T), \\ &= \mathbf{Q}\mathfrak{h}(\mathbf{P_e})\mathbf{Q}^T - \mathfrak{h}(\mathbf{P_f}), \\ &= [\beta(\mathbf{e}) - \beta(\mathbf{f})]\mathbf{1} + [\gamma(\mathbf{e}) - \gamma(\mathbf{f})]\mathbf{P_f}. \end{aligned} \quad \text{(IV.4-19)}$$

Because $\mathbf{1}$ and $\mathbf{P_f}$ are linearly independent,

$$\beta(\mathbf{e}) = \beta(\mathbf{f}) \qquad \text{and} \qquad \gamma(\mathbf{e}) = \gamma(\mathbf{f}). \tag{IV.4-20}$$

Since $\mathbf{e}$ and $\mathbf{f}$ are any two unit vectors, β and γ are constants.

Suppose now that $\mathfrak{h}$ is an affine function. Then there is a constant symmetric tensor $\mathbf{K}$ and a linear function $\mathfrak{l}$ such that

$$\mathfrak{h} = \mathbf{K} + \mathfrak{l}, \tag{IV.4-21}$$

and, by (11),

$$\mathbf{K} + \mathfrak{l}(\mathbf{QAQ}^T) = \mathbf{QKQ}^T + \mathbf{Q}\mathfrak{l}(\mathbf{A})\mathbf{Q}^T \tag{IV.4-22}$$

for all symmetric $\mathbf{A}$ and for all orthogonal $\mathbf{Q}$. Since $\mathfrak{l}$ is linear, $\mathfrak{l}(\mathbf{0}) = \mathbf{0}$. According to (22), then, the constant symmetric tensor $\mathbf{K}$ commutes with every orthogonal tensor.

Exercise IV.4.1. Prove that if $\mathbf{K}$ commutes with every orthogonal tensor, then

$$\mathbf{K} = \alpha\mathbf{1}. \tag{IV.4-23}$$

Thus the linear function $\mathfrak{l}$ in (21) must satisfy (11). The fact that it is linear allows us to conclude from (18) that

$$\begin{aligned}
\mathfrak{l}(\mathbf{A}) &= \mathfrak{l}\left(\sum_{k=1}^{n} a_k \mathbf{P}_{\mathbf{e}_k}\right), \\
&= \sum_{k=1}^{n} a_k \,\mathfrak{l}(\mathbf{P}_{\mathbf{e}_k}), \\
&= \sum_{j=1}^{n} a_k(\beta\mathbf{1} + \gamma\mathbf{P}_{\mathbf{e}_k}), \\
&= \left(\beta \sum_{k=1}^{n} a_k\right)\mathbf{1} + \gamma\left(\sum_{k=1}^{n} a_k \mathbf{P}_{\mathbf{e}_k}\right), \\
&= \beta(\operatorname{tr}\mathbf{A})\mathbf{1} + \gamma\mathbf{A},
\end{aligned} \tag{IV.4-24}$$

the numbers $a_1, a_2, \ldots, a_n$ being the latent roots of $\mathbf{A}$. Putting (23) and (24) into (21) shows that $\mathfrak{h}$ must have the form (17).

Conversely, it is plain that (17) is an isotropic affine function for every choice of α, β, and γ. $\triangle$

By combining the theorems of CAUCHY and NOLL we obtain the following

Theorem (STOKES). *The constitutive relation of a linearly viscous fluid is of the form*

$$\mathbf{T} = (-p + \lambda \operatorname{tr} \mathbf{D})\mathbf{1} + 2\mu\mathbf{D}, \tag{IV.4-25}$$

where p, λ, and μ are function of ρ. Every such relation defines a linearly viscous fluid.

The theory based on (25) is called the *Navier–Stokes Theory of Fluids*; under various hypothèses, (25) or major special cases of it were derived by NAVIER, CAUCHY, ST. VENANT, and STOKES. The coefficients λ and μ are called the *viscosities* of the fluid. In rigid motions the Navier–Stokes Theory reduces to Eulerian hydrodynamics, so the fluid it defines exhibits the phenomenon of "flow" in the sense described above, namely, in a state of rest it can sustain only hydrostatic stress. If $\lambda = \mu = 0$, the linearly viscous fluid reduces to an elastic fluid, and for this reason elastic fluids are sometimes called "inviscid" or "perfect".

A material slightly more general than any of those introduced so far in this section is defined by the case when the mapping $\mathfrak{G}$ in (IV.3-5) reduces to a function of $\mathbf{F}(\mathbf{X}, t)$ and $\dot{\mathbf{F}}(\mathbf{X}, t)$ which is affine in $\dot{\mathbf{F}}$:

$$\mathbf{T} = \mathsf{K}(\mathbf{F}, \mathbf{X})[\dot{\mathbf{F}}] = \mathsf{L}(\mathbf{F}, \mathbf{X})[\mathbf{G}], \tag{IV.4-26}$$

where the second form follows from the first by (II.11-5), and where the domain of the affine operator L, indicated by the brackets, is the space of tensors over $\mathscr{V}$. Such a material is called *linearly viscous*.

Exercise IV.4.2. Prove that (26) satisfies the principle of Material Frame-Indifference if and only if it is equivalent to

$$\mathbf{R}^{\mathsf{T}}\mathbf{T}\mathbf{R} = \mathsf{M}(\mathbf{C}, \mathbf{X})[\mathbf{R}^{\mathsf{T}}\mathbf{D}\mathbf{R}], \tag{IV.4-27}$$

$\mathsf{M}(\mathbf{C}, \mathbf{X})$ being an affine operator on the space of symmetric tensors over $\mathscr{V}$.

BOLTZMANN's accumulative theory of visco-elasticity is obtained if we suppose the mapping $\mathfrak{G}$ in (IV.3-5) to be expressible as an integral from $s = 0$ to $s = \infty$. In this case, too, the principle of frame-indifference imposes a restriction upon the class of mappings $\mathfrak{G}$, but we shall not presently carry out the reduction to frame-indifferent form, which we defer to §XIII.6 in Volume 2.

In the Boltzmann theory, as in the theory of elasticity, a further simplification is often attained at the cost of supposing that $|\mathbf{F} - \mathbf{1}|$ or some measure of the magnitude of $\mathbf{F}^t - \mathbf{1}$ be small in some sense. Approximations of this kind make it easier to solve some special problems but are more confusing than helpful in analysis of the general theory.

Exercise IV.4.3. Prove that other than a constant multiple of $\mathbf{1}$, there is no affine function $\mathfrak{g}$ in (1) that satisfies the principle of Material Frame-Indifference. (Do not confuse this condition with that of taking the restriction of $\mathfrak{g}$ to positive symmetric arguments as affine.) Interpret this result in terms of the theory of elasticity.

In this section we have defined and named some of the most important special materials of old. Also, we have illustrated the force of the Axiom of Material Frame-Indifference by showing how it serves to *delimit those functions or mappings that may enter a putative class of constitutive relations.* In the next section we shall give a more general argument of the same kind, an argument which applies to all simple materials.

5. Frame-Indifference. Reduced Constitutive Relations

According to Axiom N3, the response $\mathfrak{G}$ must be such as to make the constitutive relation (IV.3-5) satisfy the Principle of Material Frame-Indifference. Under the change of frame (II.14-3) the local deformation $\mathbf{F}$ obeys the transformation rule (II.14-6), and hence its history $\mathbf{F}^t$ obeys the rule

$$\mathbf{F}^{*t^*} = \mathbf{Q}^t\mathbf{F}^t, \tag{IV.5-1}$$

$\mathbf{Q}^t$ being the history of the orthogonal tensor function $\mathbf{Q}$ occurring in (II.14-3), while the stress tensor $\mathbf{T}$ satisfies (IV.1-1). Hence in order that Axiom N3 hold, $\mathfrak{G}$ must be such that

$$\mathfrak{G}(\mathbf{Q}^t\mathbf{F}^t) = \mathbf{Q}(t)\mathfrak{G}(\mathbf{F}^t)\mathbf{Q}(t)^{\mathrm{T}} \tag{IV.5-2}$$

for *every* orthogonal tensor history $\mathbf{Q}^t$ and for *every* invertible tensor history $\mathbf{F}^t$ in a suitable class. Here $\mathbf{Q}(t)$ is the present value of the function $\mathbf{Q}$, so that $\mathbf{Q}^t(0) = \mathbf{Q}(t)$. Conversely, if (2) is satisfied, so is Axiom N3.

Following an analysis first given by NOLL, we can solve the equation (2) for $\mathfrak{G}$, once and for all. Indeed, by the polar decomposition theorem (II.9-1) we see that $\mathbf{F}^t = \mathbf{R}^t\mathbf{U}^t$, so that (2) becomes

$$\mathbf{Q}(t)^{\mathrm{T}}\mathfrak{G}(\mathbf{Q}^t\mathbf{R}^t\mathbf{U}^t)\mathbf{Q}(t) = \mathfrak{G}(\mathbf{F}^t). \tag{IV.5-3}$$

We may now choose the orthogonal tensor history $\mathbf{Q}^t$ in such a way that $\mathbf{Q}^t(s) = \mathbf{R}^t(s)^{\mathrm{T}}$, $0 \leqq s < \infty$. Hence $\mathbf{Q}(t) = \mathbf{R}(t)^{\mathrm{T}}$, and (3) becomes

$$\mathfrak{G}(\mathbf{F}^t) = \mathbf{R}(t)\mathfrak{G}(\mathbf{U}^t)\mathbf{R}(t)^{\mathrm{T}}. \tag{IV.5-4}$$

Conversely, if (4) holds, it is easy to show that Axiom N3 is satisfied. We have proved the following

Reduction Theorem (NOLL). *Let $\mathfrak{G}$ denote a mapping of positive symmetric tensor histories onto symmetric tensors. Then every constitutive relation for a simple material is of the form*

$$\mathbf{T}(t) = \mathbf{R}(t)\mathfrak{G}(\mathbf{U}^t)\mathbf{R}(t)^{\mathrm{T}}, \tag{IV.5-5}$$

and conversely, any such mapping defines a simple material.

A constitutive equation of this kind, in which the mappings or functions occurring are not subject to any further restriction, is called a *reduced form.*

The result (5) shows that while the stretch history $\mathbf{U}^t$ of a simple material may influence its present stress in any way whatever, past rotations have no influence at all. The present rotation $\mathbf{R}$ enters (5) explicitly. Thus the reduced form enables us to dispense with considering rotation in determining the response to a motion. If we like, we may regard (4) as effecting an extension of $\mathfrak{G}$ from a domain of positive symmetric tensor histories to the full domain of invertible tensor histories. In writing it and similar formulae henceforth we shall usually omit the argument t of $\mathbf{T}$, $\mathbf{U}$, $\mathbf{R}$, *etc.*, although of course t must still appear in the notation for histories $\mathbf{U}^t$, *etc.*

The reduced form enables us also, in principle, to reduce the number of tests needed to determine the response $\mathfrak{G}$ by experiment. Indeed, consider pure stretch histories: $\mathbf{R}^t = \mathbf{1}$. If we know the stress $\mathbf{T}$ corresponding to an arbitrary homogeneous pure stretch history $\mathbf{U}^t$, we have a relation of the form $\mathbf{T} = \mathfrak{G}(\mathbf{U}^t)$. By (5) we then know $\mathbf{T}$ for all deformation histories. Alternatively, consider irrotational histories: $\mathbf{W} = \mathbf{0}$. Given any $\mathbf{U}^t$, we can determine $\mathbf{R}^t$ by integrating $(\mathrm{II}.11\text{-}13)_2$ with $\mathbf{W}$ set equal to $\mathbf{0}$. If we know the stress corresponding to an arbitrary irrotational history, by putting the corresponding values of $\mathbf{R}$ into (5) we can again determine $\mathfrak{G}$. Thus, we may characterize simple materials in either of two more economical ways: *A material is simple if and only if its response in general is determined by the restriction of its response to homogeneous pure stretch histories, or to homogeneous irrotational deformation histories.*

In the polar decomposition (II.9-1) two measures of stretch, $\mathbf{U}$ and $\mathbf{V}$, are introduced. Kinematically, there is no reason to prefer one to the other. From (4) we see that use of $\mathbf{U}$ as a measure of stretch history leads to a simple reduced form for the constitutive equations of a simple material. If we like, of course we may use $\mathbf{V}$ instead. Since $\mathbf{U}^t = (\mathbf{R}^t)^{\mathrm{T}}\mathbf{V}^t\mathbf{R}^t$, substitution into (4) shows that by using $\mathbf{V}$ we do not generally eliminate the rotation history $\mathbf{R}$. That is, use of $\mathbf{V}$ does not lead to a simple result. There are many other tensors that measure stretch just as well as **U** and **V**. In the older literature one or another of these is called a "strain" tensor, but the term "strain" has led to such confusion that we are better advised to avoid it altogether.

Exercise IV.5.1. Show that had we started from a relation of the form

$$\mathbf{T}(X, t) = \mathfrak{G}(\mathbf{F}^t, \mathbf{X}, \dot{\mathbf{x}}, \mathbf{x}, t) \tag{IV.5-6}$$

as the definition of a simple body-point, the Principle of Material Frame-Indifference would have reduced it to our actual starting point (IV.3-1). (*Cf.* the analysis of the assumption (IV.4-5) in §IV.4.)

Exercise IV.5.2. Show that all the reductions obtained in §IV.4 are in fact special cases of the reduction indicated in the preceding exercise, followed by the reduction of (IV.3-5) to (5).

There are infinitely many other reduced forms for the constitutive relation of a simple material. Since $\mathbf{C}^t = (\mathbf{U}^t)^2$, one such form is

$$\begin{aligned}\mathbf{T} &= \mathbf{R}\mathbf{U}\mathbf{U}^{-1}\mathfrak{G}(\sqrt{\mathbf{C}^t})\mathbf{U}^{-1}\mathbf{U}\mathbf{R}^{\mathrm{T}},\\ &= \mathbf{F}\mathfrak{L}(\mathbf{C}^t)\mathbf{F}^{\mathrm{T}},\end{aligned} \tag{IV.5-7}$$

$\mathfrak{L}$ being defined as follows:

$$\mathfrak{L}(\mathbf{C}^t) \equiv \mathbf{U}^{-1}\mathfrak{G}(\sqrt{\mathbf{C}^t})\mathbf{U}^{-1}. \tag{IV.5-8}$$

In §II.8 we constructed the kinematical apparatus for using the actual placement as the reference placement. It is natural to ask if the response of a simple material can be described entirely in terms of this apparatus. Of course the answer is no, but an analysis due to NOLL shows just how far we can go toward expressing the constitutive relation in terms of $\mathbf{F}_t^t$ rather than $\mathbf{F}^t$. To do so, we note from (II.8-7) and (II.9-1) that for given **X**

$$\mathbf{F}(\tau) = \mathbf{R}_t(\tau)\mathbf{U}_t(\tau)\mathbf{R}(t)\mathbf{U}(t). \tag{IV.5-9}$$

Thus

$$\mathbf{F}(\tau) = \mathbf{R}_t(\tau)\mathbf{R}(t)[\mathbf{R}(t)^{\mathrm{T}}\mathbf{U}_t(\tau)\mathbf{R}(t)]\mathbf{U}(t). \tag{IV.5-10}$$

Putting $\tau = t - s$ in (10) yields $\mathbf{F}^t = \mathbf{R}_t^t\mathbf{R}(t)[\mathbf{R}(t)^{\mathrm{T}}\mathbf{U}_t^t\mathbf{R}(t)]\mathbf{U}(t)$, so if

$$\mathbf{Q}^t(s) \equiv (\mathbf{R}_t^t(s)\mathbf{R}(t))^{\mathrm{T}}, \tag{IV.5-11}$$

we obtain

$$\mathbf{Q}^t\mathbf{F}^t = \mathbf{R}(t)^{\mathrm{T}}\mathbf{U}_t^t\mathbf{R}(t)\mathbf{U}(t). \tag{IV.5-12}$$

We may write the requirement of frame-indifference (2) in the equivalent form

$$\mathfrak{G}(\mathbf{F}^t) = \mathbf{Q}(t)^{\mathrm{T}}\mathfrak{G}(\mathbf{Q}^t\mathbf{F}^t)\mathbf{Q}(t) \qquad \forall \mathbf{Q}^t. \tag{IV.5-13}$$

With the choice of $\mathbf{Q}^t$ given by (11) we have $\mathbf{Q}(t) = \mathbf{R}(t)^{\mathrm{T}}$, so (12) and (13) yield

$$\mathbf{R}^{\mathrm{T}}\mathbf{T}\mathbf{R} = \mathfrak{G}(\mathbf{R}^{\mathrm{T}}\mathbf{U}_t^t\,\mathbf{R}\mathbf{U}). \tag{IV.5-14}$$

For applications an equivalent expression in terms of $\mathbf{C}_t^t$ and $\mathbf{C}(t)$ is more convenient:

$$\mathbf{R}^{\mathrm{T}}\mathbf{T}\mathbf{R} = \mathfrak{R}(\mathbf{R}^{\mathrm{T}}\mathbf{C}_t^t\,\mathbf{R}, \mathbf{C}). \tag{IV.5-15}$$

NOLL's reduced forms (14) and (15) show that it is not possible to express the effect of the deformation history in determining the stress entirely by reference to the present shape. While the effect of all the *past* history, $0 < s < \infty$, is accounted for in this way, a fixed reference placement is required, in general, to allow for the effect of the deformation and rotation at the present instant, as shown by the appearance of $\mathbf{R}$ and $\mathbf{C}$ in (15). Of course the relative rotation $\mathbf{R}_t$ has no effect at all. Roughly, these results show that memory effects can be accounted for entirely by use of the relative local deformation, but finite-strain effects require use of some fixed reference placement, any one we please. This is as far as we can go, in general, toward avoiding the use of some fixed reference placement when we wish to describe the response of a simple material.

We conclude this section by remarking upon an important special case. A material point is said to have a *placement at ease* $\boldsymbol{\kappa}_0$ if the stress vanishes when a neighborhood of that point has been held at rest in $\boldsymbol{\kappa}_0$ at all times, past and present. In general, of course, a material point need not have any such placement, as is shown by the case of an Eulerian fluid, defined by (IV.4-4), since usually the pressure function p is assumed to be such that $p(\rho) > 0$ unless $\rho = 0$, the exception $\rho = 0$ being excluded because it violates the condition (II.2-5). When a placement at ease $\boldsymbol{\kappa}_0$ exists, if we choose it as the reference placement $\boldsymbol{\kappa}$ we obtain

$$\mathfrak{G}(\mathbf{1}^t) = \mathbf{0}, \tag{IV.5-16}$$

$\mathbf{1}^t$ being the history up to the time t of the tensor function $\mathbf{F}$ such that $\mathbf{F}(t) = \mathbf{1}$ for all times t. By (2) we see that

$$\mathfrak{G}(\mathbf{Q}^t) = \mathbf{0}. \tag{IV.5-17}$$

Thus any rotation, constant or varying in time in any way, carries one placement at ease into another. The converse is not true, for a material point may have two distinct placements at ease that are not obtained from one another by a rotation.

In this book we shall not assume in general that any material point has a placement at ease.

6. Internal Constraints

So far, we have been assuming that the material is capable, if subjected to appropriate forces, of undergoing any smooth deformation. If the class of possible transplacements is limited *a priori* at interior points of $\chi(\mathscr{B}, t)$, the material is said to be subject to an *internal constraint*. A *simple constraint* is expressed by requiring the local deformation $\mathbf{F}$ to satisfy an equation of the form

$$\gamma(\mathbf{F}) = 0, \tag{IV.6-1}$$

where γ is a frame-indifferent scalar function.

Exercise IV.6.1. Prove that γ is frame-indifferent if and only if

$$\gamma(\mathbf{F}) = \gamma(\mathbf{U}). \tag{IV.6-2}$$

Hence a simple constraint may be written in the form

$$\lambda(\mathbf{C}) = 0, \tag{IV.6-3}$$

where λ is a scalar function. Let λ have been determined, and let f be any real function that vanishes at 0 only. Then $\mathbf{F}$ satisfies the frame-indifferent simple constraint (1) if and only if it satisfies $f(\lambda(\mathbf{C})) = 0$.

If we differentiate (3) with respect to time at a given body-point, we obtain

$$\dot{\lambda} = \partial_{\mathbf{C}} \lambda(\mathbf{C}) \cdot \dot{\mathbf{C}} = 0. \tag{IV.6-4}$$

That is, in view of $(\text{II}.11\text{-}13)_1$,

$$(\mathbf{F} \partial_{\mathbf{C}} \lambda(\mathbf{C}) \mathbf{F}^{\mathrm{T}}) \cdot \mathbf{D} = 0 \tag{IV.6-5}$$

for invertible $\mathbf{F}$ compatible with the constraint and for all symmetric $\mathbf{D}$. Conversely, if (5) holds at each instant for the body-point in question, by integration we conclude that $\lambda(\mathbf{C}) = \text{const.}$; therefore, (5) asserts that if (3) holds at one instant, it holds always. Thus (5) may be used alternatively as a general expression for a frame-indifferent simple constraint.

In all examples so far found to be of interest, for every positive $\mathbf{C}$

$$\partial_{\mathbf{C}} \lambda(\mathbf{C}) \neq \mathbf{0}, \tag{IV.6-6}$$

and we shall consider only constraints of this kind. Because $\mathbf{F}\partial_{\mathbf{C}}\lambda(\mathbf{C})\mathbf{F}^{\mathrm{T}}$ is a symmetric tensor, we may interpret (5) as requiring all $\mathbf{D}$ corresponding to $\mathbf{F}$ to lie in a certain 5-dimensional plane determined by $\mathbf{C}$.

7. Principle of Determinism for Constrained Simple Materials

Constraints, since they assert that some kinds of deformation cannot occur, must be maintained by forces. Since the constraints, by definition, are immutable, the forces maintaining them cannot be determined by the motion itself or its history. In particular, simple internal constraints must be maintained by appropriate stresses, and the constitutive equation of a constrained simple material must be such as to allow these stresses to operate, irrespective of the deformation history.

For constrained materials, accordingly, the principle of determinism must be relaxed. *A fortiori*, the necessary modification of that principle cannot be deduced from the principle itself but must be brought in through a more general axiom.

There are, presumably, many systems of forces which could effect any given constraint. The simplest of these are the ones whose power vanishes in any motion compatible with the constraint. In a constrained material stresses that do no work will therefore be assumed to remain arbitrary in the sense that *they are not determined by the history of the motion.*

Thus we have given reasons for laying down the following

Axiom N1$_{\mathrm{C}}$ (Principle of Determinism for Simple Materials Subject to Constraints). *The stress is determined by the history of the motion only to within an arbitrary tensor that does no work in any motion compatible with the constraint. That is,*

$$\mathbf{T} = \mathbf{N} + \mathfrak{G}(\mathbf{F}^t), \qquad \text{(IV.7-1)}$$

$\mathbf{N}$ *being a stress for which the stress-power vanishes in any motion satisfying the constraint. The mapping* $\mathfrak{G}$ *need be defined only for arguments* $\mathbf{F}^t$ *such as to satisfy the constraint. It is understood here that* $\mathbf{N}$ *and* $\mathfrak{G}$ *may depend upon* $\mathbf{X}$. *The tensor* $\mathbf{N}$, *since the condition defining it makes no use of a reference placement, is independent of* $\boldsymbol{\kappa}$; *of course* $\mathfrak{G}$ *generally depends upon* $\boldsymbol{\kappa}$.

$\mathbf{T} - \mathbf{N}$ is called the *determinate stress*. Axiom N1$_{\mathrm{C}}$ generalizes Axiom N1 of §IV.2 and reduces to it when no internal constraints are imposed.

The problem is now to find $\mathbf{N}$. The stress-power w of a symmetric stress tensor $\mathbf{T}$ in a motion with stretching tensor $\mathbf{D}$ is given by (III.6-13). Accordingly, we are to find the general solution $\mathbf{N}$ of the equation

$$\mathbf{N} \cdot \mathbf{D} = 0 \tag{IV.7-2}$$

if $\mathbf{D}$ is any symmetric tensor that satisfies (IV.6-5). Hence the symmetric tensor $\mathbf{N}$ must be perpendicular to every vector $\mathbf{D}$ that is perpendicular to $\mathbf{F}\partial_{\mathbf{C}}\lambda(\mathbf{C})\mathbf{F}^{\mathrm{T}}$. Thus $\mathbf{N}$ is parallel to this latter vector:

$$\mathbf{N} = q\mathbf{F}\partial_{\mathbf{C}}\lambda(\mathbf{C})\mathbf{F}^{\mathrm{T}}, \tag{IV.7-3}$$

q being an arbitrary scalar. This is the general solution of (2).

If there are k constraints $\lambda^m(\mathbf{C}) = 0$, $m = 1, 2, \ldots, k$, then

$$\mathbf{N} = \sum_{m=1}^{k} q_m \mathbf{F}\partial_{\mathbf{C}}\lambda^m(\mathbf{C})\mathbf{F}^{\mathrm{T}}. \tag{IV.7-4}$$

That is, the symmetric tensor $\mathbf{F}^{-1}\mathbf{N}(\mathbf{F}^{-1})^{\mathrm{T}}$ must lie in the span of the k symmetric tensors $\partial_{\mathbf{C}}\lambda^m(\mathbf{C})$, $m = 1, 2, \ldots, k$. If the k tensors $\partial_{\mathbf{C}}\lambda^m(\mathbf{C})$ are linearly independent, their span is a k-dimensional plane. If $k \geqq 6$, no restriction upon $\mathbf{N}$ results. Thus in a material subject to 6 or more constraints with linearly independent gradients, the stress is altogether arbitrary.

The argument given here applies at a single material point. Usually the same constraints will be laid down for all points of a body. In that case (4) will result for each, but the theory does not require that the quantities q_m in (4) for the several points be related to one another in any particular way. In order to obtain a constitutive relation leading to definite solutions of specific problems it is customary to assume that each multiplier q_m is a smooth field $q_m(\mathbf{x}, t)$ on the present shape of $\mathscr{B}$. Substitution into (1) yields the general constitutive equation for simple material subject to k simple frame-indifferent internal constraints.

The determinate stress, $\mathfrak{G}(\mathbf{F}^t)$, may be expressed in reduced forms identical with those found in §IV.4 for unconstrained materials.

We now consider some examples of internal constraints.

1. *Incompressibility.* A material is said to be *incompressible* if it can experience only isochoric motions. By $(\text{II.5-8})_3$ and $(\text{II.9-7})_9$, an appropriate constraint function for an incompressible material is

$$\lambda(\mathbf{C}) = \det \mathbf{C} - 1. \tag{IV.7-5}$$

Since

$$\mathbf{F}\partial_{\mathbf{C}}\lambda(\mathbf{C})\mathbf{F}^{\mathrm{T}} = \mathbf{F}\mathbf{C}^{-1}\mathbf{F}^{\mathrm{T}} \det \mathbf{C} = \mathbf{1}, \tag{IV.7-6}$$

(3) yields

$$\mathbf{N} = -p\mathbf{1}, \tag{IV.7-7}$$

where p is an arbitrary scalar. Thus we have verified a result asserted in effect by POINCARÉ: *In an incompressible material the stress is determinate from the motion only to within an arbitrary hydrostatic pressure.*

2. *Inextensibility.* If $\mathbf{e}_\kappa$ is a unit position vector in the reference shape $\kappa(\mathscr{B})$, $\mathbf{F}\mathbf{e}_\kappa$ is the vector $\mathbf{e}$ into which it is carried in a homogeneous transplacement with gradient $\mathbf{F}$, as we have seen in §II.12. Accordingly, for a material *inextensible* in the actual direction $\mathbf{e}$ an appropriate constraint function is

$$\lambda(\mathbf{C}) = |\mathbf{F}\mathbf{e}_\kappa|^2 - 1 = \mathbf{e}_\kappa \cdot \mathbf{C}\mathbf{e}_\kappa - 1. \tag{IV.7-8}$$

Because

$$\partial_{\mathbf{C}} \lambda(\mathbf{C}) = \mathbf{e}_\kappa \otimes \mathbf{e}_\kappa, \tag{IV.7-9}$$

(3) yields

$$\mathbf{N} = q\mathbf{F}(\mathbf{e}_\kappa \otimes \mathbf{e}_\kappa)\mathbf{F}^{\mathrm{T}} = q\mathbf{e} \otimes \mathbf{e}. \tag{IV.7-10}$$

Since $\mathbf{N}$ is an arbitrary uniaxial tension in the direction of $\mathbf{e}$, we recover a result first found by ADKINS & RIVLIN: *In a material inextensible in a certain direction, the stress is determinate only to within a uniaxial tension in that direction.*

3. *Rigidity.* A material is *rigid* if it is inextensible in every direction. By the result just established, the stress in a rigid material is determinate only to within an arbitrary tension in every direction. Therefore, the stress in a rigid material is altogether unaffected by the motion. This is not surprising in view of the fact, which we have demonstrated in §I.13, that the rigid motion of any body is determined without knowledge of what the stress may be.

Exercise IV.7.1. For incompressible materials obtain counterparts of the reduced forms (IV.5-5), (IV.5-7), (IV.5-14), and (IV.5-15). Show that the constitutive relation of an incompressible elastic material is of the form

$$\mathbf{T} = -p\mathbf{1} + \mathbf{R}\mathfrak{g}(\mathbf{U})\mathbf{R}^{\mathrm{T}}, \qquad |\det \mathbf{U}| = 1; \tag{IV.7-11}$$

of an incompressible elastic fluid,

$$\mathbf{T} = -p\mathbf{1}; \tag{IV.7-12}$$

of an incompressible linearly viscous fluid,

$$\mathbf{T} = -p\mathbf{1} + 2\mu\mathbf{D}, \qquad \operatorname{tr} \mathbf{D} = 0. \tag{IV.7-13}$$

In all three cases the hydrostatic pressure p is indeterminate in the sense that it may be assigned independently of the history of the motion. Show that, conversely, any relation having one of the above three forms defines, respectively, an incompressible elastic material, an incompressible elastic fluid, and an incompressible linearly viscous fluid.

Exercise IV.7.2 (Energy Theorem for Incompressible Fluids in Classical Hydrodynamics). Noting that all motions of an incompressible elastic fluid body $\mathscr{B}$ are mechanically perfect, and considering the case when $\mathscr{B}$ is subject to conservative body force, by use of the results of Exercise III.6.5 show that

$$K + V = \text{const.} \qquad \text{(III.6-18)}_r$$

if on $\partial\chi(\mathscr{B}, t)$ the pressure is everywhere constant or the velocity is everywhere tangential. (A more general result of this kind is given in §XII.7, below.)

8. Equations of Motion for Simple Bodies

A body $\mathscr{B}$ all of whose points are of simple material is a *simple body*. To obtain the *equation of motion* for a body, we need only substitute the constitutive relation into CAUCHY's First Law (III.6-1).

For an unconstrained body, the constitutive relation is (IV.3-5), and the equation of motion is

$$\rho\ddot{\mathbf{x}} = \operatorname{div} \mathfrak{G}(\mathbf{F}^t) + \rho\mathbf{b}. \qquad \text{(IV.8-1)}$$

We think of $\mathbf{b}$ as given—typically, as being a constant vector or $\mathbf{0}$—and then (1) becomes *a condition on the transplacement* χ_κ. In the older theories this condition is a differential equation of second order in the time and the co-ordinates, separately or jointly. In general, it is a differential-functional equation which in view of the reduced form (IV.5-5) *is never linear in the derivatives with respect to spatial co-ordinates.* The resources of analysis are far from sufficient today to approach the general solution of initial-value or boundary-value problems for such equations. Nevertheless, a great deal is known about particular solutions for particular classes of responses $\mathfrak{G}$, and the rest of this book is devoted to the proof and explanation of some of these now known theorems of rational mechanics.

We have just made it plain that a constrained body is by no means a special case of an unconstrained one. Rather, the reverse holds, and the unconstrained body emerges as special. The behavior of a constrained body is not the same as that of any corresponding unconstrained one which happens to experience a motion satisfying the constraint. For example, if an unconstrained body happens to have been subjected to an isochoric deformation history, the stress field on its present shape is determined by that history. An incompressible body, by definition, can never be subjected to anything but isochoric deformation histories, but its stress field is never completely determined by them, being always indeterminate to the extent of an arbitrary hydrostatic pressure field.

Writers on hydrodynamics are guilty of propagating not only bad English but also confusion when they refer to "incompressible flows". A flow, obviously, cannot be compressed. A flow may or may not be isochoric, and a fluid may or may not be incompressible; the behavior of an incompressible fluid in a certain, necessarily isochoric flow is generally not at all the same as that of any compressible fluid in the same isochoric flow.

A constrained body is susceptible of a smaller class of deformations than is an unconstrained one. Corresponding to this restriction are certain arbitrary stresses, arbitrary in the sense that they are not determined by the deformation history. When we seek to determine whether or not a given deformation history of a body consisting of constrained points be compatible with the axioms of mechanics and assigned body force, the presence of these arbitrary stresses gives us greater freedom than for an unconstrained body undergoing the same deformation history subject to the same body force. In this sense a single deformation history satisfying a certain internal constraint will correspond to infinitely many different stress fields, provided it correspond to any at all. Roughly, we may say that while a constrained body is susceptible, by definition, to a restricted class of deformation histories, it is easier to solve problems concerning those histories for a constrained body than for a corresponding unconstrained one. We shall frequently illustrate this evident but important fact.

The extreme case is furnished by the rigid body, where the deformations allowed are reduced to so special a class that the stress, whatever it may be, has no effect at all on the motion of the body, which can be determined by solving ordinary differential equations which express no more than the principles of linear and rotational momentum for the whole body, with no reference to what the actions of its subbodies upon one another may or may not be.

The most useful constrained body is the incompressible one. If we substitute (IV.7-7) into (IV.7-1) and then put the result into CAUCHY's First Law, we obtain the equation of motion

$$\rho(\ddot{\mathbf{x}} - \mathbf{b}) = -\operatorname{grad} p + \operatorname{div} \mathfrak{G}(\mathbf{F}^t). \tag{IV.8-2}$$

If we suppose $\mathbf{b}$ given, χ_κ must satisfy (1) for an unconstrained body, (2) for an incompressible one. In the former all fields χ_κ are eligible to compete, and few will be found successful. For the latter, only those fields χ_κ such that $\det \mathbf{F}^t = 1$ are allowed, but the scalar field p may be adjusted to aid in finding a solution. The condition upon the motion alone, of course, is now

$$\operatorname{grad}[\operatorname{div} \mathfrak{G}(\mathbf{F}^t) - \rho(\ddot{\mathbf{x}} - \mathbf{b})] = \{\operatorname{grad}[\operatorname{div} \mathfrak{G}(\mathbf{F}^t) - \rho(\ddot{\mathbf{x}} - \mathbf{b})]\}^{\mathsf{T}}, \quad \text{(IV.8-3)}$$

a differential-functional equation of order higher than that of (1).

To see the effect of this difference, let us suppose the field of body force be lamellar, so that (III.1-7) holds. If we set

$$\varphi \equiv \frac{p}{\rho} + \varpi, \quad \text{(IV.8-4)}$$

ϖ being the potential of $\mathbf{b}$, and if we suppose also that ρ is uniform, then (2) becomes

$$\ddot{\mathbf{x}} = -\operatorname{grad} \varphi + \frac{1}{\rho} \operatorname{div} \mathfrak{G}(\mathbf{F}^t). \quad \text{(IV.8-5)}$$

Thus p and ϖ enter the equation of motion only through the combination denoted by φ. Suppose, now, that for a given incompressible simple body, that is, for a given response $\mathfrak{G}$, a certain isochoric deformation history satisfy (5) with a certain pressure field p_1 and a field of body force with potential ϖ_1. Let p_2 and ϖ_2 be any scalar fields such that

$$\frac{p_2}{\rho} + \varpi_2 = \frac{p_1}{\rho} + \varpi_1. \quad \text{(IV.8-6)}$$

By (4), the equation of motion (5) is satisfied when p_1 and ϖ_1 are replaced by p_2 and ϖ_2. Thus we have the following theorem, due in principle to EULER: *If for an incompressible material of uniform density a certain isochoric deformation history satisfies the equation of motion for the body force whose potential is ϖ_1, it satisfies the equation of motion for the same material subject to the body force whose potential is ϖ_2.*

When $\mathbf{b} = \mathbf{0}$, the only forces applied to $\mathscr{B}$ from without are tractions upon the boundary $\partial\chi(\mathscr{B}, t)$. Thus we have the following corollary to EULER's theorem: *A motion of an incompressible body $\mathscr{B}$ of uniform density is possible subject to the lamellar field of body force with potential ϖ if and only if it is possible for that same body subject to surface tractions alone.*

Since no more than adjustment of the pressure field is needed to convert the solution of a problem in which there is no body force at all to one in which the lamellar body force with potential ϖ is applied, and since the pressure field is arbitrary, there is little loss in generality in supposing that $\mathbf{b} = \mathbf{0}$ when we treat problems concerning incompressible bodies.

9. Homogeneous Transplacements of Unconstrained Simple Bodies

The constitutive relation of an unconstrained simple body with respect to the reference placement $\boldsymbol{\kappa}$ is of the form

$$\mathbf{T}(\mathbf{X}, t) = \mathfrak{G}_{\boldsymbol{\kappa}}(\mathbf{F}_{\boldsymbol{\kappa}}^{t}(\mathbf{X}), \mathbf{X}). \tag{IV.9-1}$$

Therefore, as we have explained in §IV.3, the restriction of the constitutive mapping of a simple body to the histories of transplacements homogeneous with respect to $\boldsymbol{\kappa}$ determines its response to all deformation histories altogether. Thus in an ideal program of experiment we should subject a body of given material to every transplacement of the form (*cf.* §II.12)

$$\mathbf{x} = \chi_{\boldsymbol{\kappa}}(\mathbf{X}, t) = \mathbf{x}_0(t) + \mathbf{F}(t)(\mathbf{X} - \mathbf{X}_0), \qquad \det \mathbf{F}(t) \neq 0, \tag{IV.9-2}$$

and record the stresses obtained. The results would amount to a determination of the response $\mathfrak{G}_{\boldsymbol{\kappa}}$. We now ask whether such a program be possible in principle.

Can the motion (2) be produced in a body of the material defined by (1)? If the body force $\mathbf{b}$ in CAUCHY's First Law (III.6-1) is disposable, the answer is of course yes. However, while in considering the totality of dynamical processes we saw no reason to exclude any $\mathbf{b}$, it is a different matter when we come to think about particular experiments, for only very special body forces are available in the laboratory. Practically speaking, a uniform field $\mathbf{b} = \text{const.}$ is all we are likely to be able to produce, unless we call upon electromagnetic forces, the effects of which are not taken up in this book. We then ask if the homogeneous motion (2) can be produced in the body defined by (1) if suitable surface tractions be supplied. We shall approach the problem only for homogeneous bodies.

A simple body is *homogeneous* if there is a reference placement $\boldsymbol{\kappa}$, called a *homogeneous reference*, such that the response $\mathfrak{G}_{\boldsymbol{\kappa}}$ is independent of $\mathbf{X}$. That is,

$$\mathfrak{G}_{\boldsymbol{\kappa}}(\mathbf{F}^t, \mathbf{X}) = \mathfrak{G}_{\boldsymbol{\kappa}}(\mathbf{F}^t, \mathbf{Y}) \qquad \forall \mathbf{X}, \mathbf{Y} \in \boldsymbol{\kappa}(\mathscr{B}), \tag{IV.9-3}$$

and for all histories $\mathbf{F}^t$ in a suitable class.

Therefore, in a homogeneous simple body subject to a homogeneous deformation history from a homogeneous reference the stress field $\mathbf{T}$ has the same value at every place, so that

$$\operatorname{div} \mathbf{T} = \mathbf{0}. \tag{IV.9-4}$$

The question we now put is, if $\mathbf{b} = \text{const.}$, is it possible to supply boundary tractions such as to produce the homogeneous transplacement (2) with respect to a homogeneous reference of a simple body? Substitution of (4) into the equation of motion (IV.8-1) yields the condition

$$\rho \mathbf{b} = \rho \ddot{\mathbf{x}}. \tag{IV.9-5}$$

This requirement is compatible with the motion (2) if and only if

$$\ddot{\mathbf{F}} = \mathbf{0}, \qquad \ddot{\mathbf{x}}_0 = \mathbf{b}. \tag{IV.9-6}$$

Hence

$$\mathbf{F}(t) = \mathbf{F}_0(\mathbf{1} + t\mathbf{F}_1), \qquad \mathbf{x}_0(t) = \tfrac{1}{2}t^2\mathbf{b} + t\mathbf{e} + \mathbf{f}, \tag{IV.9-7}$$

$\mathbf{F}_0$ being an arbitrary invertible tensor, $\mathbf{F}_1$ an arbitrary tensor, $\mathbf{e}$ an arbitrary vector, and $\mathbf{f}$ an arbitrary fixed place.

Exercise IV.9.1 (TRUESDELL). Prove that in these motions

$$\mathbf{G} = \mathbf{F}_0\mathbf{F}_1(\mathbf{1} + t\mathbf{F}_1)^{-1}\mathbf{F}_0^{-1}. \tag{IV.9-8}$$

From (7) we see that in general the homogeneous transplacement (2) cannot be produced by surface tractions alone when $\mathbf{b} = \text{const}$. Only a special class of homogeneous transplacements, defined by (7), is possible. Therefore, the ideal program of determining $\mathfrak{G}_\kappa$ by effecting all homogeneous transplacements from $\boldsymbol{\kappa}$ is not practicable. This does not mean that no other method of determining $\mathfrak{G}_\kappa$ may be found but merely that the method of homogeneous transplacements, used to interpret the *definition* of a simple material, is not feasible for finding $\mathfrak{G}_\kappa$ by experiment without use of artificial body forces.

What is left from our analysis is a particular class of homogeneous transplacements that can be effected in *any unconstrained homogeneous simple body* by application of suitable boundary tractions. In other words, a class of *exact solutions* for all homogeneous unconstrained bodies has been found. These particular solutions generally exist only for a finite interval of time. By assumption, $\det \mathbf{F}(0) = \det \mathbf{F}_0 \neq 0$. Then $\mathbf{F}$ is invertible only so long as

$$\det(\mathbf{1} + t\mathbf{F}_1) \neq 0, \tag{IV.9-9}$$

that is, in an interval of time $[t_-, t_+]$ containing 0 and such that $-1/t$ never equals a proper number of $\mathbf{F}_1$. Since $\mathbf{F}_1$ is an arbitrary tensor, possibly singular, nothing can be said in general about its proper numbers. The possibilities that $t_- = -\infty$ or $t_+ = +\infty$ are not excluded. *E.g.*, in an isochoric transplacement of this class,

$$|\det \mathbf{F}_0| = 1, \qquad \det(\mathbf{1} + t\mathbf{F}_1) = 1, \tag{IV.9-10}$$

and the interval in which the motion exists is $]-\infty, +\infty[$.

Exercise IV.9.2. Prove that the motion is isochoric if and only if

$$|\det \mathbf{F}_0| = 1, \qquad \operatorname{tr} \mathbf{F}_1 = 0, \qquad \operatorname{tr} \mathbf{F}_1^2 = 0, \quad \det \mathbf{F}_1 = 0. \tag{IV.9-11}$$

By counterexample show that $\mathbf{F}_1^2$ need not equal $\mathbf{0}$.

An important example is furnished by the motion of *simple shearing*, which has been used traditionally to illustrate every special theory of continuum mechanics. The components of the velocity field with respect to a rectangular Cartesian system are given by (II.11-11). In a suitable pair of rectangular Cartesian systems, one on $\boldsymbol{\kappa}(\mathscr{B})$ and one on $\boldsymbol{\chi}(\mathscr{B}, t)$, the components of the transplacement are

$$\begin{aligned} x_1 &= X_1 \\ x_2 &= X_2 + \kappa t X_1, \\ x_3 &= X_3. \end{aligned} \tag{IV.9-12}$$

Then

$$[\mathbf{F}] = \begin{Vmatrix} 1 & 0 & 0 \\ \kappa t & 1 & 0 \\ 0 & 0 & 1 \end{Vmatrix}, \qquad \mathbf{b} = \mathbf{0}, \tag{IV.9-13}$$

so that

$$\mathbf{F}_0 = \mathbf{1}, \qquad [\mathbf{F}_1] = \kappa \begin{Vmatrix} 0 & 0 & 0 \\ 1 & 0 & 0 \\ 0 & 0 & 0 \end{Vmatrix}, \tag{IV.9-14}$$

and (11) is satisfied. (In fact, $\mathbf{F}_1^2 = \mathbf{0}$.) Therefore, a simple shearing is possible, subject to the action of surface tractions alone, in any homogeneous unconstrained simple body.

Another example is furnished by a homogeneous irrotational pure stretch: $\mathbf{R} = \mathbf{1}$, $\mathbf{W} = \mathbf{0}$, $\mathbf{U} = \mathbf{U}(t)$. From $(\text{II}.11\text{-}13)_2$ we see that $\mathbf{U}$ must satisfy the differential equation

$$\dot{\mathbf{U}}\mathbf{U} = \mathbf{U}\dot{\mathbf{U}}. \tag{IV.9-15}$$

Exercise IV.9.3. Prove that (15) holds if and only if $\mathbf{U}$ has an orthogonal triad of proper vectors which are constant in time. Thus

$$\mathbf{U} = \sum_{i=1}^{3} v_i(t) \mathbf{e}_i \otimes \mathbf{e}_i, \tag{IV.9-16}$$

the $\mathbf{e}_i$ being fixed mutually orthogonal unit vectors. Prove that the corresponding homogeneous pure stretch has constant acceleration if and only if $\ddot{\mathbf{x}}_0 = \text{const.}$ and the v_i are positive affine functions of t. Prove that a rectangular block with faces normal to the $\mathbf{e}_i$ is transformed by the motion into another such block at any time within the interval for which the motion exists. Prove that this motion is isochoric if and only if it reduces to a translation.

The two families of motions just exhibited are particularly interesting special cases of the class of homogeneous transplacements we have proved to be possible in any homogeneous unconstrained simple body upon which suitable boundary tractions alone are brought to bear. As we shall see presently, the corresponding class of solutions for incompressible bodies is much greater.

10. Homogeneous Transplacements of Incompressible Simple Bodies

We now determine all homogeneous transplacements that can be effected by boundary tractions and lamellar body forces acting upon homogeneous incompressible bodies. The term "homogeneous" applied to an incompressible body will be taken to mean not only that the response $\mathfrak{G}_\kappa$ does not depend upon $\mathbf{X}$ but also that ρ_κ is an assigned constant. The reference placement $\boldsymbol{\kappa}$ will then be called *homogeneous* for $\mathscr{B}$, and it is transplacements homogeneous with respect to $\boldsymbol{\kappa}$ that we shall consider.

In order that a certain transplacement be possible in all homogeneous incompressible simple bodies, subject to lamellar body force only, that transplacement must be possible in particular in homogeneous incompressible elastic fluids, the constitutive relation of which is (IV.7-12).

Exercise IV.10.1 (EULER, KELVIN). Using results given in §II.11, show that in any such transplacement

$$\mathbf{W}_2 = \mathbf{0}; \qquad \text{(II.11-31)}_r$$

equivalently,

$$(\ddot{\mathbf{F}}\mathbf{F}^{-1})^T = \ddot{\mathbf{F}}\mathbf{F}^{-1}; \qquad \text{(IV.10-1)}$$

the flow is circulation-preserving (§II.13), so there is an acceleration-potential P_2:

$$\ddot{\mathbf{x}} = -\operatorname{grad} P_2. \qquad \text{(II.11-33)}_r$$

Thus the transplacements we seek must correspond to circulation-preserving flows.

Conversely, however, an arbitrary homogeneous incompressible body cannot be made to undergo an arbitrary circulation-preserving flow through the agency of surface tractions alone. Indeed, suppose (II.11-33) holds. Then the equation of motion (IV.8-5) becomes

$$\frac{1}{\rho}\operatorname{div}\mathfrak{G}(\mathbf{F}^t) = -\operatorname{grad}\lambda, \tag{IV.10-2}$$

in which

$$\lambda \equiv P_2 - \varphi + h, \tag{IV.10-3}$$

h being a function of time only. A necessary condition for (2) to hold is

$$\operatorname{grad}\operatorname{div}\mathfrak{G}(\mathbf{F}^t) = (\operatorname{grad}\operatorname{div}\mathfrak{G}(\mathbf{F}^t))^{\mathrm{T}}, \tag{IV.10-4}$$

and this condition is also sufficient for the existence of λ in a simply connected region. If (2) and (3) hold, then by (IV.8-4) we see that

$$p = \rho(P_2 - \lambda - \varpi + h). \tag{IV.10-5}$$

Thus we have established the following

Theorem (COLEMAN & TRUESDELL). *Let a certain circulation-preserving flow be possible, subject to the action of surface tractions alone, for a certain incompressible simple body of uniform density. Then the pressure p, which is not determined by the history of the motion, is determined by the balance of linear momentum to within a function of time only.*

This theorem shows that while the constitutive relation of the body-point leaves p unrestricted, the balance of linear momentum for the whole body does come close to determining p for the class of transplacements considered, namely, those possible in a perfect fluid. Substitution of (5) and (IV.7-7) into (IV.7-1) yields

$$\mathbf{T} = -\rho(P_2 - \lambda - \varpi + h)\mathbf{1} + \mathfrak{G}(\mathbf{F}^t). \tag{IV.10-6}$$

The acceleration-potential P_2 is known if the circulation-preserving flow is known; the potential λ is known if that flow satisfies the condition of compatibility (4) for the given homogeneous incompressible body in the given motion; the potential ϖ is assigned. Thus the entire stress $\mathbf{T}$ is determined to within a hydrostatic pressure field ρh, which depends upon the time only.

The theorem we have just stated is more general than our present purpose demands, since it refers to all circulation-preserving flows. To specialize it to homogeneous transplacements, we first remark that for any homogeneous deformation history (2) is satisfied by taking $\lambda = 0$. Thus in order that a homogeneous deformation history may be effected in all homogeneous incompressible bodies by the action of suitable surface tractions alone, it is necessary and sufficient that the velocity field of that history be circulation-preserving. Therefore, we need only determine those homogeneous deformation histories that are circulation-preserving. To do so, we note that (1) becomes a second-order differential equation for $\mathbf{F}$, which is now a function of time only. If (1) holds, it is easy to verify from $(\text{II}.12\text{-}16)_2$ that the acceleration-potential P_2 is given by

$$-P_2 = (\mathbf{x} - \mathbf{x}_0) \cdot [\tfrac{1}{2}\ddot{\mathbf{F}}\mathbf{F}^{-1}(\mathbf{x} - \mathbf{x}_0) + \ddot{\mathbf{x}}_0]. \qquad \text{(IV.10-7)}$$

We have proved, then, that *a homogeneous isochoric transplacement is possible, subject to boundary tractions alone, in every homogeneous incompressible simple body if and only if* $\mathbf{F}$ *satisfies* (1). If (1) holds, we may obtain the stress by substituting (7) into (6):

$$\mathbf{T} = \rho[(\mathbf{x} - \mathbf{x}_0) \cdot (\tfrac{1}{2}\ddot{\mathbf{F}}\mathbf{F}^{-1}(\mathbf{x} - \mathbf{x}_0) + \ddot{\mathbf{x}}_0) + \lambda + \varpi - h]\mathbf{1} + \mathfrak{G}(\mathbf{F}^t). \qquad \text{(IV.10-8)}$$

Any unimodular solution $\mathbf{F}$ of (1) and any point-valued function $\mathbf{x}_0$ if put into (II.12-1) yield a possible motion, and then substitution into (8) yields the stresses required to produce it, subject to the action of the body force with potential ϖ.

If we are given a homogeneous stretch history $\mathbf{U}^t$, we may set $\mathbf{W} = \mathbf{0}$ in $(\text{II}.11\text{-}13)_2$ and integrate the resulting ordinary differential equation for $\mathbf{R}$. In this way we can determine a rotation history $\mathbf{R}^t$ such that the flow corresponding to $\mathbf{R}^t\mathbf{U}^t$ is irrotational. As we remarked in §II.13, every irrotational flow is circulation-preserving. If $\det \mathbf{U}(t) = 1$, the result demonstrated above shows that the motion just determined in principle can be produced in any homogeneous incompressible body by applying suitable boundary tractions. Consequently, the ideal experimental program proposed initially can be achieved, for homogeneous incompressible bodies, without calling upon artificial body forces, in fact without use of any body force at all.

If we substitute CAUCHY's criterion (II.11-32) into $(\text{II}.11\text{-}13)_2$, we find that

$$\dot{\mathbf{R}} = \mathbf{R}\mathbf{Y}, \qquad \text{(IV.10-9)}$$

$\mathbf{Y}$ being defined as follows:

$$\mathbf{Y} \equiv \tfrac{1}{2}(\mathbf{U}^{-1}\dot{\mathbf{U}} - \dot{\mathbf{U}}\mathbf{U}^{-1}) + \mathbf{U}^{-1}\mathbf{W}_\kappa\mathbf{U}^{-1}. \qquad \text{(IV.10-10)}$$

Suppose now a homogeneous stretch $\mathbf{U}$ and an aribitrary spin $\mathbf{W}_\kappa$ in the reference shape be given. Then $\mathbf{Y}$ is a known function of t. If $\mathbf{U}$ and $\mathbf{W}_\kappa$ are such that $\mathbf{Y}$ is continuous, the first-order linear differential equation (9) determines a unique rotation $\mathbf{R}(t)$ corresponding to any assigned initial rotation $\mathbf{R}(0)$. The homogeneous motion with $\mathbf{F}(t) = \mathbf{R}(t)\mathbf{U}(t)$ is then circulation-preserving. In view of the theorem proved earlier in this section, we may collect these last results in the following

Theorem (COLEMAN & TRUESDELL). *By applying suitable boundary tractions alone, it is possible to cause any homogeneous incompressible body to undergo any desired isochoric homogeneous stretch history* $\mathbf{U}^t$. *The corresponding rotation history* $\mathbf{R}^t$, *which is independent of the material, is obtained from the unique solution of* (9) *corresponding to assigned initial values* $\mathbf{R}(0)$ *and* $\mathbf{W}_\kappa$. *Conversely, the only homogeneous transplacements that can be effected in all homogeneous incompressible bodies by the application of boundary tractions and a lamellar body force are those in which* $\mathbf{R}$ *is determined from* $\mathbf{U}$, $\mathbf{R}(0)$, *and* $\mathbf{W}_\kappa$ *by* (9).

Putting $\mathbf{W}_\kappa = \mathbf{0}$ in the foregoing theorem shows that *irrotational histories suffice to construct homogeneous transplacements corresponding to all possible stretch histories.* Clearly pure stretch histories do not, since $\mathbf{R} = \mathbf{1}$ is not generally a solution of (9).

Exercise IV.10.2. Prove that a pure stretch is circulation-preserving if and only if

$$\dot{\mathbf{U}}\mathbf{U} - \mathbf{U}\dot{\mathbf{U}} = \text{const.}, \tag{IV.10-11}$$

generalizing (IV.9-15). Hence, in general, a homogeneous isochoric pure stretch cannot be produced in an arbitrary homogeneous incompressible simple body by the effect of boundary tractions alone. Among those special homogeneous isochoric pure stretches that can be so produced are the irrotational ones, given by (IV.9-16) with the added restriction $v_1(t)v_2(t)v_3(t) = 1$.

In Exercise IV.9.3 we have seen that the only homogeneous isochoric pure stretch possible in all unconstrained bodies subject to constant body force is a state of rest. Exercise IV.10.2 shows that any homogeneous isochoric irrotational pure stretch history can be produced by the action of surface tractions alone, or of any lamellar body force, in any homogeneous incompressible body. This result and others obtained in this section show the difference between the stress system in a compressible body that just happens to undergo an isochoric motion and that in a corresponding incompressible body undergoing the same local deformation. For the unconstrained body, change of volume is avoided because the stresses are selected in just the right way, and that way is specified uniquely by the response $\mathfrak{G}$. For the incompressible body, no system of stresses can produce anything but an isochoric motion, and corresponding to that fact there is a hydrostatic pressure which is arbitrary in the sense that it is not determined by the history of the local deformation.

When given body forces are applied, CAUCHY's First Law restricts that arbitrary pressure but does not determine it uniquely. In this sense a single isochoric deformation history if possible at all for a given incompressible body is possible subject to infinitely many different body forces and surface tractions.

Exercise IV.10.3. Suppose that $\mathfrak{G}$ be the response of a certain unconstrained simple body $\mathscr{B}$, and that the restriction of $\mathfrak{G}$ to isochoric deformation histories be the response of a certain incompressible simple body $\mathscr{B}_0$. How does the stress system required to effect a certain simple shearing in $\mathscr{B}$ differ from that required to effect just the same simple shearing in $\mathscr{B}_0$?

Internal constraints such as incompressibility reduce the class of possible motions but expand the class of stresses compatible with such motions as may take place. The theory of a constrained body is therefore essentially easier to work out. The far-reaching simplification that results from assuming the material to be incompressible was seen and exploited by RIVLIN in his pioneer researches on the non-linear continuum theories in 1946–1955. Most of the explicit solutions known today concern incompressible bodies.

11. Material Isomorphisms

Up to now we have considered the constitutive relation of a single body-point, or a single homogeneous body made up of points all having the same response relative to a given reference placement $\boldsymbol{\kappa}$. When can we say that two body-points X_1 and X_2 of $\mathscr{B}$ are of the *same* material? When it is possible to bring the portions $\mathscr{P}_1$ and $\mathscr{P}_2$ of $\mathscr{B}$ near X_1 and X_2 into reference shapes $\boldsymbol{\kappa}_1(\mathscr{P})$ and $\boldsymbol{\kappa}_2(\mathscr{P})$, respectively, such that any subsequent deformation history gives rise to exactly the same stress at the places $\mathbf{X}_1$ and $\mathbf{X}_2$ occupied by X_1 and X_2. Thus no experimental measurement of stress as determined by deformation can detect whether we started with the part $\mathscr{P}_1$ containing $\mathbf{X}_1$ in $\boldsymbol{\kappa}_1(\mathscr{B})$ or the part $\mathscr{P}_2$ containing $\mathbf{X}_2$ in $\boldsymbol{\kappa}_2(\mathscr{B})$, it being understood that $\mathbf{X}_1 = \boldsymbol{\kappa}_1(X_1)$, $\mathbf{X}_2 = \boldsymbol{\kappa}_2(X_2)$. This interpretation suggests also that we should require the densities $\rho_{\boldsymbol{\kappa}_1}$ and $\rho_{\boldsymbol{\kappa}_2}$ to be equal and uniform near X_1 and X_2, as we shall.

To render this idea formal, we erect the following

Definition (NOLL). *Let $\mathfrak{G}_{\boldsymbol{\kappa}}$ be the response of a simple material with respect to the reference placement $\boldsymbol{\kappa}$. The points X_1 and X_2 of $\mathscr{B}$ are* materially isomorphic *if there are reference placements* κ_1 *and* κ_2 *such that* $\rho_{\kappa_1} = \rho_{\kappa_2} = \text{const.}$ *near* $\mathbf{X}_1$ *and* $\mathbf{X}_2$ *and*

$$\mathfrak{G}_{\kappa_1}(\mathbf{F}^t, \mathbf{X}_1) = \mathfrak{G}_{\kappa_2}(\mathbf{F}^t, \mathbf{X}_2) \tag{IV.11-1}$$

for every history $\mathbf{F}^t$ *in a suitable class.*

This definition embodies the idea just stated informally, for the value of the left-hand side is the stress at the place occupied by X_1 when the material points constituting $\mathscr{B}$ have been subjected to a history of deformation $\mathbf{F}^t$ with respect to $\kappa_1(\mathscr{B})$, while the right-hand side is the stress at the place occupied by X_2 when the material points constituting $\mathscr{B}$ have been subjected to just the same history of deformation with respect to $\kappa_2(\mathscr{B})$. Since (1) must hold for all $\mathbf{F}^t$, we can bring the parts $\mathscr{P}_1$ and $\mathscr{P}_2$ of the body that contain X_1 and X_2, respectively, into shapes indistinguishable by any measurement described by the theory of simple bodies.

If each body-point is materially isomorphic to every other one, then every sufficiently small part of $\mathscr{B}$ has just the same properties as every other sufficiently small part, and we say that the body is *uniform*. Now this means that the responses of $\mathscr{B}$ at X_1 and X_2 are the same with respect to suitable reference placements κ_1 and κ_2, that the responses of $\mathscr{B}$ at X_2 and X_3 are the same with respect to suitable reference placements κ_2' and κ_3', *etc.* There need be no single reference placement κ such that all the material points making up $\mathscr{B}$ have one and the same response: $\mathfrak{G}_\kappa(\cdot, \mathbf{X}) = \mathfrak{G}_\kappa(\cdot, \mathbf{Y})\ \forall \mathbf{X}, \mathbf{Y} \in \kappa(\mathscr{B})$. In order to demonstrate the isomorphism of each pair of body-points it may be necessary (in imagination, of course) to cut the body into small pieces and bring each piece separately into an appropriate shape before beginning the experiment. These small pieces need not fit together to form a shape of all of $\mathscr{B}$.

If the isomorphism of all the points of a uniform body may be demonstrated by use of a single reference placement κ, the body is *homogeneous*. The response $\mathfrak{G}_\kappa$ with respect to this particular κ is independent of $\mathbf{X}$, and ρ_κ is likewise independent of $\mathbf{X}$, so that the definition of "homogeneous" in terms of the concept of material isomorphism is equivalent to the one we have introduced already, in §IV.9.

While every homogeneous body is uniform, the converse is false. Uniform but non-homogeneous bodies seem to correspond in some cases to what in physics are called bodies with "defects" and "dislocations". In this book, henceforth, we shall consider only homogeneous bodies.[1]

[1] The general theory and solutions of particular problems concerning non-homogeneous uniform simple bodies are presented in the book by W. NOLL & R. A. TOUPIN & C.-C. WANG, *Continuum Theory of Inhomogeneities in Simple Bodies*, Berlin and New York, Springer-Verlag, 1968, and also in Chapters V and VI of the book by WANG & TRUESDELL, *Introduction to Rational Elasticity*, Leyden, Noordhoff, 1973.

The concept of material isomorphism is of far greater use than merely to define homogeneity, as we shall see in the next section.

12. The Peer Group

Trivially, every point X of $\mathscr{B}$ is materially isomorphic to itself, but there may be also non-trivial isomorphisms of X with itself. We shall analyse this possibility by the aid of an arbitrarily selected reference placement $\boldsymbol{\kappa}_1$, and since we shall consider now a single body-point X, we shall drop it from the notation. Thus (IV.11-1) yields the condition

$$\mathfrak{G}_{\kappa_1}(\mathbf{F}^t) = \mathfrak{G}_{\kappa_2}(\mathbf{F}^t). \tag{IV.12-1}$$

If we can find a $\boldsymbol{\kappa}_2$ distinct from $\boldsymbol{\kappa}_1$ such that (1) holds for all $\mathbf{F}^t$, we shall have shown that the response of the given body-point X is just the same in deformations with respect to two distinct reference placements. That is, in terms of the ideal experiments we sometimes invoke so as to visualize the assertions of the theory, no experiment on the part of $\mathscr{B}$ near X can distinguish $\boldsymbol{\kappa}_2(\mathscr{B})$ from $\boldsymbol{\kappa}_1(\mathscr{B})$. Thus the reference placements $\boldsymbol{\kappa}_1$ and $\boldsymbol{\kappa}_2$ are *peers* at X.

From their very definition, the maps that transform one peer into another form a group, which is defined by properties of $\mathfrak{G}_\kappa$. Thus the group depends both on the constitutive relation at X and on some one, any one, of the placements $\boldsymbol{\kappa}$ that the members of the group interconvert. A different choice of $\boldsymbol{\kappa}$ will lead in general to a different group of maps. While it might seem natural to call the group defined by $\mathfrak{G}_\kappa$ the *peer group* of the material with respect to $\boldsymbol{\kappa}$ at X, that name, or the equivalent name *isotropy group*[1], is commonly applied to the group $\mathscr{g}_\kappa$ of local deformations defined by such maps. That these likewise form a group, this time under tensor multiplication, is obvious from the rule of composition (II.7-5) for gradients. Since the $\boldsymbol{\kappa}$ interconverted by elements of the peer group have the same density, by $(\text{II}.5\text{-}8)_3$ the gradient $\mathbf{P}$ of such a map is unimodular: $\det \mathbf{P} = \pm 1$. Thus the peer group $\mathscr{g}_\kappa$ of $\boldsymbol{\kappa}$ at X is a subgroup of the unimodular group $\mathscr{u}$:

$$\mathscr{g}_\kappa \subset \mathscr{u}. \tag{IV.12-2}$$

[1] The term "isotropy group", used by NOLL in introducing these groups, is misleading here because it derives from the concept of turning, while the elements of the peer group need not all be rotations; "symmetry", while closer to the popular speech of physicists, would be equally misleading because it derives from the concept of distance, which is irrevelant in material response. The term "peer" is intended to suggest its root meaning, which is "equal in status before the law", the "law" being here the constitutive relation of the material.

It is the group of gradients of *all maps that carry* $\boldsymbol{\kappa}$ *into its peers*, namely, the other reference placements indistinguishable from $\boldsymbol{\kappa}$ by measurements of stress arising from deformation of parts of $\boldsymbol{\kappa}(\mathscr{B})$.

By substituting (IV.3-3) into (1), we find that the elements of the peer group $g_{\boldsymbol{\kappa}}$ are unimodular tensors $\mathbf{H}$ such that for all histories $\mathbf{F}^t$ in the domain of $\mathfrak{G}_{\boldsymbol{\kappa}}$

$$\mathfrak{G}_{\boldsymbol{\kappa}}(\mathbf{F}^t\mathbf{H}) = \mathfrak{G}_{\boldsymbol{\kappa}}(\mathbf{F}^t), \tag{IV.12-3}$$

and conversely, any such $\mathbf{H}$ is an element of $g_{\boldsymbol{\kappa}}$.

Exercise IV.12.1. Directly from (3), verify the fact that the collection of solutions $\mathbf{H}$ forms a group.

As a part of the definition of the peer group we have required that its members be unimodular. We have done so in favor of the intended application rather than for any mathematical objection to more general isomorphisms. By considering in (3) the case of the rest history $\mathbf{F}(t) \equiv \mathbf{1}$, we see that if $\mathbf{H} \in g_{\boldsymbol{\kappa}}$ and $n = 1, 2, 3, \ldots$, then $\mathfrak{G}_{\boldsymbol{\kappa}}((\mathbf{H}^n)^t) = \mathfrak{G}_{\boldsymbol{\kappa}}(\mathbf{1}^t)$; here $(\mathbf{H}^n)^t$ denotes the history of the constant tensor $\mathbf{H}^n$, and $\mathbf{1}^t$ denotes the history of $\mathbf{1}$. If $|\det \mathbf{H}| < 1$, this result and (II.5-4) show that we can find a placement which has arbitrarily large density and in which a part of the body can be held at rest indefinitely under just the same stress as that required for equilibrium in $\boldsymbol{\kappa}$. If $|\det \mathbf{H}| > 1$, the same can be said for a placement with arbitrary small density. Such a material would be a strange one. In particular, no Eulerian fluid with invertible pressure function is of this kind. In this book we merely leave out of account any $\mathbf{H}$ that satisfies (3) and is not unimodular, but the foregoing remarks would lend support to *requiring*, as part of the definition of simple material, that $\mathfrak{G}_{\boldsymbol{\kappa}}$ allow no solutions $\mathbf{H}$ of (3) that are not unimodular.[1]

The members $\mathbf{H}$ of $g_{\boldsymbol{\kappa}}$ need not be orthogonal, but they may be. Since $\mathbf{1} \in g_{\boldsymbol{\kappa}}$ for every material point and every reference placement $\boldsymbol{\kappa}$, at least one member of $g_{\boldsymbol{\kappa}}$ is orthogonal. If an orthogonal tensor $\mathbf{Q} \in g_{\boldsymbol{\kappa}}$, then also $\mathbf{Q}^{\mathrm{T}} \in g_{\boldsymbol{\kappa}}$ since $g_{\boldsymbol{\kappa}}$ is a group; also, if $\mathbf{F}^t$ runs over all invertible tensor histories, so does $\mathbf{QF}^t$. Thus, when $\mathbf{H} = \mathbf{Q}^{\mathrm{T}}$, (3) is equivalent to

$$\mathfrak{G}_{\boldsymbol{\kappa}}(\mathbf{QF}^t\mathbf{Q}^{\mathrm{T}}) = \mathfrak{G}_{\boldsymbol{\kappa}}(\mathbf{QF}^t). \tag{IV.12-4}$$

[1] In a theory of thermomechanics it is possible to define peer groups and to prove that in order to satisfy certain reasonable requirements they must be subgroups of u, as has been shown by M. E. GURTIN & W. O. WILLIAMS, "On the inclusion of the complete symmetry group in the unimodular group," *Archive for Rational Mechanics and Analysis* **23** (1966/7), 163–172 (1966).

In the condition (IV.5-2), which expresses the principle of Material Frame-Indifference, we select the particular history $\mathbf{Q}^t = \mathbf{Q}(t) = \mathbf{Q}$ and obtain

$$\mathfrak{G}_\kappa(\mathbf{Q}\mathbf{F}^t) = \mathbf{Q}\mathfrak{G}_\kappa(\mathbf{F}^t)\mathbf{Q}^T. \tag{IV.12-5}$$

This relation holds for all $\mathbf{F}^t$ and for all orthogonal tensors $\mathbf{Q}$, while (4) holds only for those $\mathbf{Q}$ that belong to g_κ. Combining the two relations yields

$$\mathfrak{G}_\kappa(\mathbf{Q}\mathbf{F}^t\mathbf{Q}^T) = \mathbf{Q}\mathfrak{G}_\kappa(\mathbf{F}^t)\mathbf{Q}^T \tag{IV.12-6}$$

as a necessary condition to be satisfied by all orthogonal members of g_κ.

Exercise IV.12.2. Prove that conversely, if $\mathbf{Q}$ satisfies (6), then $\mathbf{Q} \in g_\kappa$.

Thus (6) is *a necessary and sufficient condition for the orthogonal tensor* $\mathbf{Q}$ *to belong to the peer group.*

From (6) we see that $-\mathbf{1} \in g_\kappa$ for all materials and all κ. Since $-\mathbf{1}$ is a central inversion, it does not correspond to any deformation that could be effected physically but merely expresses the invariance of material properties under reflections of the reference placement.[1] Since $-\mathbf{1} \in g_\kappa$ and g_κ is a group, $-\mathbf{H} \in g_\kappa \Leftrightarrow \mathbf{H} \in g_\kappa$. Thus g_κ can be expressed as the direct product of the trivial group consisting in $\mathbf{1}$ and $-\mathbf{1}$ alone and a group g_κ^+ all of whose members have determinant $+1$:

$$g_\kappa = \{\mathbf{1}, -\mathbf{1}\} \otimes g_\kappa^+, \tag{IV.12-7}$$

and it is only the elements of g_κ^+ that can be interpreted in terms not only of change of reference placement but also as gradients of transplacements that map one shape of a given body onto another. These are the transplacements that cannot be distinguished from one another by experiment,[2] but it is formally more convenient to retain the trivial central inversions and so operate with g_κ itself. We have shown, then, that $\{\mathbf{1}, -\mathbf{1}\}$ is the smallest possible peer group:

$$\{\mathbf{1}, -\mathbf{1}\} \subset g_\kappa \subset \mathscr{u}. \tag{IV.12-8}$$

The foregoing constructions and results in these precise, abstract forms were introduced by NOLL, generalizing earlier and more special notions.

[1] The reader should not extrapolate this result to other theories such as those of heat conduction and electromagnetism; in them there is no such invariance, because the local deformation $\mathbf{F}$ is not the only independent variable in the constitutive relations.

[2] Again the reader must be warned that while this fact expresses a proved theorem of the theory presented in this book, nothing of the sort holds for the peer groups that can be defined by parallel constructions in other theories such as optics.

Any subgroup of the unimodular group that includes $\{\mathbf{1}, -\mathbf{1}\}$ may be the peer group of a material point. Corresponding to any assigned unimodular subgroup $\mathscr{g}$, it is possible to construct infinitely many responses $\mathfrak{G}$; more specifically, it is possible to write $\mathfrak{G}$ in a reduced form such as to be frame-indifferent and to include automatically *all* materials having an assigned peer group, and these only.[1] In the following sections we shall consider only such $\mathscr{g}$ as are notable or lead to especially simple representations for $\mathfrak{G}$. In particular, we shall use the ideas and apparatus just given so as to define the concepts of "fluid", "solid", and "isotropic".

13. Comparison of Peer Groups with Respect to Different Reference Placements

The peer group $\mathscr{g}_{\boldsymbol{\kappa}}$ at a material point depends, as does the response $\mathfrak{G}_{\boldsymbol{\kappa}}$ of the material, upon the choice of reference placement $\boldsymbol{\kappa}$. Since $\mathfrak{G}_{\boldsymbol{\kappa}_1}$ determines $\mathfrak{G}_{\boldsymbol{\kappa}_2}$ for all $\boldsymbol{\kappa}_2$, the same should be true of $\mathscr{g}_{\boldsymbol{\kappa}_1}$ and $\mathscr{g}_{\boldsymbol{\kappa}_2}$. This is so, and either group determines the other through a rule found by NOLL:

$$\mathscr{g}_{\boldsymbol{\kappa}_2} = \mathbf{P}\mathscr{g}_{\boldsymbol{\kappa}_1}\mathbf{P}^{-1}. \tag{IV.13-1}$$

To prove this rule, we simply apply (IV.3-3) to each member of (IV.12-3) after replacing $\boldsymbol{\kappa}$ therein by $\boldsymbol{\kappa}_1$:

$$\mathfrak{G}_{\boldsymbol{\kappa}_2}(\mathbf{F}^t\mathbf{H}\mathbf{P}^{-1}) = \mathfrak{G}_{\boldsymbol{\kappa}_2}(\mathbf{F}^t\mathbf{P}^{-1}); \tag{IV.13-2}$$

here $\mathbf{P} \equiv \nabla\boldsymbol{\lambda}$ and $\boldsymbol{\lambda} \equiv \boldsymbol{\kappa}_2 \circ \boldsymbol{\kappa}_1^{-1}$. As $\mathbf{F}^t$ runs over all invertible tensor histories, so does $\mathbf{F}^t\mathbf{P}^{-1}$ for any assigned invertible tensor $\mathbf{P}$. Hence (2) is equivalent to

$$\mathfrak{G}_{\boldsymbol{\kappa}_2}(\mathbf{F}^t\mathbf{P}\mathbf{H}\mathbf{P}^{-1}) = \mathfrak{G}_{\boldsymbol{\kappa}_2}(\mathbf{F}^t), \tag{IV.13-3}$$

which is of the form (IV.12-3) with $\boldsymbol{\kappa}$ replaced by $\boldsymbol{\kappa}_2$. Since $\mathbf{PHP}^{-1}$ is unimodular if $\mathbf{H}$ is, every solution $\mathbf{H}$ of (IV.12-3) corresponds to a unimodular solution $\mathbf{PHP}^{-1}$ of (3), and conversely. The rule (1) is an abbreviated statement of this fact.

It is a trivial consequence of (IV.12-1) that *if* $\boldsymbol{\kappa}_1$ *and* $\boldsymbol{\kappa}_2$ *are peers, they have the same peer groups.*

While the members of $\mathscr{g}_{\boldsymbol{\kappa}_1}$ and $\mathscr{g}_{\boldsymbol{\kappa}_2}$ are unimodular tensors, the reference placements $\boldsymbol{\kappa}_1$ and $\boldsymbol{\kappa}_2$ themselves need not have the same density. In particular, if we let $\boldsymbol{\kappa}_2$ be obtained from $\boldsymbol{\kappa}_1$ by a dilatation, then $\mathbf{P} = K\mathbf{1}$ and $K \neq 0$, so $\mathbf{P}^{-1} = K^{-1}\mathbf{1}$, and thus (1) yields $\mathscr{g}_{\boldsymbol{\kappa}_2} = \mathscr{g}_{\boldsymbol{\kappa}_1}$. *Thus the peer group is unaltered by a dilatation.*

[1] C.-C. WANG, "On a general representation theorem for constitutive relations," *Archive for Rational Mechanics and Analysis* **33**, 1–25 (1969).

Whatever be $\mathscr{g}_{\kappa_1}$, (1) shows that for some choice of κ_2 we may expect to obtain a different peer group $\mathscr{g}_{\kappa_2}$. Thus the concept of peerage is a relative one, depending upon the choice of reference placement. It is possible, however, that $\mathscr{g}_{\kappa_1} = \mathscr{g}_{\kappa_2}$ for all choices of κ_1 and κ_2. In that case we shall say that the material at X is *egalitarian*: No deformation can lessen or enlarge its peer group. A glance at (1) reveals two groups corresponding to egalitarian materials:

$$\mathscr{g} = \{\mathbf{1}, -\mathbf{1}\} \qquad \text{or} \qquad \mathscr{g} = \mathscr{u}. \tag{IV.13-4}$$

According to a theorem of group theory,[1] the proper unimodular group $\mathscr{u}^+$ is "simple", which means that the equation

$$\mathscr{g} = \mathbf{P}\mathscr{g}\mathbf{P}^{-1} \qquad \forall \mathbf{P} \in \mathscr{u} \tag{IV.13-5}$$

has no solutions $\mathscr{g}$ that can be peer groups other than the trivial ones given by (4). *Thus the groups* (4) *correspond to the only possible egalitarian materials.*[2] In §IV.16 we shall see an important consequence of this fact. Here we remark merely that the fact itself is not surprising and could easily be conjectured from experience in mechanics, since it asserts that only for the two extremes of response can no deformation either create new peers or unseat any of the old. At one extreme, all placements are peers; at the other, no placement has any peers but the two trivial ones.

14. Isotropic Materials

A homogeneous body is isotropic if it can be brought into a shape, no rotations of which can be detected by experiment. Isotropy is an example, and the most important one, of *material symmetry*. To consider material symmetries, we fix attention upon the peer groups of a single material point X. In this section and the next two we shall use the phrase "a material

[1] My inquiries have not led to a simple, direct proof. The result is a consequence of more powerful theorems of group theory presented by J. J. ROTMAN, *The Theory of Groups*, 2nd ed., Rockleigh, New Jersey, Allyn & Bacon, 1973. The projective special linear group $PSL(n, K) \equiv SL(n, K)/Z_0$. Here K is an arbitrary field; $SL(n, K)$ is the multiplicative group of proper unimodular $n \times n$ matrices over K; and Z_0 is its center, that is, the group of all elements that commute with every element. ROTMAN's Theorem 8.25 asserts that $PSL(m, K)$ is simple if $m \geqq 3$; his Theorem 8.13, that the center of $SL(3, R)$ is the unit matrix. Thus $PSL(3, R) = SL(3, R)/Z_0 = SL(3, R)$, so $SL(3, R)$ is simple.

[2] Again the reader should be warned not to expect that the result proved here for the mechanics of simple materials can be extended to other theories in which a peer group may be defined. For example, in optics there are four groups that correspond to egalitarian materials: not only those given by (4) but also $\{\mathbf{1}\}$ and $\mathscr{u}^+$.

is ..." instead of "a material point is ...". Since in the rest of the book we consider only homogeneous bodies, so all the material points that constitute a body must have the same material symmetry, we could just as well write in each case "a body is ...". The letter $\mathscr{o}$ will denote the full orthogonal group.

Definition (CAUCHY, NOLL). *A material is* isotropic *if there is a reference placement* $\boldsymbol{\kappa}$ *such that*

$$\mathscr{g}_{\boldsymbol{\kappa}} \supset \mathscr{o}. \tag{IV.14-1}$$

Such a placement $\boldsymbol{\kappa}$ is called *undistorted*. According to this definition, every orthogonal deformation of an undistorted placement carries it into a peer. From NOLL's rule (IV.13-1) we see that for other placements $\boldsymbol{\kappa}'$ the peer groups $\mathscr{g}_{\boldsymbol{\kappa}'}$ need not contain $\mathscr{o}$. That is, rotations of $\boldsymbol{\kappa}'$ generally can be detected by experiment, though rotations of an undistorted placement $\boldsymbol{\kappa}$ cannot. Of course, that same rule shows us that *an orthogonal deformation carries one undistorted placement of an isotropic material into another*, a fact which merely reflects the definition of "isotropic material", besides showing that an isotropic material has infinitely many undistorted placements.

For an isotropic material (IV.12-6) changes from an equation to be solved for certain $\mathbf{Q}$ into an identity satisfied by all $\mathbf{Q}$, and likewise (IV.12-3) is satisfied by all orthogonal $\mathbf{H}$. By this latter equation, then, the value of $\mathbf{T}$ is unchanged if we replace $\mathbf{F}^t$ by $\mathbf{F}^t\mathbf{Q}$, where $\mathbf{Q}$ is any constant orthogonal tensor. In particular, we may replace $\mathbf{F}^t$ by $\mathbf{F}^t\mathbf{R}^{\mathrm{T}}$ without changing the value of $\mathbf{T}$. We apply this transformation to the constitutive relation in the reduced form (IV.5-15). In this transformation $\mathbf{F}$ is replaced by $\mathbf{F}\mathbf{R}^{\mathrm{T}}$; that is, $\mathbf{RU}$ is replaced by $\mathbf{RUR}^{\mathrm{T}}$, and since $\mathbf{RUR}^{\mathrm{T}}$ is positive and symmetric, the polar decomposition theorem shows that $\mathbf{R}$ is replaced by $\mathbf{1}$ and $\mathbf{U}$ is replaced by $\mathbf{RUR}^{\mathrm{T}}$. Then $\mathbf{R}^{\mathrm{T}}\mathbf{C}_t^t\mathbf{R}$ is replaced by $\mathbf{C}_t^t$, and $\mathbf{C}$ is replaced by $\mathbf{RCR}^{\mathrm{T}}$, which by virtue of $(\mathrm{II}.9\text{-}5)_5$ is $\mathbf{B}$. Thus (IV.5-15) yields as NOLL's *form of the constitutive relation for isotropic materials*:

$$\mathbf{T} = \mathfrak{K}(\mathbf{C}_t^t; \mathbf{B}), \tag{IV.14-2}$$

in which, as was to be expected, the rotation does not appear at all.

According to (IV.12-6), moreover, if $\mathbf{F}^t$ is replaced by $\mathbf{Q}\mathbf{F}^t\mathbf{Q}^{\mathrm{T}}$, for any $\mathbf{Q}$, the stress $\mathbf{T}$ is replaced by $\mathbf{QTQ}^{\mathrm{T}}$. In this replacement $\mathbf{C}_t^t$ and $\mathbf{B}$ are replaced by $\mathbf{Q}\mathbf{C}_t^t(s)\mathbf{Q}^{\mathrm{T}}$ and $\mathbf{QBQ}^{\mathrm{T}}$, as is easily verified from (II.9-10) and (II.9-5). Thus the mapping $\mathfrak{K}$ in (2) must satisfy the identity

$$\mathfrak{K}(\mathbf{Q}\mathbf{C}_t^t\mathbf{Q}^{\mathrm{T}}; \mathbf{QBQ}^{\mathrm{T}}) = \mathbf{Q}\mathfrak{K}(\mathbf{C}_t^t; \mathbf{B})\mathbf{Q}^{\mathrm{T}}, \tag{IV.14-3}$$

for every orthogonal tensor $\mathbf{Q}$, for every positive symmetric tensor history $\mathbf{C}_t^t$, and for every positive symmetric tensor $\mathbf{B}$.

A mapping satisfying this requirement for all $\mathbf{Q}$ is called *isotropic*. Thus the concept of isotropic mapping generalizes that of isotropic function defined by (IV.4-9). Conversely, if (3) is satisfied by $\mathfrak{K}$, (2) gives the constitutive equation of an isotropic simple material, referred to an undistorted placement. If a reference placement that is not undistorted is used, the constitutive relation of an isotropic material cannot have the form (2) and generally shows no recognizable simplicity.

Exercise IV.14.1. Verify the steps in the foregoing proof of (2) and (3).

While (1) embodies a natural concept of isotropy, it seems more general than in fact it is. According to a theorem of group theory, the orthogonal group is maximal in the unimodular group.[1] That is, if $\mathscr{g}$ is a group such that $\mathscr{o} \subset \mathscr{g} \subset \mathscr{u}$, then

$$\text{either} \quad \mathscr{g} = \mathscr{o} \qquad \text{or} \quad \mathscr{g} = \mathscr{u}. \tag{IV.14-4}$$

Thus the peer group of an isotropic material in an undistorted placement is either the orthogonal group or the unimodular group.

15. Solids

In ordinary experience we commonly think of a body as being "solid" if after changing its form we can discern a difference in the way it responds to further deformation. A solid, then, has some placement, any non-rigid transplacement of which is detectable by some subsequent experiment. Thus, still considering a particular material point, we lay down the following formal

Definition (NOLL). *A material is* solid *if there is a reference placement* $\boldsymbol{\kappa}$ *such that*

$$\mathscr{g}_{\boldsymbol{\kappa}} \subset \mathscr{o}. \tag{IV.15-1}$$

Such a placement $\boldsymbol{\kappa}$ is called *undistorted*. According to this definition no non-orthogonal deformation belongs to the peer group $\mathscr{g}_{\boldsymbol{\kappa}}$ corresponding to an undistorted $\boldsymbol{\kappa}$.

A material for which $\mathscr{g} = \{\mathbf{1}, -\mathbf{1}\}$ is a solid. Such a material, which is called *triclinic*, furnishes an example of a *crystalline solid* in the classical sense. All the classical crystallographic groups, provided they be extended so as to include $-\mathbf{1}$, correspond to solids. So also do the groups defining "transversely isotropic" and "orthotropic" materials, and many others.

[1] *E.g.* W. NOLL, "Proof of the maximality of the orthogonal group in the unimodular group," *Archive for Rational Mechanics and Analysis* **18**, 100–102 (1965), reprinted in NOLL's *Foundations of Mechanics and Thermodynamics*, Berlin and New York, Springer-Verlag, 1974.

For solids, no particularly simple form of the constitutive relation has been found.

An *isotropic solid*, of course, is a material that is both solid and isotropic. Both of these qualities have been defined in terms of the existence of special reference placements, both of which have been called "undistorted". Denoting by $\boldsymbol{\kappa}$ the one used to define "isotropic" and by $\bar{\boldsymbol{\kappa}}$ the one used to define "solid", we thus assume that

$$\mathscr{g}_{\boldsymbol{\kappa}} \supset \mathscr{o}, \qquad \mathscr{g}_{\bar{\boldsymbol{\kappa}}} \subset \mathscr{o}. \tag{IV.15-2}$$

The relation between any such pair of reference placements is laid bare in the following

Theorem 1.

$$\mathscr{g}_{\boldsymbol{\kappa}} = \mathscr{g}_{\bar{\boldsymbol{\kappa}}} = \mathscr{o}. \tag{IV.15-3}$$

Proof. According to the last statement in §IV.14, either $\mathscr{g}_{\boldsymbol{\kappa}} = \mathscr{o}$ or $\mathscr{g}_{\boldsymbol{\kappa}} = \mathscr{u}$. If $\mathscr{g}_{\boldsymbol{\kappa}} = \mathscr{u}$, then by (IV.13-1) $\mathscr{g}_{\bar{\boldsymbol{\kappa}}} = \mathscr{u}$; since $(2)_2$ contradicts this conclusion, we are left with the former alternative, $\mathscr{g}_{\boldsymbol{\kappa}} = \mathscr{o}$. Thus $\boldsymbol{\kappa}$ is an undistorted placement of the solid.

If $\boldsymbol{\lambda} \equiv \bar{\boldsymbol{\kappa}} \circ \boldsymbol{\kappa}^{-1}$ and $\mathbf{P} \equiv \nabla\boldsymbol{\lambda}$, NOLL's rule (IV.13-1) assures us that there is a $\boldsymbol{\lambda}$ such that $\mathscr{g}_{\bar{\boldsymbol{\kappa}}} = \mathbf{P}\mathscr{o}\mathbf{P}^{-1}$. If we can find an orthogonal tensor $\mathbf{R}$ such that $\mathscr{g}_{\bar{\boldsymbol{\kappa}}} = \mathbf{R}\mathscr{o}\mathbf{R}^{-1}$, then we shall have proved the theorem, since the only orthogonal conjugate of $\mathscr{o}$ is $\mathscr{o}$ itself. That such an $\mathbf{R}$ exists, is a consequence of a more general result which is stated in the following exercise. △

Exercise IV.15.1 (COLEMAN & NOLL). Let $\boldsymbol{\kappa}$ and $\boldsymbol{\kappa}^*$ be two undistorted placements of a solid, and let $\mathbf{P} = \nabla(\boldsymbol{\kappa}^* \circ \boldsymbol{\kappa}^{-1})$. If the polar decomposition of $\mathbf{P}$ is $\mathbf{P} = \mathbf{R}_0\mathbf{U}_0$, and if $\mathbf{Q}^*$ and $\mathbf{Q}$ are elements of $\mathscr{g}_{\boldsymbol{\kappa}^*}$ and $\mathscr{g}_{\boldsymbol{\kappa}}$ that correspond to one another through NOLL's rule, prove that

$$\mathbf{Q}^* = \mathbf{R}_0\mathbf{Q}\mathbf{R}_0^{\mathrm{T}}, \qquad \mathbf{U}_0 = \mathbf{Q}^{\mathrm{T}}\mathbf{U}_0\mathbf{Q}. \tag{IV.15-4}$$

Hence

$$\mathscr{g}_{\boldsymbol{\kappa}^*} = \mathbf{R}_0\,\mathscr{g}_{\boldsymbol{\kappa}}\,\mathbf{R}_0^{-1}. \tag{IV.15-5}$$

That is, $\mathscr{g}_{\boldsymbol{\kappa}^*}$ is an orthogonal conjugate of $\mathscr{g}_{\boldsymbol{\kappa}}$.

Returning to the consideration of a solid in general, we remark that its peer group with respect to an undistorted placement may be any subgroup of the orthogonal group that contains $-\mathbf{1}$.

However, only certain particular kinds of anisotropy attract much notice. These are the ones corresponding to the 32 crystal classes, which are defined by optical symmetries, and to two further types which correspond to certain manufactured products. In order to define these particular symmetries, we let $\mathbf{R}_{\mathbf{a}}^{\varphi}$ denote a right-handed rotation of angle φ about an axis in the direction of the vector $\mathbf{a}$; we let $(\mathbf{i}, \mathbf{j}, \mathbf{k})$ be an orthonormal basis, and we set $\mathbf{p} \equiv \sqrt{\tfrac{1}{3}}(\mathbf{i} + \mathbf{j} + \mathbf{k})$. In view of (IV.12-7), it suffices to specify $\mathscr{g}^+$, which is a group of rotations.

A material such that $\mathscr{g}^+$ consists in $\mathbf{1}$ and all rotations $\mathbf{R}_\mathbf{k}^\varphi$ for a fixed $\mathbf{k}$ and all angles φ is called *transversely isotropic* with respect to $\mathbf{k}$.

The 32 crystal classes reduce to 11 in the context of the present, purely mechanical theory. Definitions of these, along with the standard crystallographic names, are given in the following table, summarizing results obtained by COLEMAN & NOLL. The directions of the particular unit vectors **i**, **j**, **k** are called the *crystallographic axes.*

	Crystal class	Generators[1] of $\mathscr{g}^+$	Order of $\mathscr{g}$
1.	*Triclinic system* all classes	$\mathbf{1}$	2
2.	*Monoclinic system* all classes	$\mathbf{R}_\mathbf{k}^\pi$	4
3.	*Rhombic system* all classes	$\mathbf{R}_\mathbf{i}^\pi$, $\mathbf{R}_\mathbf{j}^\pi$	8
4.	*Tetragonal system* tetragonal-disphenoidal tetragonal-pyramidal tetragonal-dipyramidal	$\mathbf{R}_\mathbf{k}^{\pi/2}$	8
5.	tetragonal-scalenohedral ditetragonal-pyramidal tetragonal-trapezohedral ditetragonal-dipyramidal	$\mathbf{R}_\mathbf{k}^{\pi/2}$, $\mathbf{R}_\mathbf{i}^\pi$	16
6.	*Cubic system* tetartoidal diploidal	$\mathbf{R}_\mathbf{i}^\pi$, $\mathbf{R}_\mathbf{j}^\pi$, $\mathbf{R}_\mathbf{p}^{2\pi/3}$	24
7.	hextetrahedral gyroidal hexoctahedral	$\mathbf{R}_\mathbf{i}^{\pi/2}$, $\mathbf{R}_\mathbf{j}^{\pi/2}$, $\mathbf{R}_\mathbf{k}^{\pi/2}$	48
8.	*Hexagonal system* trigonal-pyramidal rhombohedral	$\mathbf{R}_\mathbf{k}^{2\pi/3}$	6
9.	ditrigonal-pyramidal trigonal-trapezohedral hexagonal-scalenohedral	$\mathbf{R}_\mathbf{i}^\pi$, $\mathbf{R}_\mathbf{k}^{2\pi/3}$	12
10.	trigonal-dipyramidal hexagonal-pyramidal hexagonal-dipyramidal	$\mathbf{R}_\mathbf{k}^{\pi/3}$	12
11.	ditrigonal-dipyramidal dihexagonal-pyramidal hexagonal-trapezohedral dihexagonal-dipyramidal	$\mathbf{R}_\mathbf{i}^\pi$, $\mathbf{R}_\mathbf{k}^{\pi/3}$	24

[1] The members of a set of elements of a group $\mathscr{g}$ are called *the generators* of $\mathscr{g}$ if products of their powers exhaust $\mathscr{g}$.

Finally, a material is called *orthotropic* if $\mathscr{g}$ contains the reflections $-\mathbf{R}_\mathbf{i}^\pi$, $-\mathbf{R}_\mathbf{j}^\pi$, $-\mathbf{R}_\mathbf{k}^\pi$. Since $\mathbf{R}_\mathbf{i}^\pi\mathbf{R}_\mathbf{j}^\pi = \mathbf{R}_\mathbf{k}^\pi$ and $(\mathbf{R}_\mathbf{i}^{\pi/2})^2 = \mathbf{R}_\mathbf{i}^\pi$, the materials belonging to the classes numbered 3, 5, 6, and 7 in the table are orthotropic.

In this book we shall not have occasion to treat crystals or other materials of special symmetry, except, of course, isotropic materials. The definitions just given are included only so as to help the student understand the meanings of the terms, should he encounter them elsewhere.

Traditionally the use of these classical "point groups" is motivated by CAUCHY's theory of stress in a lattice of mass-points.[1] ERICKSEN[2] has pointed out that the arguments used apply only when $|\mathbf{F} - \mathbf{1}|$ is small. The theory of point lattices if taken seriously for local deformations of any magnitude suggests that the peer groups of crystals should contain some non-orthogonal tensors, no matter what reference placement be used. Thus a crystal lattice is not a solid in the sense defined by (1) and used throughout this book.

Returning to the consideration of solids as defined by (1), we note first that only certain particular placements will be undistorted. Indeed, if $\boldsymbol{\kappa}$ is undistorted and if $\mathbf{P} \equiv \nabla(\boldsymbol{\kappa}^* \circ \boldsymbol{\kappa}^{-1})$, by NOLL's rule we have

$$\mathscr{g}_{\boldsymbol{\kappa}^*} = \mathbf{P}\mathscr{g}_{\boldsymbol{\kappa}}\mathbf{P}^{-1}, \qquad \mathscr{g}_{\boldsymbol{\kappa}} \subset \mathscr{o}. \tag{IV.15-6}$$

Now if $\mathbf{Q}$ is orthogonal, $\mathbf{PQP}^{-1}$ generally fails to be orthogonal. Thus *not all placements are undistorted.*

Exercise IV.15.2. By applying (5), show that the peer groups corresponding to different undistorted placements of a particular solid are not generally the same, and that the undistorted placements of an anisotropic solid generally fail to be peers.

Thus we may set ourselves the following task: to find the largest class of mappings $\boldsymbol{\lambda}$ that carry in an undistorted placement $\boldsymbol{\kappa}$ defined by a given group $\mathscr{g}_{\boldsymbol{\kappa}}$ into places in another undistorted placement.

For the largest and smallest possible peer groups, the answer is easy to get. First, if $\mathscr{g}_{\boldsymbol{\kappa}} = \{\mathbf{1}, -\mathbf{1}\}$, then, as shown in §IV.13 all placements are undistorted, so any $\boldsymbol{\lambda}$ has the property sought. The second case is settled by

Theorem 2. *A transplacement of an isotropic solid maps one undistorted shape onto another if and only if it is conformal.*[3]

[1] An outline of this theory is given by A. E. H. LOVE in Note B, "The notion of stress," in his *A Treatise on the Mathematical Theory of Elasticity*, Cambridge, Cambridge University Press, 2nd–4th editions, 1906/1927, variously reprinted.

[2] J. L. ERICKSEN, "Nonlinear elasticity of diatomic crystals," *International Journal of Solids and Structures* **6**, 951–952 (1970), and Chapter IV of "Special topics in elastostatics," *Advances in Applied Mechanics*, in press.

[3] A transplacement $\boldsymbol{\lambda}$ is conformal if it preserves the angles between material curves; equivalently, there is an orthogonal tensor $\mathbf{R}$ such that $\nabla\boldsymbol{\lambda} = K\mathbf{R}$ and $K \neq 0$.

The sufficiency of the condition follows trivially from a theorem established in §IV.13. Necessity is a consequence, as we shall see presently, of the following more general

Theorem 3 (COLEMAN & NOLL). *Let* $\boldsymbol{\kappa}$ *be an undistorted placement of a solid body* $\mathscr{B}$, *and set* $\boldsymbol{\lambda} \equiv \boldsymbol{\kappa}^* \circ \boldsymbol{\kappa}^{-1}$, *so* $\boldsymbol{\lambda}$ *maps* $\boldsymbol{\kappa}(\mathscr{B})$ *onto* $\boldsymbol{\kappa}^*(\mathscr{B})$. *Then* $\boldsymbol{\kappa}^*$ *is undistorted if and only if the proper spaces of the right stretch tensor* $\mathbf{U}_0$ *of* $\nabla\boldsymbol{\lambda}$ *are invariant under all the rotations in the peer group* $g_{\boldsymbol{\kappa}}$.

Proof. By $(4)_2$, every member $\mathbf{Q}$ of $g_{\boldsymbol{\kappa}}$ commutes with $\mathbf{U}_0$. According to a theorem of algebra,[1] $\mathbf{Q}$ satisfies this condition if and only if it leaves the proper spaces of $\mathbf{U}_0$ invariant.

Exercise IV.15.3. Prove the statement of sufficiency in Theorem 2 as a corollary of Theorem 3. △

We turn now to the use of Theorem 3 so as to complete the proof of necessity in Theorem 2. By Theorem 1 we know that if $\boldsymbol{\kappa}$ is undistorted, $g_{\boldsymbol{\kappa}} = o$. If $\boldsymbol{\lambda}$ carries $\boldsymbol{\kappa}$ into another undistorted placement $\boldsymbol{\kappa}^*$, then by Theorem 3 every orthogonal tensor must leave invariant the proper spaces of $\mathbf{U}_0$. Therefore, the proper space of $\mathbf{U}_0$ can be nothing but $\mathscr{V}$ itself. Hence $\mathbf{U}_0$ has only one proper number, so that $\mathbf{U}_0 = K\mathbf{1}$, and consequently $\nabla\boldsymbol{\lambda} = K\mathbf{R}$. △

Theorem 3 itself may be used to determine the most general form of $\mathbf{U}_0$ compatible with a given $g_{\boldsymbol{\kappa}}$ in cases other than the two already disposed of: isotropic or triclinic solids. The results for the crystalline solids are shown in the following table, due to COLEMAN & NOLL. The numbers in parentheses refer to the definitions of the special kinds of aeolotropy in the table printed above in this section.

Type of aeolotropy	Restrictions on $\mathbf{U}_0$
Triclinic system (1)	no restriction
Monoclinic system (2)	$\mathbf{k}$ is a proper vector of $\mathbf{U}_0$
Rhombic system (3)	$\mathbf{i}, \mathbf{j}, \mathbf{k}$ are proper vectors of $\mathbf{U}_0$
Tetragonal system (4, 5) Hexagonal system (8, 9, 10, 11) Transverse isotropy	$\mathbf{U}_0 = A\mathbf{1} + B\mathbf{k} \otimes \mathbf{k}$
Cubic system (6, 7)	$\mathbf{U}_0 = A\mathbf{1}$

[1] This theorem is a corollary of Theorem 2, §43, and Theorem 3, §79, of P. R. HALMOS, *Finite-Dimensional Vector Spaces*, 2nd ed., Princeton, Van Nostrand 1958.

16. Fluids

There are various physical notions concerned with fluids. One is that a fluid is a substance which can flow. "Flow" itself is a vague term. One meaning of "flow" is simply deformation under stress, which does not distinguish a fluid from any other non-rigid material. Another is that steady velocity results from constant stress, which seems to be special and to apply only with difficulty and to particular flows. Another is inability to support shear stress when in equilibrium. Formally, within the theory of simple materials, such a definition would yield

$$\mathbf{T} = -p(\rho)\mathbf{1} + \mathfrak{F}(\mathbf{F}^t), \tag{IV.16-1}$$

where $\mathfrak{F}(\mathbf{1}^t) = \mathbf{0}$, $\mathbf{1}^t$ being the history whose value is always $\mathbf{1}$. Since the material so defined may have any peer group whatever, including one of those already considered to define a solid, this definition does not lend itself to a criterion in terms of common response.

Exercise IV.16.1. Show that the relation $\mathbf{T} = -K(3 - \operatorname{tr} \mathbf{U})\mathbf{1}$ defines an isotropic elastic solid which has infinitely many placements at ease and never experiences non-vanishing shear stress, no matter how it be deformed.

Finally, a fluid is regarded as a material having "no preferred configuration". In terms of peer groups we may realize this somewhat vague idea by the following

Definition. *A fluid is a non-solid egalitarian material.*

In §IV.13 we have shown that for an egalitarian material either $\mathscr{g} = \{\mathbf{1}, -\mathbf{1}\}$ or $\mathscr{g} = \mathscr{u}$. The former case corresponds to a solid, according to the definition given in §IV.15. Thus we have the following

Theorem. *A material is a fluid if and only if*

$$\mathscr{g}_{\boldsymbol{\kappa}} = \mathscr{u} \tag{IV.16-2}$$

for all $\boldsymbol{\kappa}$.

From this theorem, some preceding ones, and the definitions, we read off the following trivial but important corollaries:

1. *Every fluid is isotropic.*
2. *Every placement of a fluid is undistorted.*

3. *A material is egalitarian if and only if it is either a fluid or a triclinic solid.*
4. *The only isotropic materials are fluids and isotropic solids.*

The condition (2) was laid down as the definition of a fluid by NOLL, who derived thereupon the following

Fundamental Theorem on Fluids. *Every fluid has a constitutive relation of the form*

$$\mathbf{T} = \mathfrak{K}(\mathbf{C}_t^t; \rho), \tag{IV.16-3}$$

and

$$\mathfrak{K}(\mathbf{Q}\mathbf{C}_t^t\mathbf{Q}^{\mathsf{T}}; \rho) = \mathbf{Q}\mathfrak{K}(\mathbf{C}_t^t; \rho)\mathbf{Q}^{\mathsf{T}} \tag{IV.16-4}$$

for all orthogonal $\mathbf{Q}$ *and all arguments* $\mathbf{C}_t^t$, ρ *that lie in the domain of* $\mathfrak{K}$. *Every such isotropic mapping of positive symmetric tensor histories into symmetric tensors defines a fluid. Furthermore,*

$$\mathfrak{K}(\mathbf{1}^t; \rho) = -p(\rho)\mathbf{1}. \tag{IV.16-5}$$

This last result states that all fluids obey in equilibrium the laws of Eulerian hydrostatics, according to which the stress is a hydrostatic pressure which depends on the density alone. In particular, a fluid exhibits the phenomenon of "flow" in one of the common senses, namely, that it cannot support any shear stress when it has been at rest for all times, past and present, in any placement whatever. As we have shown at the beginning of this section, the converse is false: A material capable of "flow" in this sense may have any peer group whatever.

Proof of Noll's theorem. Since a fluid is isotropic and every placement is undistorted, we may apply (IV.14-2) for any reference placement $\boldsymbol{\kappa}$. Since for a fluid $\mathbf{T}$ cannot be changed by a static deformation from one placement to another with the same density, the dependence upon $\mathbf{B}(t)$ in (IV.14-2) must reduce to dependence on det $\mathbf{B}(t)$, or, what is the same thing, dependence on ρ, and this establishes the necessity of (3). Furthermore, $\mathfrak{K}$ must satisfy (IV.14-3), which now reduces to (4). For the particular case of the rest history, $\mathbf{C}_t^t = \mathbf{1}^t$, so (4) yields

$$\mathbf{T} = \mathfrak{K}(\mathbf{1}^t; \rho) = \mathbf{Q}\mathfrak{K}(\mathbf{1}^t, \rho)\mathbf{Q}^{\mathsf{T}} = \mathbf{Q}\mathbf{T}\mathbf{Q}^{\mathsf{T}} \tag{IV.16-6}$$

Thus in a fluid which has always been at rest $\mathbf{T}$ commutes with every orthogonal tensor. The result of Exercise IV.4-1 establishes the necessity of (5).

Exercise IV.16.2. Show that (3) and (4) imply that $\mathscr{g}_{\kappa} = \mathscr{u}$ for every $\boldsymbol{\kappa}$.

This exercise completes the proof of NOLL's theorem. △

We may express the foregoing result also as follows: The constitutive relation of a fluid is of the form

$$\mathbf{T} = -p(\rho)\mathbf{1} + \mathfrak{C}(\mathbf{C}_t^t - \mathbf{1}; \rho); \qquad \text{(IV.16-7)}$$

the mapping $\mathfrak{C}$ is isotropic, and it vanishes when its argument is the history $\mathbf{0}^t$ whose value is always $\mathbf{0}$. Conversely, every relation of this form defines a fluid.

A trivial corollary of the foregoing, which may be proved in several other ways, is that *any relation of the form* (IV.4-4) *defines an elastic fluid.* While in hydrodynamics it is customary to impose the condition that $p(\rho) > 0$ for all ρ, or at least the weaker requirement that $p(\rho) > 0$ for all but a discrete set of values of ρ, this condition does not follow from any general principle of mechanics. Since hydrostatic tensions of some magnitude have been produced, with extreme pains, in certain very quiet laboratories, perhaps the condition $p(\rho) > 0$ should be regarded as expressing stability rather than a constitutive restriction.

A fluid may react to its entire deformation history, yet its reaction cannot be different for different placements with the same density. A fluid reconciles these two seemingly contradictory qualities—ability to remember all its past and inability to regard one placement as different from another—by reacting to the past only insofar as it may differ from the present, which may be ever changing.

17. Fluid Crystals

To exhaust the possible types of simple materials, any material that is not a solid we shall call a *fluid crystal.* For a fluid crystal, then, $\mathscr{g}_{\kappa} \not\subset \mathscr{o}$ no matter what be the reference placement $\boldsymbol{\kappa}$. Thus the peer group with respect to every placement has some non-orthogonal elements. That is, there is always some non-rigid transplacement that no experiment can detect. In this regard a fluid crystal is like a fluid, for which all changes of shape without change of density are undetectable. Since it is impossible that $\mathscr{g}_{\kappa} \supset \mathscr{o}$ unless the fluid crystal be in fact a fluid, for a non-isotropic fluid crystal there are also some non-identical rotations that are detectable. In this property an anisotropic fluid resembles an anisotropic solid.

The definitions and results in the preceding section show that *a fluid crystal is a fluid if and only if it is isotropic.*

In this book we shall not go any further into the theory of fluid crystals.[1]

18. Motions with Constant Principal Relative Stretch Histories

Continuum mechanics, even the mechanics of simple materials, covers so vast a range of possible behavior that little can be learned from it without descending to special cases. In this complexity continuum mechanics mirrors nature itself, for only by specifying particular features of a phenomenon can we so much as name it, let alone describe it. In the mechanics of simple materials two kinds of specialization are fruitful:

1. of the material
2. of the motions to which the body is subjected.

We have given examples of the former in the immediately preceding sections. The constitutive relations of fluids and isotropic solids are simpler than the general one, and we can expect the solution of problems for these two classes of materials to be relatively easier than for anisotropic solids or fluid crystals. The continuum mechanics of the last century carried this kind of specialization much further and restricted attention to materials specified by one or two constants. As a result, the solution of wide classes of boundary-value problems became easy—deceptively so, since only rarely can the properties of natural bodies be condensed adequately into one or two numbers fit to be tabulated in a handbook.

We have given a specimen of the second simplification in §§IV.9 and IV.10, where we have seen that we may determine, once and for all, all

[1] The peer groups of certain fluid crystals have been defined and interpreted by B. D. COLEMAN, "Simple liquid crystals," *Archive for Rational Mechanics and Analysis* **20**, 41–58 (1965), and C.-C. WANG, "A general theory of subfluids," *ibid.* **20**, 1–40 (1965).

Fluid crystals as defined here are not to be confused with the "liquid crystals" of ERICKSEN and others; liquid crystals do not fit into the framework established and studied in this book, although they are simple materials in the more general sense introduced in NOLL's paper of 1972, which is cited at the end of this chapter. Surveys of the extensive literature on liquid crystals are available:

1. *Static Theory*. J. L. ERICKSEN, "Equilibrium theory of liquid crystals," *Advances in Liquid Crystals*, Vol. 2, ed. G. BROWN, New York, Academic Press, 1976.
2. ***Dynamic Theory***. **Э. Л. Аэро и А. Н. Булыгин, "Гидромеханика Жидких Кристаллов,"** pp. 106–214 of *Гидромеханика* (Итоги науки и техники), Том 7, Moscow, 1973.

homogeneous deformations that can be produced in an arbitrary homogeneous simple body by bringing to bear suitable tractions upon the boundary. We now define and analyse certain particular motions in which the effects of material memory, which for a simple material may indeed be various and complicated in a general motion, are given little chance to manifest themselves, because there is little to remember.

Consider, for example, the constitutive equation of a simple fluid:

$$\mathbf{T} = \mathfrak{R}(\mathbf{C}_t^t; \rho). \qquad \text{(IV.16-3)}_r$$

In the particular case when $\rho = \text{const.}$ and $\mathbf{C}_t^t(s)$ is the same function of s for all t, the stress becomes constant in time for a given particle. The body may have experienced deformation for all past time, but as each body-point looks backward, so to speak, it sees the entire sequence of past deformations referred to its present placement remain unchanged.

More generally, since the Principle of Material Frame-Indifference (§IV.5) forbids past rotations to enter the constitutive relation and renders explicit the effect of present rotation, we should be able to simplify the constitutive relation almost as much in the somewhat more general case when for some orthogonal tensor $\mathbf{Q}(t)$

$$\mathbf{C}_t^t(s) = \mathbf{Q}(t)\mathbf{C}_0^0(s)\mathbf{Q}(t)^{\mathsf{T}}, \qquad 0 \leqq s < \infty. \qquad \text{(IV.18-1)}$$

Here, of course, $\mathbf{C}_0^0$ denotes $\mathbf{C}_t^t$ when $t = 0$, and $\mathbf{Q}(0) = \mathbf{1}$. Such motions were introduced and called *substantially stagnant* by COLEMAN. In them, an observer situate upon the moving particle may choose his frame in such a way as to see behind him always the same deformation history referred to the present placement. The proper numbers of $\mathbf{C}_t^t(s)$ are the same as those of $\mathbf{C}_0^0(s)$, although the principal axes of the one tensor may rotate arbitrarily with respect to those of the other. Thus, while the principal relative stretches $v_{(t)i}$ may vary with t, they do so in such a way that their histories up to the time t remain unchanged:

$$v_{(t)k}^t = v_{(0)k}^0, \qquad k = 1, 2, 3, \quad -\infty < t < \infty. \qquad \text{(IV.18-2)}$$

By (2), substantially stagnant motions may be defined alternatively as those having *constant principal relative stretch histories*.

Since the definition of this property makes no use of a fixed reference placement, it pertains to the motion itself rather than to any of its embodiments as a transplacement. Moreover, in view of a result obtained in Exercise I.11.2, this property is a frame-indifferent one.

We turn now to the pure kinematics of motions with constant principal relative stretch histories. All such motions are characterized by the following

Fundamental Theorem (NOLL). *A motion has constant principal relative stretch histories if and only if there are an orthogonal tensor* $\mathbf{Q}(t)$, *a scalar* κ, *and a constant tensor* $\mathbf{N}_0$ *such that*

$$\begin{aligned}\mathbf{F}_0(\tau) &= \mathbf{Q}(\tau)e^{\tau\kappa\mathbf{N}_0},\\ \mathbf{Q}(0) &= \mathbf{1}, \qquad |\mathbf{N}_0| = 1.\end{aligned} \tag{IV.18-3}$$

Proof. We begin from the hypothesis (1) and set

$$\mathbf{H}(s) = \mathbf{C}_0(-s) = \mathbf{Q}(t)^{\mathrm{T}}\mathbf{C}_t(t-s)\mathbf{Q}(t). \tag{IV.18-4}$$

By (II.8-8), $\mathbf{F}_t(\tau) = \mathbf{F}_0(\tau)\mathbf{F}_0(t)^{-1}$, so

$$\begin{aligned}\mathbf{Q}(t)\mathbf{H}(s)\mathbf{Q}(t)^{\mathrm{T}} &= \mathbf{C}_t(t-s),\\ &= [\mathbf{F}_0(t)^{\mathrm{T}}]^{-1}\mathbf{C}_0(t-s)\mathbf{F}_0(t)^{-1},\\ &= [\mathbf{F}_0(t)^{\mathrm{T}}]^{-1}\mathbf{H}(s-t)\mathbf{F}_0(t)^{-1}.\end{aligned} \tag{IV.18-5}$$

If we set

$$\mathbf{E}(t) \equiv \mathbf{Q}(t)^{\mathrm{T}}\mathbf{F}_0(t), \tag{IV.18-6}$$

then (5) assumes the form of a difference equation:

$$\mathbf{H}(s-t) = \mathbf{E}(t)^{\mathrm{T}}\mathbf{H}(s)\mathbf{E}(t). \tag{IV.18-7}$$

To obtain a necessary condition for a solution $\mathbf{H}(s)$, we differentiate[1] (7) with respect to t and put $t = 0$, obtaining the first-order linear differential equation

$$-\dot{\mathbf{H}}(s) = \mathbf{M}^{\mathrm{T}}\mathbf{H}(s) + \mathbf{H}(s)\mathbf{M}; \tag{IV.18-8}$$

here $\mathbf{M} \equiv \dot{\mathbf{E}}(0)$, and the dot denotes differentiation with respect to s. The unique solution of (8) such that $\mathbf{H}(0) = \mathbf{1}$ is easily seen to be

$$\mathbf{H}(s) = e^{-s\mathbf{M}^{\mathrm{T}}}e^{-s\mathbf{M}}. \tag{IV.18-9}$$

Since histories are defined only when $s \geqq 0$, this result has been derived only for that domain. Nevertheless, the difference equation (7) serves to define $\mathbf{H}(s)$ for negative s as well and shows that $\mathbf{H}$ is analytic. Since the right-hand side of (9) is analytic, the principle of analytic continuation shows that (9) gives the unique solution for all s, when $\mathbf{E}(t)$ is assigned. If we substitute (9) back into (7), by putting $s = 0$ we obtain

$$\mathbf{E}(t)e^{-t\mathbf{M}}[\mathbf{E}(t)e^{-t\mathbf{M}}]^{\mathrm{T}} = \mathbf{1}. \tag{IV.18-10}$$

[1] That the assertion of the theorem remains true even if $\mathbf{H}$ is merely continuous and $\mathbf{E}$ is completely arbitrary has been shown by W. NOLL, "The representation of monotonous processes by exponentials," *Indiana University Mathematics Journal* **25**, 209–214 (1976).

Hence $\mathbf{E}(t)e^{-t\mathbf{M}}$ is an orthogonal tensor, say $\overline{\mathbf{Q}}(t)$. By (6), then,

$$\mathbf{F}_0(t) = \mathbf{Q}(t)\overline{\mathbf{Q}}(t)e^{t\mathbf{M}}. \tag{IV.18-11}$$

The form asserted by NOLL's theorem holds trivially if $\mathbf{M} = \mathbf{0}$; if $\mathbf{M} \neq \mathbf{0}$, it follows if we set

$$\kappa \equiv |\mathbf{M}|, \qquad \mathbf{N}_0 = \frac{1}{\kappa}\mathbf{M}. \tag{IV.18-12}$$

(The proof reveals that the tensor $\mathbf{Q}(t)$ occurring in the result (3) is not generally the same orthogonal tensor as that occurring in the hypothesis (1).) Conversely, if (3) holds, an easy calculation shows that the motion is one with constant relative principal stretch histories. △

Exercise IV.18.1 (NOLL). Prove that in a motion with constant principal relative stretch histories

$$\mathbf{F}_t(\tau) = \mathbf{Q}(\tau)\mathbf{Q}(t)^{\mathsf{T}}e^{(\tau-t)\kappa\mathbf{N}} = \mathbf{Q}(\tau)e^{(\tau-t)\kappa\mathbf{N}_0}\mathbf{Q}(t)^{\mathsf{T}}, \tag{IV.18-13}$$

$\mathbf{N}$ being defined as follows:

$$\mathbf{N} = \mathbf{Q}(t)\mathbf{N}_0\,\mathbf{Q}(t)^{\mathsf{T}}, \qquad |\mathbf{N}| = 1; \tag{IV.18-14}$$

conversely, if $\mathbf{F}_t(\tau)$ has the form (13), any motion to which it corresponds is a motion with constant principal relative stretch histories. In such a motion

$$\begin{aligned}
\mathbf{C}_t^t(s) &= e^{-s\kappa\mathbf{N}^{\mathsf{T}}}e^{-s\kappa\mathbf{N}},\\
\mathbf{G} &= \kappa\mathbf{N} + \dot{\mathbf{Q}}(t)\mathbf{Q}(t)^{\mathsf{T}},\\
\mathbf{A}_1 &= \dot{\mathbf{C}}_t^t(0) = \kappa(\mathbf{N} + \mathbf{N}^{\mathsf{T}}),\\
\mathbf{A}_2 &= \ddot{\mathbf{C}}_t^t(0) = \kappa(\mathbf{N}^{\mathsf{T}}\mathbf{A}_1 + \mathbf{A}_1\mathbf{N}) = \kappa^2(2\mathbf{N}^{\mathsf{T}}\mathbf{N} + \mathbf{N}^2 + (\mathbf{N}^{\mathsf{T}})^2),\\
\mathbf{A}_3 &= \kappa(\mathbf{N}^{\mathsf{T}}\mathbf{A}_2 + \mathbf{A}_2\mathbf{N}), \quad \ldots\\
\mathbf{A}_k &= \kappa(\mathbf{N}^{\mathsf{T}}\mathbf{A}_{k-1} + \mathbf{A}_{k-1}\mathbf{N}),
\end{aligned} \tag{IV.18-15}$$

the notations being those of §II.11. A motion with constant principal relative stretch histories is isochoric if and only if

$$\operatorname{tr}\mathbf{N}_0 = 0, \tag{IV.18-16}$$

and of course also $\operatorname{tr}\mathbf{N} = 0$.

With the aid of these results we are able to see easily the extremely special nature of motions with constant principal relative stretch histories, which is expressed by the following

Corollary (WANG). *The relative deformation history $\mathbf{C}_t^t$ of a motion with constant principal relative stretch histories is determined uniquely by its first three Rivlin–Ericksen tensors.* That is, if three tensors $\mathbf{A}_1(t)$, $\mathbf{A}_2(t)$, and $\mathbf{A}_3(t)$

are given, they can be the first three Rivlin–Ericksen tensors corresponding to at most one relative deformation history $\mathbf{C}_t^t$ satisfying the defining condition (1).

The proof rests upon a simple lemma. Let **S** be a symmetric tensor and **W** a skew tensor in 3-dimensional space. Without loss of generality we can take the matrices of these tensors as having the forms

$$[\mathbf{S}] = \begin{Vmatrix} a & 0 & 0 \\ 0 & b & 0 \\ 0 & 0 & c \end{Vmatrix}, \qquad [\mathbf{W}] = \begin{Vmatrix} 0 & x & y \\ -x & 0 & z \\ -y & -z & 0 \end{Vmatrix}. \tag{IV.18-17}$$

Then

$$[\mathbf{SW} - \mathbf{WS}] = \begin{Vmatrix} 0 & (a-b)x & (a-c)y \\ (a-b)x & 0 & (b-c)z \\ (a-c)y & (b-c)z & 0 \end{Vmatrix}. \tag{IV.18-18}$$

Hence **S** and **W** commute if and only if

$$(a-b)x = 0, \qquad (a-c)y = 0, \qquad (b-c)z = 0. \tag{IV.18-19}$$

Consequently, if **S** has 3 proper numbers, it commutes with no other skew tensor than **0**. If $a = b \neq c$, **S** commutes with **W** if and only if $y = z = 0$. If $a = b = c$, **S** commutes with all **W**.

WANG's corollary may now be proved in stages. If two motions with constant principal relative stretch histories can correspond to $\mathbf{A}_1$ and $\mathbf{A}_2$, then because of $(15)_{4,6}$ there are tensors **M** and $\overline{\mathbf{M}}$ such that

$$\begin{aligned} \mathbf{M} + \mathbf{M}^{\mathrm{T}} &= \overline{\mathbf{M}} + \overline{\mathbf{M}}^{\mathrm{T}}, \\ \mathbf{M}^{\mathrm{T}}\mathbf{A}_1 + \mathbf{A}_1\mathbf{M} &= \overline{\mathbf{M}}^{\mathrm{T}}\mathbf{A}_1 + \mathbf{A}_1\overline{\mathbf{M}}. \end{aligned} \tag{IV.18-20}$$

The first of these equations asserts that $\mathbf{M} - \overline{\mathbf{M}}$ is skew; the second, that $\mathbf{M} - \overline{\mathbf{M}}$ commutes with $\mathbf{A}_1$. If $\mathbf{A}_1$ has 3 proper numbers, the lemma shows that $\mathbf{M} - \overline{\mathbf{M}} = \mathbf{0}$.

Suppose now that $\mathbf{A}_1$ has 2 proper numbers. Then relative to a suitable orthonormal basis

$$[\mathbf{A}_1] = \begin{Vmatrix} a & 0 & 0 \\ 0 & a & 0 \\ 0 & 0 & b \end{Vmatrix}, \qquad a \neq b. \tag{IV.18-21}$$

Case 1. Assume that, relative to this same basis,

$$[\mathbf{A}_2] = \begin{Vmatrix} u & 0 & 0 \\ 0 & u & 0 \\ 0 & 0 & v \end{Vmatrix}. \tag{IV.18-22}$$

The most general $\mathbf{M}$ compatible with $(15)_4$ and (21) is given by

$$\kappa[\mathbf{M}] = \begin{Vmatrix} \frac{1}{2}a & x & y \\ -x & \frac{1}{2}a & z \\ -y & -z & \frac{1}{2}b \end{Vmatrix}. \qquad \text{(IV.18-23)}$$

By (21) and (22)

$$\kappa[\mathbf{M}^{\mathrm{T}}\mathbf{A}_1 + \mathbf{A}_1\mathbf{M}] = \begin{Vmatrix} a^2 & 0 & (a-b)y \\ 0 & a^2 & (a-b)z \\ (a-b)y & (a-b)z & b^2 \end{Vmatrix}. \qquad \text{(IV.18-24)}$$

Since $a \neq b$, it follows from $(15)_6$ and (22) that

$$u = a^2, \qquad v = b^2, \qquad y = 0, \qquad z = 0. \qquad \text{(IV.18-25)}$$

Exercise IV.18.2. From (23) and (25) show that $\mathbf{M}$ commutes with $\mathbf{M}^{\mathrm{T}}$, and hence by $(15)_{1,\,3}$ conclude that

$$\mathbf{C}_t^t(s) = e^{-s\mathbf{A}_1}. \qquad \text{(IV.18-26)}$$

Case 2. Still on the supposition that $\mathbf{A}_1$ is of the form (21), but regardless of whether (22) does or does not hold, we assume that two motions with constant principal relative stretch history can correspond to $\mathbf{A}_1$, $\mathbf{A}_2$, and $\mathbf{A}_3$. Then again, there are tensors $\mathbf{M}$ and $\overline{\mathbf{M}}$ such as to satisfy (20). Since $\mathbf{M} - \overline{\mathbf{M}}$ is a skew tensor that commutes with $\mathbf{A}_1$ as given by (21), the lemma shows that

$$[\mathbf{M} - \overline{\mathbf{M}}] = \begin{Vmatrix} 0 & x & 0 \\ -x & 0 & 0 \\ 0 & 0 & 0 \end{Vmatrix}. \qquad \text{(IV.18-27)}$$

But also by $(15)_8$

$$\mathbf{M}^{\mathrm{T}}\mathbf{A}_2 + \mathbf{A}_2\mathbf{M} = \overline{\mathbf{M}}^{\mathrm{T}}\mathbf{A}_2 + \mathbf{A}_2\overline{\mathbf{M}}, \qquad \text{(IV.18-28)}$$

so that $\mathbf{M} - \overline{\mathbf{M}}$ commutes with $\mathbf{A}_2$.

Exercise IV.18.3. Now assuming that (22) does not hold, prove that $\mathbf{M} = \overline{\mathbf{M}}$ in Case 2. Finally, if $\mathbf{A}_1 = \alpha\mathbf{1}$, prove that (26) holds, and hence complete the argument for WANG's corollary. △

Accordingly, then, three given tensors $\mathbf{A}_1(t)$, $\mathbf{A}_2(t)$, and $\mathbf{A}_3(t)$ can be the Rivlin–Ericksen tensors corresponding to *at most one* $\mathbf{C}_t^t(s)$ belonging to a motion with constant principal relative stretch histories. In general, on the contrary, three symmetric tensors taken arbitrarily will fail to be the first three Rivlin–Ericksen tensors of any motion at all, let alone one with constant principal relative stretch histories, since they will fail to satisfy conditions of compatibility[1] expressing the fact that they derive from a velocity field in a region. We shall not take up these conditions since our interest lies in simplifying a constitutive relation when the motion is known to be one with constant principal relative stretch histories.

While NOLL's theorem clearly is independent of the dimension of the space, WANG's corollary rests heavily on use of the dimension 3.

NOLL's theorem (3) suggests an invariant classification of all motions with constant principal relative stretch histories into three mutually exclusive types:

1. $\mathbf{N}_0^2 = \mathbf{0}$. These motions are called *viscometric flows*.
2. $\mathbf{N}_0^3 = \mathbf{0}$ but $\mathbf{N}_0^2 \neq \mathbf{0}$.
3. $\mathbf{N}_0$ is not nilpotent.

In types 1 and 2, since $\text{tr}\,\mathbf{N}_0 = 0$, the motion is isochoric.

There are interesting examples of all three types, but the simplest, the viscometric flows, are used most in applications. We shall study them further in the next chapter.

Exercise IV.18.4. The relative local deformation $\mathbf{F}_t$ of a viscometric flow has the form

$$\begin{gathered}\mathbf{F}_t(\tau) = \mathbf{Q}(\tau)\mathbf{Q}(t)^{\mathrm{T}}[\mathbf{1} + (\tau - t)\kappa\mathbf{Q}(t)\mathbf{N}_0\,\mathbf{Q}(t)^{\mathrm{T}}], \\ \mathbf{N}_0 = \text{const.}, \qquad |\mathbf{N}_0| = 1, \qquad \mathbf{N}_0^2 = \mathbf{0}, \qquad \kappa = \text{a scalar field.}\end{gathered} \tag{IV.18-29}$$

Conversely, any relative local deformation of this form corresponds to a viscometric flow.

Exercise IV.18.5. Prove that in any motion with constant principal relative stretch histories

$$\mathbf{A}_2 - \mathbf{A}_1^2 = \kappa^2(\mathbf{N}^{\mathrm{T}}\mathbf{N} - \mathbf{N}\mathbf{N}^{\mathrm{T}}) \tag{IV.18-30}$$

and hence that

$$\text{tr}\,\mathbf{A}_1^2 = \text{tr}\,\mathbf{A}_2 = 2\kappa^2(1 + \text{tr}\,\mathbf{N}^2). \tag{IV.18-31}$$

Thus in a viscometric flow

$$\kappa^2 = \tfrac{1}{2}\,\text{tr}\,\mathbf{A}_1^2 = \tfrac{1}{2}\,\text{tr}\,\mathbf{A}_2. \tag{IV.18-32}$$

[1] Necessary and sufficient conditions are obtained by R. R. HUILGOL, "Sur les conditions nécessaires et suffisantes pour les mouvements à histoire de déformation constante," *Comptes Rendus Hebdomédaires de l'Académie des Sciences, Paris,* **282**, A67–A69 (1976), and "Algorithms for motions with constant stretch history," *Rheologica Acta* **15**, 120–129 (1976).

Exercise IV.18.6. (NOLL, COLEMAN & NOLL). Prove that the motion defined by (II.11-11) is viscometric; that the motion whose Cartesian velocity components are

$$\dot{x}_1 = 0, \qquad \dot{x}_2 = \kappa x_1, \qquad \dot{x}_3 = \lambda x_1 + \nu x_2, \tag{IV.18-33}$$

belongs to NOLL's second class if $\kappa\nu \neq 0$; that the motion whose Cartesian velocity components are

$$\dot{x}_k = a_k x_k = \text{const.}, \quad k = 1, 2, 3, \tag{IV.18-34}$$

belongs to NOLL's third class if at least one of the a_k does not vanish. Show that (34) gives rise to a homogeneous transplacement of any shape the body occupies. Prove that if $a_1 + a_2 + a_3 = 0$ the motions corresponding to (34) are possible, subject to surface tractions alone, in any incompressible simple body of uniform density.

19. Reduction of the Constitutive Relation for a Simple Material in a Motion with Constant Principal Relative Stretch Histories

In view of WANG's corollary, any information that can be determined from $\mathbf{C}_t^t$ in a motion with constant principal relative stretch histories can be determined also from $\mathbf{A}_1(t)$, $\mathbf{A}_2(t)$, $\mathbf{A}_3(t)$. Therefore, any *functional* of $\mathbf{C}_t^t$ equals, in these motions, a *function* of $\mathbf{A}_1(t)$, $\mathbf{A}_2(t)$, $\mathbf{A}_3(t)$. Consequently the general constitutive relation (IV.5-15) may be replaced, as far as motions with constant principal relative stretch histories are concerned, by

$$\mathbf{R}^{\mathrm{T}}\mathbf{T}\mathbf{R} = \mathfrak{f}(\mathbf{R}^{\mathrm{T}}\mathbf{A}_1(t)\mathbf{R},\ \mathbf{R}^{\mathrm{T}}\mathbf{A}_2(t)\mathbf{R},\ \mathbf{R}^{\mathrm{T}}\mathbf{A}_3(t)\mathbf{R},\ \mathbf{C}(t)), \tag{IV.19-1}$$

$\mathfrak{f}$ being a function. A material whose constitutive relation is (1) is called *a material of differential type of complexity 3*. By (1), then, we have the following

Theorem (WANG). *In the class of motions with constant principal relative stretch histories a simple material cannot be distinguished from some material of differential type of complexity 3.*

In other words, *no experiment based on interpretation of results for motions with constant principal relative stretch histories can distinguish a general simple material from a differential material of complexity 3.* As we shall see in the next chapter, the special flows most commonly used to describe the properties of natural fluids are of the kind considered here and hence are of very limited service in exploring the physical properties of those fluids.

An isotropic material of differential type is called a *Rivlin–Ericksen material.* For it, (1) becomes

$$\mathbf{T}(t) = \mathfrak{f}(\mathbf{A}_1(t), \mathbf{A}_2(t), \mathbf{A}_3(t), \mathbf{B}(t)), \qquad \text{(IV.19-2)}$$

and when the isotropic material is a fluid,

$$\mathbf{T}(t) = -p(\rho)\mathbf{1} + \mathfrak{f}(\mathbf{A}_1(t), \mathbf{A}_2(t), \mathbf{A}_3(t), \rho); \qquad \text{(IV.19-3)}$$

the functions $\mathfrak{f}$, in the two cases, are isotropic in the sense that for all symmetric $\mathbf{A}_1$, $\mathbf{A}_2$, $\mathbf{A}_3$, $\mathbf{B}$, and for all orthogonal $\mathbf{Q}$

$$\mathfrak{f}(\mathbf{Q}\mathbf{A}_1\mathbf{Q}^T, \mathbf{Q}\mathbf{A}_2\mathbf{Q}^T, \mathbf{Q}\mathbf{A}_3\mathbf{Q}^T, \mathbf{Q}\mathbf{B}\mathbf{Q}^T \text{ or } \rho) = \mathbf{Q}\mathfrak{f}(\mathbf{A}_1, \mathbf{A}_2, \mathbf{A}_3, \mathbf{B} \text{ or } \rho)\mathbf{Q}^T, \qquad \text{(IV.19-4)}$$

this being the functional equation to which (IV.14-3) reduces in the present instance. Moreover, for a fluid $\mathfrak{f}(\mathbf{0}, \mathbf{0}, \mathbf{0}, \rho) = \mathbf{0}$.

The results in §IV.7 enable the reader to write down at once the constitutive relations for incompressible Rivlin–Ericksen materials.

These reductions just given may be interpreted in two ways. On the one hand, they enable us to solve easily various special problems concerned with motions having constant principal relative stretch histories. However complicated may be in general the response of a material, in these particular motions we need consider only a simple special constitutive equation. On the other hand, they show that observation of this class of flows is insufficient to tell us much about a material, since most of the complexities of material response are prevented from manifesting themselves.

In §VI.1 we shall discuss materials of the differential type in somewhat more detail, but in the next chapter we shall exploit the present results so as the obtain specific solutions for viscometric flows of simple fluids.

In a viscometric flow, by definition, $\mathbf{N}_0^2 = \mathbf{0}$, and hence by (IV.18-14) and (IV.18-15)

$$\mathbf{A}_3 = \mathbf{A}_4 = \cdots = \mathbf{0}. \qquad \text{(IV.19-5)}$$

Therefore, *in a viscometric flow a simple fluid cannot be distinguished from some Rivlin–Ericksen fluid of complexity 2.*

For a motion with constant principal relative stretch histories we see from $(\text{IV.18-15})_1$ that

$$\mathbf{R}^T\mathbf{C}_t^t(s)\mathbf{R} = \exp[-s\kappa(\mathbf{R}^T\mathbf{N}\mathbf{R})^T]\exp[-s\kappa\mathbf{R}^T\mathbf{N}\mathbf{R}]. \qquad \text{(IV.19-6)}$$

Hence any quantity determined by $\mathbf{R}^T\mathbf{C}_t^t\mathbf{R}$ in general is determined here by $\kappa\mathbf{R}^T\mathbf{N}\mathbf{R}$. Referring to the constitutive relation of a simple material as expressed by (IV.5-15), we may set

$$\mathfrak{f}(\kappa\mathbf{N}, \mathbf{C}) \equiv \mathfrak{K}(\mathbf{R}^T\mathbf{C}_t^t\mathbf{R}, \mathbf{C}) \qquad \text{(IV.19-7)}$$

and so obtain

$$\mathbf{R}^{\mathsf{T}}\mathbf{T}\mathbf{R} = \mathfrak{f}(\kappa\mathbf{R}^{\mathsf{T}}\mathbf{N}\mathbf{R}, \mathbf{C}) \tag{IV.19-8}$$

as an expression for that relation when restricted to motions with constant principal relative stretch histories. The student will see at once the simpler forms to which (8) reduces for isotropic solids and fluids. For an incompressible fluid the corresponding result is

$$\mathbf{T} = -p\mathbf{1} + \mathfrak{f}(\kappa\mathbf{N}), \tag{IV.19-9}$$

the function $\mathfrak{f}$ being subject to the requirement that

$$\mathfrak{f}(\kappa\mathbf{Q}\mathbf{N}\mathbf{Q}^{\mathsf{T}}) = \mathbf{Q}\mathfrak{f}(\kappa\mathbf{N})\mathbf{Q}^{\mathsf{T}} \tag{IV.19-10}$$

for all orthogonal tensors $\mathbf{Q}$.

The relations (9) and (10) provide the starting point for the analysis in the next chapter.

General References

§§26–30 of NFTM.

W. Noll, "A mathematical theory of the mechanical behavior of continuous media," *Archive for Rational Mechanics and Analysis* **2**, 197–226 (1958). Reprinted in *Continuum Mechanics II, The Rational Mechanics of Materials*, ed. C. Truesdell, New York, Gordon and Breach, 1965, and in *Continuum Theory of Inhomogeneities in Simple Bodies*, New York, Springer-Verlag, 1968, and in *The Foundations of Mechanics and Thermodynamics*, New York, Heidelberg, & Berlin, Springer-Verlag, 1974.

B. D. Coleman & W. Noll, "Material symmetry and thermostatic inequalities in finite elastic deformations," *Archive for Rational Mechanics and Analysis* **15**, 87–111 (1964), reprinted in Noll's *The Foundations of Mechanics and Thermodynamics.*

W. Noll, "A new mathematical theory of simple materials," *Archive for Rational Mechanics and Analysis* **48**, 1–50 (1972), reprinted with the preceding.

W. Noll, "Lectures on the foundations of continuum mechanics and thermodynamics," *Archive for Rational Mechanics and Analysis* **52**, 62–92 (1973), reprinted with the preceding.

Appendix I

General Scheme of Notation

The kinds of letters are listed in the order of their appearance in the book. Departures from this scheme occur here and there within single sections.

Script majuscules: $\mathscr{A}, \mathscr{B}, \mathscr{C}, \ldots, \mathscr{X}, \mathscr{Y}, \mathscr{Z}$, denote bodies, sets, regions of space, curves, and surfaces.

Lightfaced italics, both majuscule and minuscule, stand for scalars and scalar-valued functions: $A, B, C, \ldots, X, Y, Z, a, b, c, \ldots, x, y, z$. Included are the components of vectors and tensors with respect to particular bases.

Exception: X usually stands for a body-point.

Special letter: t always denotes the time.

Note also the uses of o and O explained below in §C1 of Appendix II.

Boldfaced roman minuscules stand for vectors and vector-valued functions: $\mathbf{a}, \mathbf{b}, \ldots, \mathbf{u}, \mathbf{v}$, except that $\mathbf{x}, \mathbf{y}, \mathbf{z}$ are places.

Special letter: $\mathbf{n}$ always denotes an oriented unit normal to a surface.

Boldfaced Greek minuscules $\boldsymbol{\varphi}, \boldsymbol{\lambda}, \boldsymbol{\tau}$, etc., in Chapter I denote mappings other than functions of vectors.

Special letters:

$\boldsymbol{\chi}$ always denotes the motion of a body (§I.7).

$\boldsymbol{\kappa}$ always denotes a reference placement of a body (§II.3).

$\boldsymbol{\chi}_{\boldsymbol{\kappa}}$ always denotes the transplacement of a body from the reference placement $\boldsymbol{\kappa}$ to the actual placement (§II.3).

Boldfaced majuscules **A**, **B**, ..., **U**, **V**, **W** denote linear transformations (second-order tensors) over finite-dimensional (usually 3-dimensional) vector spaces.

Exception: **X** is always a place in a reference shape.

Special letters:

Q and **R** are always orthogonal.

W is always skew.

F is always an invertible tensor which can be interpreted as a local deformation.

Lightfaced Greek minuscules are used for three different kinds of quantities:

1. For angles, rates of change of angles, and other pure rates.
2. For scalar potentials.
3. For scalar moduli or scalar-valued material functions of a real variable.

Exceptions:

ρ is always the mass-density (§II.2).

δ and ε are scalars which can be chosen arbitrarily small.

Fraktur letters, both majuscule and minuscule, denote constitutive mappings (responses). Lightfaced $\mathfrak{a}, \mathfrak{b}, \mathfrak{c}, \ldots, \mathfrak{A}, \mathfrak{B}, \mathfrak{C}, \ldots$ are used if the values of the mappings are scalars; boldfaced $\boldsymbol{\mathfrak{a}}, \boldsymbol{\mathfrak{b}}, \boldsymbol{\mathfrak{c}}, \ldots, \boldsymbol{\mathfrak{A}}, \boldsymbol{\mathfrak{B}}, \boldsymbol{\mathfrak{C}}, \ldots$, if the values are vectors or tensors.

Script minuscules denote groups of tensors.

Special letters:

$\mathscr{o}$ is the full orthogonal group.

$\mathscr{u}$ is the full unimodular group.

$\mathscr{g}$ is a subgroup of $\mathscr{u}$.

Lightfaced Greek majuscules Θ, Φ, ..., are used for certain angles in a reference shape.

Boldfaced gothic majuscules A, B, C, L denote third-order or fourth-order tensors over a 3-dimensional vector space.

Astronomical symbols ♆, ♃, ♂, *etc.*, stand for quantities of arbitrary tensorial order; scalars, vectors, *etc.*

Special symbol: $\oint$ denotes a framing (§I.6).

Indices:

The uses of subscripts and superscripts are standard. A few examples will suffice, but the list is far from exhaustive.

If $\mathbf{a}$, $\mathbf{A}$, and A are vectors and tensors denoted as above, then their components with respect to a basis are denoted by adjoined indices, for example a_k, A_{mp}, A_{qrsu}. The particular basis is always specified. In the case of curvilinear co-ordinates, the usual notations of contravariant and covariant components such as a^k and a_k are employed once in a while. Physical components, which are components with respect to an orthonormal basis having the co-ordinate directions, are denoted by indices following the letter at middle height: $T^{\langle rr \rangle}$, $T^{\langle \theta\theta \rangle}$, etc.

Greek minuscule indices refer to co-ordinate systems in the reference shape $\boldsymbol{\kappa}(\mathscr{B})$. For example, F^k_α is the component of $\mathbf{F}$ corresponding to x^k and X^α. Both systems of co-ordinates may be curvilinear if so desired.

A boldfaced subscript kappa, for example $\chi_{\boldsymbol{\kappa}}$ and $\mathbf{A}_{\boldsymbol{\kappa}}$, reminds the reader that the reference placement $\boldsymbol{\kappa}$ is being used.

Roman indices are labels. Examples: min and max in the obvious senses, c for "convected" or "constant", T for "transpose", *etc.*

After ∂, a subscript indicates the variable on which ∂ operates; for example, ∂_θ is the partial derivative with respect to θ.

A superimposed dot always indicates a time derivative in some sense. For example, $\dot{\chi}$ is the velocity field over a body, and $\dot{\mathbf{x}}$ is the velocity field over the present shape of that body.

General relations:

$A \Rightarrow B$	Proposition A implies Proposition B.
$A \Leftrightarrow B$	A holds if and only if B holds.
$\mathscr{A} \curlyvee \mathscr{B}$, $\mathscr{A} \curlywedge \mathscr{B}$	join and meet, respectively, of the bodies $\mathscr{A}$ and $\mathscr{B}$
$\mathscr{A} \prec \mathscr{B}$, $\mathscr{B} \succ \mathscr{A}$	$\mathscr{A}$ is a part of $\mathscr{B}$.
$\mathscr{A} \cup \mathscr{B}$, $\mathscr{A} \cap \mathscr{B}$	union and intersection, respectively, of the sets $\mathscr{A}$ and $\mathscr{B}$.
$\mathscr{A} \subset \mathscr{B}$, $\mathscr{B} \supset \mathscr{A}$	$\mathscr{A}$ is a subset of $\mathscr{B}$
$x \in \mathscr{A}$	x is an element of the set $\mathscr{A}$.
$f: \mathscr{A} \to \mathscr{B}$	f maps the set $\mathscr{A}$ into or onto the set $\mathscr{B}$.
$f: x \mapsto y$	f maps the element x onto the element y; that is, $f(x) = y$.
$f \circ g$	composition of the mappings g and f; that is, $(f \circ g)(x) = f(g(x))$.
$\forall x \in \mathscr{A}$	for every x that is an element of the set $\mathscr{A}$
$\{x : x \in \mathscr{A}\}$	the set of all x that are elements of $\mathscr{A}$
$\{x_1, x_2, \ldots, x_n\}$	the set consisting in the elements $x_1, x_2, \ldots, x_n$

Operations on vectors and tensors: see Appendix II.

Appendix II

Some Definitions and Theorems of Algebra, Geometry, and Calculus[1]

A. Algebra

1. Vector Spaces, Bases

All *vector spaces* $\mathscr{V}$, $\mathscr{V}'$ *etc.*, that are recognized explicitly in this book are *finite-dimensional*, usually *3-dimensional*, and their field of scalars is the real field. Their elements are denoted by bold-faced minuscule letters $\mathbf{a}$, $\mathbf{b}$, ..., $\mathbf{u}$, $\mathbf{v}$, $\mathbf{w}$, and their scalars by light-faced italics a, b, ..., A, B, ... The zero vector is $\mathbf{0}$.

The set of vectors $\mathbf{u}_1$, $\mathbf{u}_2$, ..., $\mathbf{u}_m$ is a *linearly dependent* set if scalars a^1, a^2, ..., a^m, not all zero, can be found such that

$$a^1\mathbf{u}_1 + a^2\mathbf{u}_2 + \cdots + a^m\mathbf{u}_m = \mathbf{0}.$$

Otherwise the set is *linearly independent*. The expression on the left-hand side is called a *linear combination* of the vectors $\mathbf{u}_1$, $\mathbf{u}_2$, ..., $\mathbf{u}_m$. The set of values of all linear combinations of $\mathbf{u}_1$, $\mathbf{u}_2$, ..., $\mathbf{u}_m$ is a *subspace*, which the vectors $\mathbf{u}_1$, $\mathbf{u}_2$, ..., $\mathbf{u}_m$ are said to *span*.

The dimension n of the space is the number of vectors in the largest linearly independent set that spans the space. The dimension of a subspace

[1] The material listed here is drawn largely from unpublished notes by W. NOLL.

of an n-dimensional vector space is at most n. (Of course infinite-dimensional vector spaces whose elements are functions occur implicitly in this book, but they are not treated formally.)

Any indexed set $\mathbf{e}_1, \mathbf{e}_2, \ldots, \mathbf{e}_n$ of n linearly independent vectors spans an n-dimensional vector space and hence is called a *basis* in it. If $\mathbf{u}$ is any vector, then

$$\mathbf{u} = u^k \mathbf{e}_k,$$

where $u^1, u^2, \ldots, u^n$ are uniquely determined scalars. In this expression, and subsequently, diagonally repeated indices are to be summed from 1 to n. The n scalars u^k are the *components* of $\mathbf{u}$ relative to the basis $\mathbf{e}_1, \mathbf{e}_2, \ldots, \mathbf{e}_n$.

If $\bar{\mathbf{e}}_1, \bar{\mathbf{e}}_2, \ldots, \bar{\mathbf{e}}_n$ is another basis, of course the vector $\mathbf{u}$ has components relative to it:

$$\mathbf{u} = \bar{u}^k \bar{\mathbf{e}}_k.$$

Also $\bar{\mathbf{e}}_p$ has components, say A_p^q, relative to $(\mathbf{e}_1, \mathbf{e}_2, \ldots, \mathbf{e}_n)$:

$$\bar{\mathbf{e}}_p = A_p^q \mathbf{e}_q, \qquad p = 1, 2, \ldots, n.$$

Likewise

$$\mathbf{e}_m = \bar{A}_m^q \bar{\mathbf{e}}_q, \qquad m = 1, 2, \ldots, n,$$

and hence, since the vectors of both bases are linearly independent,

$$A_k^q \bar{A}_q^m = \delta_k^m \equiv \begin{cases} 1 & \text{if } m = k, \\ 0 & \text{if } m \neq k. \end{cases}$$

Hence the components of any vector $\mathbf{u}$ relative to the two bases are determined from each other as follows:

$$\bar{u}^q = \bar{A}_m^q u^m, \qquad u^p = A_q^p \bar{u}^q.$$

Persons who prefer numerical to geometrical treatments may use this *transformation law* for components to define vectors. They may choose to specify a vector by prescribing its components relative to some one basis, and then use the transformation law to calculate its components relative to every other basis. Alternatively, they may start with lists of numbers $u^1, u^2, \ldots, u^n$ and $\bar{u}^1, \bar{u}^2, \ldots, \bar{u}^n, \ldots$, associated with various bases and then say that these lists do or do not constitute components of one and the same vector relative to the respective bases according as they are or are not related by the transformation law.

2. *Linear Mappings*

A mapping $\mathbf{L}$ of a vector space $\mathscr{V}$ into a vector space $\mathscr{V}'$ is *linear* if

$$\mathbf{L}(a\mathbf{u} + b\mathbf{v}) = a\mathbf{L}(\mathbf{u}) + b\mathbf{L}(\mathbf{v})$$

for all $\mathbf{u}$ and $\mathbf{v}$ in $\mathscr{V}$ and all scalars a and b. The *scalar multiple* $a\mathbf{L}$ of $\mathbf{L}$ by a and the *sum* $\mathbf{L} + \mathbf{M}$ of such mappings $\mathbf{L}$ and $\mathbf{M}$ are defined as follows:

$$(a\mathbf{L})(\mathbf{u}) \equiv a(\mathbf{L}(\mathbf{u})),$$
$$(\mathbf{L} + \mathbf{M})(\mathbf{u} + \mathbf{v}) \equiv \mathbf{L}(\mathbf{u} + \mathbf{v}) + \mathbf{M}(\mathbf{u} + \mathbf{v}).$$

It is easy to show that $a\mathbf{L}$ and $\mathbf{L} + \mathbf{M}$ are themselves linear mappings of $\mathscr{V}$ into $\mathscr{V}'$.

The *nullspace* of a linear mapping $\mathbf{L}$ is the set of vectors that $\mathbf{L}$ maps onto $\mathbf{0}$. The *range of* $\mathbf{L}$ is the set of all values $\mathbf{L}(\mathbf{u})$. These sets are subspaces of $\mathscr{V}$ and $\mathscr{V}'$, respectively, and

$$\dim \text{Nullspace } \mathbf{L} + \dim \text{Range } \mathbf{L} = \dim \mathscr{V}.$$

If $\dim \mathscr{V} = \dim \mathscr{V}'$, the linear mapping $\mathbf{L}$ may have an *inverse* $\mathbf{L}^{-1}$; such an $\mathbf{L}$ is *invertible*. If $\dim \mathscr{V} = \dim \mathscr{V}'$, any of the following statements is a necessary and sufficient condition that $\mathbf{L}$ have an inverse: $\mathbf{L}$ is one-to-one, $\mathbf{L}$ maps $\mathscr{V}$ onto $\mathscr{V}'$, the nullspace of $\mathbf{L}$ contains no vector but $\mathbf{0}$. The inverse, if it exists, is itself a linear invertible mapping. Of course $(\mathbf{L}^{-1})^{-1} = \mathbf{L}$.

3. Tensors

A linear mapping of a vector space into itself is called a *tensor* (of second order). For tensors the result of mapping a vector is written like a multiplication: $\mathbf{L}\mathbf{u} \equiv \mathbf{L}(\mathbf{u})$.

The mapping whose value for every vector is $\mathbf{0}$ is a tensor; it is called the *zero tensor* and denoted by $\mathbf{0}$. The identity mapping is a tensor; it is called the *unit tensor* and denoted by $\mathbf{1}$. Thus for all vectors $\mathbf{u}$

$$\mathbf{0}\mathbf{u} = \mathbf{0}, \qquad \mathbf{1}\mathbf{u} = \mathbf{u}.$$

The tensor that transforms every vector into its opposite is called the *central inversion* and denoted by $-\mathbf{1}$:

$$(-\mathbf{1})\mathbf{v} = -\mathbf{v}.$$

If $\mathbf{L}$ and $\mathbf{M}$ are tensors, so is their composition, which we denote by $\mathbf{M}\mathbf{L}$ and call the *product* of $\mathbf{L}$ by $\mathbf{M}$. The set of all tensors forms an *algebra* under the operations denoted by $a\mathbf{L}$, $\mathbf{L} + \mathbf{M}$, and $\mathbf{L}\mathbf{M}$. Generally $\mathbf{L}\mathbf{M} \neq \mathbf{M}\mathbf{L}$. If $\mathbf{L}\mathbf{M} = \mathbf{M}\mathbf{L}$, the tensors $\mathbf{L}$ and $\mathbf{M}$ *commute*. As is usual for algebras, $-\mathbf{L}$ is written for $(-\mathbf{1})\mathbf{L}$, which is the same as $(-1)\mathbf{L}$. Of course $\mathbf{0}\mathbf{L} = 0\mathbf{L} = \mathbf{0}$, but if $\mathbf{M}\mathbf{L} = \mathbf{0}$, neither $\mathbf{M}$ nor $\mathbf{L}$ need be $\mathbf{0}$.

The *powers* of a tensor $\mathbf{L}$ are defined as follows:

$$\mathbf{L}^0 \equiv \mathbf{1}, \qquad \mathbf{L}^1 \equiv \mathbf{L}, \qquad \mathbf{L}^2 \equiv \mathbf{LL}, \qquad etc.;$$

these obey the usual *rules of exponentiation*:

$$\mathbf{L}^m\mathbf{L}^n = \mathbf{L}^{m+n} = \mathbf{L}^n\mathbf{L}^m, \qquad (a\mathbf{L})^m = a^m\mathbf{L}^m,$$
$$(\mathbf{L}^m)^n = \mathbf{L}^{mn},$$

if $m \geqq 0$ and $n \geqq 0$.

If $\mathbf{L}^m = \mathbf{0}$ for some positive integer m but $\mathbf{L}^p \neq \mathbf{0}$ if $0 < p < m$, the tensor $\mathbf{L}$ is *nilpotent* of *order m*. Nilpotent tensors of orders 1, 2, 3, ..., n exist, but not of any greater order. That is, if $\mathbf{L}^n \neq \mathbf{0}$, then $\mathbf{L}$ is not nilpotent.

If $\mathbf{L}$ is invertible,

$$\mathbf{L}\mathbf{L}^{-1} = \mathbf{L}^{-1}\mathbf{L} = \mathbf{1}.$$

Also if there is a tensor $\mathbf{M}$ such that $\mathbf{LM} = \mathbf{ML} = \mathbf{1}$, then $\mathbf{L}$ is invertible, and $\mathbf{M} = \mathbf{L}^{-1}$. Clearly, $\mathbf{1}^{-1} = \mathbf{1}, (-\mathbf{1})^{-1} = -\mathbf{1}$. The tensor $\mathbf{0}$ is not invertible, nor is any nilpotent tensor. A tensor that is not invertible is sometimes called *singular*.

The product of two invertible tensors is invertible, so is the multiple of an invertible tensor by any scalar other than 0, and $(a\mathbf{L})^{-1} = a^{-1}\mathbf{L}^{-1}$. If $\mathbf{L}$ is invertible, then

$$(\mathbf{L}^{-1})^n = (\mathbf{L}^n)^{-1},$$

and the rules of exponentiation extend to negative powers.

If $\mathbf{e}_1, \mathbf{e}_2, \ldots, \mathbf{e}_n$ is a basis of the vector space, the conditions

$$\mathbf{L}\mathbf{e}_k = L^q{}_k \mathbf{e}_q$$

define unique scalars $L^q{}_k$, which are called the *components* of $\mathbf{L}$ relative to the basis. The matrix $\|L^q{}_k\|$ of the components $L^q{}_k$ is called the *matrix* of $\mathbf{L}$ relative to $\mathbf{e}_1, \mathbf{e}_2, \ldots, \mathbf{e}_n$ and is denoted also by $[\mathbf{L}]$. That is,

$$[\mathbf{L}] \equiv \|L^q{}_k\| \equiv \begin{Vmatrix} L^1{}_1 & L^1{}_2 & \cdots & L^1{}_n \\ L^2{}_1 & & & \\ \vdots & & & \\ L^n{}_1 & L^n{}_2 & \cdots & L^n{}_n \end{Vmatrix}.$$

Of course $[a\mathbf{L}] = a[\mathbf{L}]$. The matrix of the product $\mathbf{LM}$ is the product of the matrix of $\mathbf{M}$ by the matrix of $\mathbf{L}$. The components of $\mathbf{LM}$ are $L^p{}_k M^k{}_q$.

No matter what the choice of basis,

$$[\mathbf{0}] = \begin{Vmatrix} 0 & 0 & \cdots & 0 \\ 0 & 0 & \cdots & 0 \\ \vdots & & & \vdots \\ 0 & 0 & \cdots & 0 \end{Vmatrix}, \qquad [\mathbf{1}] = \begin{Vmatrix} 1 & 0 & \cdots & 0 \\ 0 & 1 & \cdots & 0 \\ \vdots & & & \vdots \\ 0 & 0 & \cdots & 1 \end{Vmatrix},$$

$$[-\mathbf{1}] = \begin{Vmatrix} -1 & 0 & \cdots & 0 \\ 0 & -1 & \cdots & 0 \\ \vdots & & & \vdots \\ 0 & 0 & \cdots & -1 \end{Vmatrix}.$$

If the bases $\mathbf{e}_1, \mathbf{e}_2, \ldots, \mathbf{e}_n$ and $\bar{\mathbf{e}}_1, \bar{\mathbf{e}}_2, \ldots, \bar{\mathbf{e}}_n$ are related as in §IIA.1, the components $L^m{}_k$ and $\bar{L}^p{}_q$ of $\mathbf{L}$ relative to them are determined uniquely in terms of each other by the following transformation law:

$$\bar{L}^p{}_q = \bar{A}^p_r A^s_q L^r{}_s, \qquad L^r{}_s = \bar{A}^r_k A^m_s \bar{L}^k{}_m,$$

where the scalars A^q_p and $\bar{A}^k_m$ are the coefficients defining the change of basis, introduced in §A1.

The set of all tensors over a vector space of dimension n is itself a vector space of dimension n^2 under the operations of addition and scalar multiplication already introduced. The vector $\mathbf{0}$ is simply the tensor $\mathbf{0}$.

Over the vector space of dimension n^2 so obtained, we may consider linear transformations in just the same way as before. If M is such a tensor, its *components* $M^h{}_f{}^k{}_g$ relative to the basis $\mathbf{e}_1, \mathbf{e}_2, \ldots, \mathbf{e}_m$ can be determined from definitions already given. Under change of basis those components transform as follows:

$$\bar{M}^p{}_q{}^r{}_s = \bar{A}^p_h A^f_q \bar{A}^r_k A^g_s M^h{}_f{}^k{}_g.$$

Rules of this kind may be used, alternatively, to define tensors of higher order in terms of their components.

Tensors of order higher than two are used in this book only in certain special contexts. The considerations concerning them may be understood either abstractly or in terms of components, as the reader prefers.

For the result of operating with a fourth-order tensor K upon a second-order tensor $\mathbf{L}$ we use the special notation $\mathsf{K}[\mathbf{L}]$. The components of $\mathsf{K}[\mathbf{L}]$ are $K^k{}_m{}^p{}_q L_p{}^q$.

4. Determinant of a Tensor

The determinant of the matrix whose components are $L^p{}_q$ may be defined as follows:

$$\det \|L^p{}_q\| \equiv \varepsilon^{k_1 k_2 \ldots k_n} L^1{}_{k_1} L^2{}_{k_2} \cdots L^n{}_{k_n},$$

$\varepsilon^{k_1 k_2 \dots k_q}$ having the value 1 if $k_1, k_2, \dots, k_q$ are obtained from 1, 2, …, q by an even permutation; -1, if by an odd permutation; and 0 otherwise. If $\mathbf{L}$ is a tensor, the determinants of its matrices of components $\|L^p{}_q\|$ all have a common value, irrespective of the choice of basis $\mathbf{e}_1, \mathbf{e}_2, \dots, \mathbf{e}_n$ used to define those components. This common value det $\mathbf{L}$ is called the *determinant* of the tensor:

$$\det \mathbf{L} \equiv \det \|L^p{}_q\|.$$

It follows that

$$\det(\mathbf{LM}) = (\det \mathbf{L})(\det \mathbf{M}) = \det(\mathbf{ML}), \qquad \det(a\mathbf{L}) = a^n \det \mathbf{L},$$

and of course $\det \mathbf{1} = 1$, $\det(-\mathbf{1}) = (-1)^n$.

A tensor $\mathbf{L}$ is invertible if and only if $\det \mathbf{L} \neq 0$. The invertible tensors constitute a group under multiplication.

A tensor $\mathbf{L}$ such that $\det \mathbf{L} = \pm 1$ is *unimodular*. The unimodular tensors constitute a subgroup of the group of invertible tensors. This group is called the *unimodular group* $\mathscr{U}$.

5. *Inner-Product Spaces*

The vector spaces encountered in this book are endowed with an *inner product*, denoted by a dot. The *magnitude* $|\mathbf{u}|$ of a vector is defined in terms of the inner product:

$$|\mathbf{u}| \equiv \sqrt{\mathbf{u} \cdot \mathbf{u}}.$$

The elements $\mathbf{u}$ and $\mathbf{v}$ are *orthogonal* or *perpendicular* if

$$\mathbf{u} \cdot \mathbf{v} = 0.$$

The only vector orthogonal to all vectors is $\mathbf{0}$. In fact, if $\mathbf{u}$ is such that $\mathbf{u} \cdot \mathbf{v}$ is bounded above for all $\mathbf{v}$, then $\mathbf{u} = \mathbf{0}$. Also

$$|\mathbf{u} \cdot \mathbf{v}| \leqq |\mathbf{u}||\mathbf{v}|,$$

$$|\mathbf{u} + \mathbf{v}| \leqq |\mathbf{u}| + |\mathbf{v}|.$$

In the former inequality, which is called *Cauchy's inequality*, the sign $=$ is valid if and only if $\mathbf{u}$ and $\mathbf{v}$ are linearly dependent.

The set of all vectors perpendicular to a given set of vectors forms a subspace. It is called the *orthogonal complement* of the subspace spanned by the given set. The vector space itself is the *direct sum* of any of its subspaces and the orthogonal complement of that subspace. This statement means that if $\mathbf{u}$ is any vector, it can be expressed as the sum of a uniquely determined vector from any desired subspace and another vector, also uniquely determined, from the orthogonal complement of that subspace.

If g is a linear scalar-valued function of vectors whose domain is the whole vector space, there is one and only one vector $\mathbf{f}$ such that

$$g(\mathbf{u}) = \mathbf{f} \cdot \mathbf{u}.$$

This statement is the *representation theorem* for linear scalar-valued functions.

If $\mathbf{e}_1, \mathbf{e}_2, \ldots, \mathbf{e}_n$ is a basis, another basis $\mathbf{e}^1, \mathbf{e}^2, \ldots, \mathbf{e}^n$ is determined uniquely by the conditions

$$\mathbf{e}_q \cdot \mathbf{e}^k = \delta_q^k.$$

The basis $\mathbf{e}^1, \mathbf{e}^2, \ldots, \mathbf{e}^n$ is *reciprocal* to the original one. The conditions

$$\mathbf{u} = u^p \mathbf{e}_p \qquad \text{and} \qquad \mathbf{u} = u_r \mathbf{e}^r$$

are equivalent, respectively, to

$$u^k = \mathbf{e}^k \cdot \mathbf{u}, \qquad u_k = \mathbf{e}_k \cdot \mathbf{u}.$$

The components u^k are called *contravariant*, while the components u_k are called *covariant*, both relative to the basis $\mathbf{e}_1, \mathbf{e}_2, \ldots, \mathbf{e}_n$. Expressions for $\mathbf{u} \cdot \mathbf{v}$ in terms of components are

$$\mathbf{u} \cdot \mathbf{v} = u^q v_q = u_p v^p.$$

A basis $\mathbf{e}_1, \mathbf{e}_2, \ldots, \mathbf{e}_n$ is *orthonormal* if

$$\mathbf{e}_q \cdot \mathbf{e}_k = \delta_{qk} \equiv \begin{cases} 1 & \text{if } q = k \\ 0 & \text{if } q \neq k. \end{cases}$$

A basis is orthonormal if and only if it is its own reciprocal. Corresponding contravariant and covariant components relative to an orthonormal basis equal one another, so when an orthonormal basis is used, one speaks simply of "components."

A familiar example of an n-dimensional inner-product space is the *Cartesian space* $\mathscr{R}_n$, the vectors of which are lists of n real numbers v_k:

$$\mathbf{v} \equiv (v_1, v_2, \ldots, v_n).$$

The *standard basis* is defined as follows:

$$\mathbf{e}_k \equiv (0, 0, \ldots, 0, 1, 0, \ldots, 0),$$

the 1 being the kth entry. Every inner-product space of dimension n is isomorphic to $\mathscr{R}_n$, but usually a conceptual argument is clearer if it does not have recourse to this fact.

6. *Tensor Products. Tensors of Orders Greater than 2*

If **a** and **b** are vectors, their *tensor product* is the tensor $\mathbf{a} \otimes \mathbf{b}$ such that

$$(\mathbf{a} \otimes \mathbf{b})\mathbf{u} = (\mathbf{u} \cdot \mathbf{b})\mathbf{a} \qquad \forall \mathbf{u} \in \mathscr{V}.$$

In components,

$$(\mathbf{a} \otimes \mathbf{b})^k{}_m = a^k b_m.$$

If $\mathbf{e}_1, \mathbf{e}_2, \ldots, \mathbf{e}_n$ and $\mathbf{f}_1, \mathbf{f}_2, \ldots, \mathbf{f}_n$ are bases of the vector space, the set of tensor products

$$\mathbf{e}_q \otimes \mathbf{f}_k$$

form a basis for the space of tensors:

$$\mathbf{L} = L^{qk}\mathbf{e}_q \otimes \mathbf{f}_k.$$

The scalars L^{qk} are the *contravariant components* of **L** with respect to the basis. Commonly $\mathbf{f}_k$ is chosen for $\mathbf{e}_k$. Then

$$\mathbf{L} = L^{qk}\mathbf{e}_q \otimes \mathbf{e}_k = L^q{}_k\mathbf{e}_q \otimes \mathbf{e}^k = L_{qk}\mathbf{e}^q \otimes \mathbf{e}^k = L_q{}^k\mathbf{e}^q \otimes \mathbf{e}_k.$$

The scalars $L^r{}_s$ here are the same as those denoted previously by the same symbol and called simply "components" of **L** relative to $\mathbf{e}_1, \mathbf{e}_2, \ldots, \mathbf{e}_n$. They are called also *mixed components* relative to that basis and its reciprocal. In the same terms, the $L_u{}^v$ are the mixed components relative to the basis $\mathbf{e}^1, \mathbf{e}^2, \ldots, \mathbf{e}^n$ and its reciprocal. The scalars L^{rw} and L_{hm} are the *contravariant* and *covariant* components, respectively, of **L** relative to the basis $\mathbf{e}_1, \mathbf{e}_2, \ldots, \mathbf{e}_n$. If the basis is orthonormal, then $L^{qk} = L^q{}_k = L_k{}^q = L_{qk}$.

We note that

$$\mathbf{1} = \mathbf{e}_k \otimes \mathbf{e}^k = \mathbf{e}^k \otimes \mathbf{e}_k.$$

The introduction of tensor products affords another method of defining tensors. For example, if we use the symbol $\mathbf{a} \otimes \mathbf{b} \otimes \mathbf{c}$ to denote that linear mapping of the given vector space into (second-order) tensors such that

$$(\mathbf{a} \otimes \mathbf{b} \otimes \mathbf{c})\mathbf{d} = (\mathbf{c} \cdot \mathbf{d})(\mathbf{a} \otimes \mathbf{b}),$$

we can prove that the products $\mathbf{e}_k \otimes \mathbf{e}_p \otimes \mathbf{e}_q$ of elements of a basis for the given vector space form a basis for the set of such transformations. That is, if N is any linear mapping of the given vector space into (second-order) tensors, it may be expressed in the form

$$\mathsf{N} = N^{kpq}\mathbf{e}_k \otimes \mathbf{e}_p \otimes \mathbf{e}_q,$$

and its *contravariant components* N^{qrk} obey the transformation law

$$\overline{N}^{qrk} = \overline{A}^q_u \overline{A}^r_m \overline{A}^k_p N^{ump}.$$

The tensors so defined are of *third* order, but the method serves to define tensors of any order.

7. *Transposition. Symmetric and Skew Tensors*

If $B(\mathbf{u}, \mathbf{v})$ is a scalar-valued bilinear function defined for all vectors $\mathbf{u}$ and $\mathbf{v}$, there is a unique tensor $\mathbf{L}$ such that

$$B(\mathbf{u}, \mathbf{v}) = \mathbf{u} \cdot \mathbf{L}\mathbf{v}.$$

This statement is the *representation theorem for bilinear functions.* If $\mathbf{L}$ is determined in this way by B, we can determine another tensor $\mathbf{L}^{\mathrm{T}}$, called the *transpose* of $\mathbf{L}$, by the requirement that

$$B(\mathbf{v}, \mathbf{u}) = \mathbf{u} \cdot \mathbf{L}^{\mathrm{T}}\mathbf{v}.$$

Then

$$\begin{aligned} (\mathbf{L} + \mathbf{M})^{\mathrm{T}} &= \mathbf{L}^{\mathrm{T}} + \mathbf{M}^{\mathrm{T}}, \\ (\mathbf{L}\mathbf{M})^{\mathrm{T}} &= \mathbf{M}^{\mathrm{T}}\mathbf{L}^{\mathrm{T}}, \\ (\mathbf{L}^{\mathrm{T}})^{\mathrm{T}} &= \mathbf{L}, \\ (\mathbf{a} \otimes \mathbf{b})^{\mathrm{T}} &= \mathbf{b} \otimes \mathbf{a}. \end{aligned}$$

If $\mathbf{L}$ is invertible, then so is $\mathbf{L}^{\mathrm{T}}$, and

$$(\mathbf{L}^{\mathrm{T}})^{-1} = (\mathbf{L}^{-1})^{\mathrm{T}}.$$

In mixed components,

$$(L^{\mathrm{T}})^{q}{}_{k} = L_{k}{}^{q}.$$

In terms of matrices, $[\mathbf{L}^{\mathrm{T}}] = [\mathbf{L}]^{\mathrm{T}}$ if the components are taken relative to an orthonormal basis, but otherwise in general $[\mathbf{L}^{\mathrm{T}}] \neq [\mathbf{L}]^{\mathrm{T}}$. Of course

$$(L^{\mathrm{T}})^{kq} = L^{qk}, \qquad (L^{\mathrm{T}})_{qk} = L_{kq};$$

that is, the matrices of contravariant and covariant components of $\mathbf{L}^{\mathrm{T}}$ are the transposes of the respective matrices of $\mathbf{L}$.

Tensors $\mathbf{S}$ and $\mathbf{W}$ such that

$$\mathbf{S} = \mathbf{S}^{\mathrm{T}}, \qquad \mathbf{W} = -\mathbf{W}^{\mathrm{T}},$$

are called *symmetric* and *skew*, respectively. The conditions are expressed as follows in terms of components:

$$\begin{aligned} S_{qk} &= S_{kq}, & S^{qk} &= S^{kq}, & S_{q}{}^{k} &= S^{k}{}_{q}, \\ W_{qk} &= -W_{kq}, & W^{qk} &= -W^{kq}, & W_{q}{}^{k} &= -W^{k}{}_{q}. \end{aligned}$$

The set of all symmetric tensors is a $\frac{1}{2}n(n+1)$-dimensional subspace of the space of tensors; the set of skew tensors, a $\frac{1}{2}n(n-1)$-dimensional subspace. Bases for these two subspaces are formed by the following sets of products of the vectors of a basis $\mathbf{e}_1, \mathbf{e}_2, \ldots, \mathbf{e}_n$:

$$\mathbf{e}_k \otimes \mathbf{e}_m + \mathbf{e}_m \otimes \mathbf{e}_k, \qquad k \leqq m,$$

and

$$\mathbf{e}_k \wedge \mathbf{e}_m, \qquad k < m,$$

the *wedge product* or *exterior product* being the skew tensor defined as follows:

$$\mathbf{a} \wedge \mathbf{b} \equiv \mathbf{a} \otimes \mathbf{b} - \mathbf{b} \otimes \mathbf{a}.$$

Any tensor $\mathbf{L}$ has a unique representation as the sum of a symmetric and a skew tensor:

$$\mathbf{L} = \mathbf{S} + \mathbf{W}, \qquad \text{where} \quad \mathbf{S} = \tfrac{1}{2}(\mathbf{L} + \mathbf{L}^{\mathrm{T}}), \ \mathbf{W} = \tfrac{1}{2}(\mathbf{L} - \mathbf{L}^{\mathrm{T}}).$$

If $\mathbf{S}$ is either symmetric or skew, $\mathbf{S}^2$ and $\mathbf{S}$ have the same nullspace. If $\mathbf{S}$ and $\mathbf{U}$ are either both symmetric or both skew, $(\mathbf{SU})^{\mathrm{T}} = \mathbf{US}$. Hence $\mathbf{S}$ and $\mathbf{U}$ commute if and only if $\mathbf{SU}$ is symmetric. If $\mathbf{W}$ is skew and dim $\mathscr{V}$ is odd, det $\mathbf{W} = 0$.

8. Orthogonal Tensors

A mapping $\mathbf{Q}$ of an inner-product space onto itself is *orthogonal* if it preserves the inner product:

$$\mathbf{Q}(\mathbf{u}) \cdot \mathbf{Q}(\mathbf{v}) = \mathbf{u} \cdot \mathbf{v}.$$

This condition is satisfied if and only if $\mathbf{Q}$ is an invertible tensor such that

$$\mathbf{Q}^{-1} = \mathbf{Q}^{\mathrm{T}}.$$

Hence

$$\det \mathbf{Q} = \pm 1.$$

If det $\mathbf{Q} = 1$, the orthogonal tensor $\mathbf{Q}$ is called *proper*, or equivalently, a *rotation*. The central inversion $-\mathbf{1}$ is orthogonal; it is a rotation if and only if n is even. If n is odd, either $\mathbf{Q}$ or $-\mathbf{Q}$ is a rotation, while the other is the product of a rotation by the central inversion.

The orthogonal tensors constitute a proper subgroup of $\mathscr{u}$ called the (full) *orthogonal group* $\mathscr{o}$, and the rotations form a proper subgroup of the orthogonal group.

Some special properties of orthogonal tensors over a 3-dimensional space are listed below in §14 of this appendix.

9. Trace, Inner Product of Tensors

The *trace* tr **A** of the tensor **A** is defined uniquely by the following two requirements: It is a scalar-valued linear function whose domain is the set of all tensors, and

$$\operatorname{tr}(\mathbf{u} \otimes \mathbf{v}) = \mathbf{u} \cdot \mathbf{v}.$$

Hence

$$\operatorname{tr} \mathbf{A} = A^k{}_k = A_k{}^k = \operatorname{tr} \mathbf{A}^{\mathrm{T}};$$

that is, the trace of a tensor is the trace of the matrix of either of its arrays of mixed components. Of course $\operatorname{tr} \mathbf{1} = n$, $\operatorname{tr} \mathbf{0} = 0$. If **W** is skew, $\operatorname{tr} \mathbf{W} = 0$.

The *inner product* $\mathbf{A} \cdot \mathbf{B}$ of the tensors **A** and **B** is defined as follows:

$$\mathbf{A} \cdot \mathbf{B} \equiv \operatorname{tr}(\mathbf{A}\mathbf{B}^{\mathrm{T}}) = \mathbf{B} \cdot \mathbf{A}.$$

With this definition the set of all tensors **A**, regarded as a vector space of dimension n^2, becomes an inner-product space. The *magnitude* $|\mathbf{A}|$ of the tensor **A** is defined from the inner product in the usual way:

$$|\mathbf{A}| \equiv \sqrt{\mathbf{A} \cdot \mathbf{A}} = \sqrt{\operatorname{tr} \mathbf{A}\mathbf{A}^{\mathrm{T}}}.$$

In components,

$$\mathbf{A} \cdot \mathbf{B} = A^k{}_q B^q{}_k,$$
$$|\mathbf{A}| = \sqrt{A^k{}_q A^q{}_k}.$$

10. Invariant Subspaces, Projections, Proper Vectors, Proper Numbers

If the tensor **A** maps a certain subspace into itself, that subspace is *invariant* under **A**. Every tensor **A** has invariant subspaces, namely, the whole vector space, the subspace $\{\mathbf{0}\}$, the range of **A**, and the nullspace of **A**. In addition **A** may have other invariant subspaces.

A tensor **E** is called a *projection* if it is idempotent: $\mathbf{E}^2 = \mathbf{E}$. A projection is called a *perpendicular projection* if also it is symmetric: $\mathbf{E}^{\mathrm{T}} = \mathbf{E}$. If **E** is a projection, there is a basis relative to which

$$[\mathbf{E}] = \left\| \begin{array}{cccccccc} 1 & 0 & 0 & & \cdots & & & 0 \\ 0 & 1 & & & & & & \vdots \\ & & \ddots & & & & & \\ & & & 1 & 0 & 0 & & 0 \\ \vdots & & & 0 & 0 & & & \\ & & & \vdots & & 0 & & \vdots \\ & & & & & & \ddots & \\ 0 & & \cdots & 0 & & & & 0 \end{array} \right\|,$$

the number of 1s being equal to the dimension of the invariant subspace of **E**. If **E** is a perpendicular projection, the basis may be chosen to be an orthonormal one.

If **e** is a unit vector, any vector **v** has a unique decomposition as the sum of a vector $\mathbf{P_e v}$ parallel to **e** and another $\mathbf{P^e v}$ perpendicular to it:

$$\mathbf{v} = \mathbf{P_e v} + \mathbf{P^e v}, \qquad \mathbf{e} \cdot \mathbf{P^e v} = 0.$$

Both $\mathbf{P_e}$ and $\mathbf{P^e}$ are perpendicular projections, and

$$\mathbf{P_e} = \mathbf{e} \otimes \mathbf{e}, \qquad \mathbf{P^e} = \mathbf{1} - \mathbf{e} \otimes \mathbf{e}.$$

$\mathbf{P_e}$ is the *projection onto the span* of **e**; $\mathbf{P^e}$ is the *projection onto the plane normal to* **e**. The *reflection* $\mathbf{R_e}$ *across the plane normal to* **e** is the orthogonal tensor defined as follows:

$$\mathbf{R_e} \equiv -\mathbf{P_e} + \mathbf{P^e} = \mathbf{1} - 2\mathbf{e} \otimes \mathbf{e}.$$

Therefore, **v** is parallel to **e** if and only if $\mathbf{R_e v} = -\mathbf{v}$; perpendicular to **e** if and only if $\mathbf{R_e v} = \mathbf{v}$.

If x is any scalar, the nullspace of $\mathbf{A} - x\mathbf{1}$ is an invariant subspace of **A**. It is called the *proper space* of **A** corresponding to x, and its dimension is the *multiplicity* of x for **A**. The scalar x is a *proper number* of **A** if any one, and hence all, of the following equivalent conditions holds:

1. There is a non-zero vector **u** such that

$$\mathbf{Au} = x\mathbf{u}.$$

2. The proper subspace of **A** corresponding to x contains a non-zero vector.
3. The multiplicity of x for **A** is not 0.

The elements of the proper space are the *proper vectors* corresponding to that proper number. A proper number is *simple* if its multiplicity is 1; that is, if its proper space is 1-dimensional. The set of all proper numbers of **A** is called the *spectrum* of **A**. By definition, the scalars constituting the spectrum are distinct.

The *characteristic equation* of **A** is

$$x^n - I_1 x^{n-1} + I_2 x^{n-2} - \cdots \pm I_n = 0,$$

the signs being alternately $-$ and $+$, and the *principal invariants* I_k of **A** being defined as follows:

$$I_k \equiv \frac{1}{k!} \delta^{s_1 s_2 \ldots s_k}_{m_1 m_2 \ldots m_k} A^{m_1}{}_{s_1} A^{m_2}{}_{s_2} \cdots A^{m_k}{}_{s_k}, \qquad k = 1, 2, \ldots, n.$$

Here $\delta^{s_1 s_2 \ldots s_k}_{m_1 m_2 \ldots m_k}$ is defined as 0 if any superscript or subscript is repeated, or if the subscripts fail to be the same numbers as the superscripts; otherwise, as ± 1 according as an even or odd permutation is needed to bring the subscripts into the same order as the superscripts. While this definition of the I_k seems to depend upon a basis, the value of I_k so obtained is the same for all bases. Clearly

$$I_1 = \text{tr}\,\mathbf{A},$$
$$I_n = \det \mathbf{A}.$$

There are uniquely determined complex numbers $a_1, a_2, \ldots, a_n$ such that

$$x^n - I_1 x^{n-1} + \cdots \pm I_n = \prod_{k=1}^{n} (x - a_k).$$

Thus the characteristic equation of **A**, regarded as an equation in the complex field, has exactly n roots $a_1, a_2, \ldots, a_n$. They are called the *latent roots* of **A**. If q of the numbers $a_1, a_2, \ldots, a_n$ are equal, their common value is said to be a latent root of *algebraic multiplicity* q. Since the principal invariants of **A** are real, its non-real roots occur in complex-conjugate pairs. The value of I_k is the sum of the products of the latent roots a_j taken k at a time. Thus, for example, if $n = 3$,

$$I_1 = a_1 + a_2 + a_3,$$
$$I_2 = a_2 a_3 + a_3 a_1 + a_1 a_2,$$
$$I_3 = a_1 a_2 a_3.$$

Every proper number of **A** is a latent root, and every real latent root is a proper number. If $\mathbf{A}\mathbf{A}^{\mathrm{T}} = \mathbf{A}^{\mathrm{T}}\mathbf{A}$, the multiplicity of a proper number of **A** is the same as its algebraic multiplicity as a latent root. Such is the case, therefore, for tensors that are symmetric, skew, or orthogonal. If n is odd, **A** has at least one proper number, but if n is even, possibly **A** has no proper numbers at all.

The *Hamilton–Cayley Theorem* states that the tensor **A** satisfies an equation having the same form as the characteristic equation:

$$\mathbf{A}^n - I_1 \mathbf{A}^{n-1} + \cdots \pm I_n \mathbf{1} = \mathbf{0}.$$

It is possible, of course, that **A** may satisfy also a polynomial equation of degree less than n.

11. Spectral Decomposition of Symmetric Tensors

Every symmetric tensor **S** has at least one proper number. In fact, the least and greatest proper numbers of **S** are the least and greatest values, respectively, of $\mathbf{u} \cdot \mathbf{S}\mathbf{u}$ as $\mathbf{u}$ ranges over all unit vectors. Every latent root of a symmetric tensor is real and hence is a proper number. The proper spaces of a symmetric tensor are mutually orthogonal. Any vector may be expressed as a linear combination of vectors, each of which belongs to one (and of course, if it is not the vector $\mathbf{0}$, to only one) of the proper spaces of $\mathbf{S}$.

If $s_1, s_2, \ldots, s_p$ are the proper numbers of $\mathbf{S}$, then there is a unique set of perpendicular projections $\mathbf{E}_1, \mathbf{E}_2, \ldots, \mathbf{E}_p$ such that

$$\mathbf{E}_k \mathbf{E}_l = \mathbf{0} \qquad \text{if} \quad k \neq l,$$

$$\sum_{k=1}^{p} \mathbf{E}_k = \mathbf{1},$$

and

$$\mathbf{S} = \sum_{k=1}^{p} s_k \mathbf{E}_k .$$

Hence there is at least one orthonormal basis $(\mathbf{e}_1, \mathbf{e}_2, \ldots, \mathbf{e}_n)$, each member of which is a proper vector of $\mathbf{S}$:

$$\mathbf{S}\mathbf{e}_q = s_q \mathbf{e}_q, \qquad q = 1, 2, \ldots, n,$$

s_q being a proper number of $\mathbf{S}$, repeated a number of times equal to its multiplicity. Such a basis is called *principal.* The matrix of components of $\mathbf{S}$ relative to this basis is diagonal. Indeed,

$$[\mathbf{S}] = \begin{Vmatrix} s_1 & 0 & \cdots & 0 \\ 0 & s_2 & & \\ \vdots & & \ddots & \\ 0 & \cdots & & s_n \end{Vmatrix},$$

where again each proper number occurs a number of times equal to its multiplicity. With the same convention of multiplicity,

$$\mathbf{S} = \sum_{k=1}^{n} s_k \mathbf{P}_{\mathbf{e}_k} .$$

In order that two symmetric tensors $\mathbf{S}$ and $\mathbf{T}$ have the same proper numbers, each with the same multiplicity, it is necessary and sufficient that there be an orthogonal tensor $\mathbf{Q}$ such that

$$\mathbf{T} = \mathbf{Q}\mathbf{S}\mathbf{Q}^{\mathsf{T}}.$$

If this condition holds, the proper spaces of $\mathbf{T}$ are the images under $\mathbf{Q}$ of the proper spaces of $\mathbf{S}$. An orthogonal tensor $\mathbf{Q}$ commutes with the symmetric tensor $\mathbf{S}$ if and only if the proper spaces of $\mathbf{S}$ are invariant subspaces of $\mathbf{Q}$.

12. Positive Tensors

A tensor $\mathbf{S}$ is *positive* if

$$\mathbf{u} \cdot \mathbf{S}\mathbf{u} > 0 \qquad \text{unless} \qquad \mathbf{u} = \mathbf{0};$$

non-negative if

$$\mathbf{u} \cdot \mathbf{S}\mathbf{u} \geqq 0 \qquad \text{for all vectors } \mathbf{u}.$$

If $-\mathbf{S}$ is positive, $\mathbf{S}$ is *negative*; if $-\mathbf{S}$ is non-negative, $\mathbf{S}$ is *non-positive*. We abbreviate these terms by the notations $\mathbf{S} > 0$, $\mathbf{S} \geqq 0$, $\mathbf{S} < 0$, $\mathbf{S} \leqq 0$, respectively. If $\mathbf{L}$ is any tensor, $\mathbf{L}\mathbf{L}^{\mathrm{T}} \geqq 0$ and $\mathbf{L}^{\mathrm{T}}\mathbf{L} \geqq 0$. If $\mathbf{L}$ is invertible, $\mathbf{L}\mathbf{L}^{\mathrm{T}} > 0$ and $\mathbf{L}^{\mathrm{T}}\mathbf{L} > 0$. If $\mathbf{L}$ is symmetric, $\mathbf{L} > 0$ if and only if all of its proper numbers are positive, and $\mathbf{L} \geqq 0$ if and only if all of its proper numbers are non-negative.

If $\mathbf{S} \geqq 0$, there is a unique non-negative $\mathbf{T}$ such that $\mathbf{T}^2 = \mathbf{S}$. We denote this tensor by $\sqrt{\mathbf{S}}$ and call it the *square root* of $\mathbf{S}$. The proper numbers of $\sqrt{\mathbf{S}}$ are the non-negative square roots of those of $\mathbf{S}$.

13. Polar Decomposition

If $\mathbf{L}$ is an invertible tensor, then there are unique positive symmetric tensors $\mathbf{S}$ and $\mathbf{T}$ and a unique orthogonal tensor $\mathbf{Q}$ such that

$$\mathbf{L} = \mathbf{Q}\mathbf{S} = \mathbf{T}\mathbf{Q}.$$

$\mathbf{T}$ and $\mathbf{S}$ determine each other as follows if $\mathbf{Q}$ is known:

$$\mathbf{T} = \mathbf{Q}\mathbf{S}\mathbf{Q}^{\mathrm{T}}.$$

14. Structure of Orthogonal Tensors over a 3-Dimensional Vector Space

In this book we need to analyse orthogonal tensors only when $\dim \mathscr{V} = 3$. Then the central inversion $-\mathbf{1}$ is orthogonal but not a rotation. Every orthogonal tensor $\mathbf{Q}$ is either a rotation $\mathbf{R}$ or the product $-\mathbf{R}$ of a rotation $\mathbf{R}$ by $-\mathbf{1}$. Thus the structure of orthogonal tensors is determined by the structure of rotations.

The latent roots of $\mathbf{R}$ are 1, $e^{i\theta}$, $e^{-i\theta}$; the angle θ, which is rendered unique by the requirement that $0 \leqq \theta < 2\pi$, is the *angle of rotation*.

If $\mathbf{R} \neq \mathbf{1}$, the proper space that corresponds to the proper number 1 is 1-dimensional. It is called the *axis* of the rotation $\mathbf{R}$. If $\mathbf{e}$ denotes either of the two unit vectors in the axis, there is a right-handed orthonormal basis $\mathbf{e}_1$, $\mathbf{e}_2$, $\mathbf{e}$ such that

$$\mathbf{R}\mathbf{e}_1 = \cos\theta\,\mathbf{e}_1 + \sin\theta\,\mathbf{e}_2,$$
$$\mathbf{R}\mathbf{e}_2 = -\sin\theta\,\mathbf{e}_1 + \cos\theta\,\mathbf{e}_2,$$
$$\mathbf{R}\mathbf{e} = \mathbf{e},$$

If $\mathbf{R} = \mathbf{1}$, then $\theta = 0$, and the preceding formula holds for every basis $\mathbf{e}_1$, $\mathbf{e}_2$, $\mathbf{e}$. The matrix of $\mathbf{R}$ with respect to the basis $\mathbf{e}_1$, $\mathbf{e}_2$, $\mathbf{e}$ is given by

$$[\mathbf{R}] = \begin{Vmatrix} \cos\theta & -\sin\theta & 0 \\ \sin\theta & \cos\theta & 0 \\ 0 & 0 & 1 \end{Vmatrix}.$$

In general co-ordinates

$$R^k{}_m = \cos\theta\,\delta^k_m + (1 - \cos\theta)e^k e_m - \sin\theta\,\varepsilon^k{}_{mp}\,e^p.$$

If $\mathbf{Q} = -\mathbf{R}$, the axis of $\mathbf{R}$ is called the *axis* of $\mathbf{Q}$, and the angle of rotation of $\mathbf{R}$ is called the *angle of rotation* of $\mathbf{Q}$.

$\mathbf{R} = \mathbf{R}^{\mathrm{T}}$ if and only if $\theta = 0$ or π.

15. Structure of Skew Tensors over a 3-Dimensional Vector Space. 3-Dimensional Vector Algebra[1]

In this book we need to analyse skew tensors only if $\dim \mathscr{V} = 3$. Then the nullspace of a non-null skew tensor $\mathbf{W}$, namely, the subspace of vectors $\mathbf{n}$ such that

$$\mathbf{W}\mathbf{n} = \mathbf{0},$$

is 1-dimensional; it is called the *axis* of $\mathbf{W}$. If $\mathbf{W} \neq \mathbf{0}$, its only proper number is 0, and its axis is its only invariant subspace. Two non-null skew tensors have the same axis if and only if they are proportional to one another.

Let $\mathbf{n}$ be one of the two unit vectors lying on the axis of a non-null skew tensor $\mathbf{W}$, and let $\mathbf{e}$ be normal to $\mathbf{n}$. Then of the two unit vectors normal to the span of $\mathbf{e}$ and $\mathbf{n}$ we may choose one, say $\mathbf{f}$, such that

$$\mathbf{W} = \tfrac{1}{2}|\mathbf{W}|\mathbf{e} \wedge \mathbf{f}.$$

Equivalently, there is an orthonormal basis such that

$$[\mathbf{W}] = \begin{Vmatrix} 0 & \tfrac{1}{2}|\mathbf{W}| & 0 \\ -\tfrac{1}{2}|\mathbf{W}| & 0 & 0 \\ 0 & 0 & 0 \end{Vmatrix}.$$

[1] *Cf.* §§7.16, 8.15, and 8.16 of W. H. GREUB, *Linear Algebra*, 3rd ed., Berlin and New York, Springer-Verlag, 1967.

Two orthonormal bases $\{\mathbf{e}_k\}$ and $\{\bar{\mathbf{e}}_k\}$ are said to have *the same orientation* if they are obtainable from one another by a rotation: $\bar{\mathbf{e}}_k = \mathbf{R}\mathbf{e}_k$, $k = 1, 2, 3$. Since every orthogonal transformation is either a rotation or the negative of one, there are exactly two distinct classes of bases having the same orientation. In 3-dimensional vector algebra one of these is set down and fixed. Of the two possible isomorphisms between skew tensors and vectors, one is specified by use of a particular orthonormal basis. "The Gibbsian cross" $\mathbf{T}_\times$ of a tensor $\mathbf{T}$ is the vector defined as follows in terms of components with respect to any such basis:

$$T_{\times 3} \equiv T_{12} - T_{21}, \qquad T_{\times 1} \equiv T_{23} - T_{32}, \qquad T_{\times 2} \equiv T_{31} - T_{13}.$$

If $\mathbf{S}$ is skew, then

$$S_{\times 3} = 2S_{12}, \qquad S_{\times 1} = 2S_{23}, \qquad S_{\times 2} = 2S_{31}.$$

The *cross product* $\mathbf{u} \times \mathbf{w}$ of two vectors $\mathbf{u}$ and $\mathbf{w}$ is defined thus:

$$\mathbf{u} \times \mathbf{w} \equiv \tfrac{1}{2}(\mathbf{u} \wedge \mathbf{w})_\times,$$

so that

$$(\mathbf{u} \times \mathbf{w})_3 = u_1 w_2 - u_2 w_1, \qquad \text{etc.}$$

and

$$|\mathbf{u} \wedge \mathbf{w}| = \sqrt{2}\,|\mathbf{u} \times \mathbf{w}|.$$

Also

$$2\mathbf{S}\mathbf{u} = -\mathbf{S}_\times \times \mathbf{u}, \qquad \mathbf{S} \cdot (\mathbf{u} \wedge \mathbf{v}) = -\mathbf{S}_\times \cdot (\mathbf{u} \times \mathbf{v}),$$

and if $\mathbf{S}$ and $\mathbf{T}$ are both skew,

$$\mathbf{S} \cdot \mathbf{T} = \tfrac{1}{2}\mathbf{S}_\times \cdot \mathbf{T}_\times, \qquad (\mathbf{S}\mathbf{T})_\times = \tfrac{1}{2}(\mathbf{S}\mathbf{T} - \mathbf{T}\mathbf{S})_\times = -\tfrac{1}{4}\mathbf{S}_\times \times \mathbf{T}_\times.$$

The first of these relations shows that $\mathbf{S} \cdot \mathbf{T} = 0$ if and only if the nullspaces of $\mathbf{S}$ and $\mathbf{T}$ are perpendicular; the second shows that $\mathbf{S}$ and $\mathbf{T}$ commute if and only if the nullspace of one of them contains the nullspace of the other.

B. Geometry

1. Euclidean Point Spaces

While in this book there are allusions to rather general manifolds, the only specific geometry employed is that of Euclidean space.

A set $\mathscr{E}$ of elements $\mathbf{x}$, $\mathbf{y}$ is an n-dimensional *Euclidean point space* or *Euclidean manifold* if it is endowed with a structure defined in reference to an n-dimensional inner-product space $\mathscr{V}$ by the following axioms:

1. Each vector $\mathbf{u} \in \mathscr{V}$ maps $\mathscr{E}$ onto itself
$$\mathbf{u}(\mathbf{x}) \in \mathscr{E} \qquad \forall \mathbf{x} \in \mathscr{E}.$$
2. The composition of the mappings $\mathbf{u}$ and $\mathbf{v}$ is their vector sum:
$$(\mathbf{u} + \mathbf{v})(\mathbf{x}) = \mathbf{u}(\mathbf{v}(\mathbf{x})) \qquad \forall \mathbf{x} \in \mathscr{E}.$$
3. For given $\mathbf{x}$ and $\mathbf{y}$ there is exactly one vector $\mathbf{u}$ such that
$$\mathbf{u}(\mathbf{x}) = \mathbf{y}.$$

The elements of $\mathscr{E}$ are called *points*. The vector space $\mathscr{V}$ is the *translation space* of $\mathscr{E}$, and its elements are called *translations* of $\mathscr{E}$. The translations may be visualized as arrows; if the butt of the arrow $\mathbf{u}$ is put at $\mathbf{x}$, its point distinguishes $\mathbf{u}(\mathbf{x})$. Thus we say that $\mathbf{u}$ *translates* $\mathbf{x}$ into $\mathbf{y}$; of course $-\mathbf{u}$ translates $\mathbf{y}$ into $\mathbf{x}$.

The second axiom, in view of the first, asserts that the result of applying first the translation $\mathbf{v}$ and then the translation $\mathbf{u}$ is the same as the result of applying $\mathbf{u} + \mathbf{v}$ to start with; thus it expresses the *axiom of resultant displacements*, familiar from elementary geometry and mechanics, and it suggests the notation

$$\mathbf{x} + \mathbf{u} \equiv \mathbf{u}(\mathbf{x}).$$

Thus we use the plus sign to denote not only addition of vectors to each other but also addition of vectors to points. Axiom 3 enables us to extend the interpretation by writing $\mathbf{y} - \mathbf{x}$ for the unique vector $\mathbf{u}$ that maps $\mathbf{x}$ into $\mathbf{y}$:

$$\mathbf{y} - \mathbf{x} \equiv \mathbf{u}.$$

Thus the difference of points is defined, and

$$\mathbf{x} + (\mathbf{y} - \mathbf{x}) = \mathbf{y}.$$

Let $\mathscr{U}$ be a subspace of $\mathscr{V}$. By a *flat* of $\mathscr{E}$ *parallel* to $\mathscr{U}$ we shall mean a non-empty set $\mathscr{F}$ of points such that if the difference $\mathbf{x} - \mathbf{y} \in \mathscr{U}$, then either both $\mathbf{x}$ and $\mathbf{y}$ belong to $\mathscr{F}$, or neither does. Two flats of $\mathscr{E}$ that are both parallel to the same subspace of $\mathscr{V}$ are *parallel* to each other. The *dimension* of $\mathscr{F}$ is the dimension of the $\mathscr{U}$ that defines it. A 1-dimensional flat is a *straight line*, while a 2-dimensional flat is a *plane*, and an $(n-1)$-dimensional flat is a *hyperplane*. Of course, if $n = 3$, planes and hyperplanes are the same thing.

Every point $\mathbf{y}$ of $\mathscr{F}$ can be expressed as

$$\mathbf{y} = \mathbf{y}_0 + \mathbf{u},$$

where $\mathbf{y}_0$ is an arbitrarily selected point of $\mathscr{F}$, and $\mathbf{u}$ is some vector in $\mathscr{U}$. In particular, an equation for a straight line is

$$\mathbf{y} = \mathbf{y}_0 + s\mathbf{e},$$

where $\mathbf{e}$ is some vector and where s runs from $-\infty$ to ∞.

2. *Distance, Congruence*

The *distance* between the points $\mathbf{x}$ and $\mathbf{y}$ is the magnitude of the vector $\mathbf{u}$ that translates $\mathbf{x}$ into $\mathbf{y}$, that is,

$$|\mathbf{y} - \mathbf{x}|.$$

It is easy to see that this function of pairs of points satisfies the axioms of a metric, and in particular that it obeys the *triangle axiom*:

$$|\mathbf{x} - \mathbf{y}| + |\mathbf{y} - \mathbf{z}| \geqq |\mathbf{x} - \mathbf{z}|.$$

A mapping $\boldsymbol{\alpha}$ of $\mathscr{E}$ onto itself is called a *congruence* if it preserves distances. The *representation theorem* for congruences asserts that to each congruence $\boldsymbol{\alpha}$ of $\mathscr{E}$ corresponds a unique orthogonal tensor $\mathbf{Q}$ over $\mathscr{V}$ such that

$$\boldsymbol{\alpha}(\mathbf{x}) = \boldsymbol{\alpha}(\mathbf{x}_0) + \mathbf{Q}(\mathbf{x} - \mathbf{x}_0)$$

for each pair of points $\mathbf{x}_0$ and $\mathbf{x}$. Thus each congruence may be regarded as the succession, in either order, of a translation of an arbitrarily selected point and a uniquely determined orthogonal transformation of the vectors that translate that point into the other points of space.

3. *Topology, Figures*

The topology of a Euclidean space is defined in the standard way by the metric $|\mathbf{x} - \mathbf{y}|$. The definitions of spheres, cubes, parallelepipeds, open and closed sets, neighborhoods, interiors, closures, and boundaries as those concepts are used in this book may be found in any text on elementary analysis.

C. Calculus

1. *Limits, Orders*

Euclidean point space has a metric; likewise, the magnitude of a difference of vectors or tensors, $|\mathbf{u} - \mathbf{v}|$ or $|\mathbf{T} - \mathbf{S}|$, serves as a metric. In terms of the topologies defined by these metrics, standard procedure defines continuity, convergence, limits, boundedness, compactness, *etc.*, in the respective spaces. Standard theorems of calculus, such as those on subsequences, Cauchy's criterion, covering theorems, the theorem of the maximum and minimum on a compact set, are easy to extend to $\mathscr{E}$ and to vector spaces.

The *order symbols* O and o are defined as follows for scalar-valued functions of a scalar variable.

If there is a constant K such that

$$|f(x)| < K|g(x)|$$

when x is sufficiently near to a, we write

$$f = O(g) \qquad \text{as} \quad x \to a.$$

If

$$\frac{f(x)}{g(x)} \to 0 \qquad \text{as} \quad x \to a,$$

we write

$$f = o(g) \qquad \text{as} \quad x \to a.$$

For example, $O(1)$ stands for a function that is bounded near a, and $o(1)$ stands for a function that tends to 0 as $x \to a$.

The statement that f is continuous at $x = a$ may be put as follows:

$$f(x) = f(a) + o(1) \qquad \text{as} \quad x \to a.$$

These definitions are easily extended to scalar-valued functions of points, vectors, or tensors. Similar definitions hold for vector-valued and tensor-valued functions. For them, we write $\mathbf{o}$ and $\mathbf{O}$ instead of o and O.

2. *Differentiation*

If $\mathbf{f}$ is a point-valued or vector-valued function of a real variable t, its *derivative* $\dot{\mathbf{f}}(t)$ at t is defined as follows: If there is a vector $\mathbf{g}$ such that

$$\mathbf{f}(t + s) = \mathbf{f}(t) + s\mathbf{g} + \mathbf{o}(s) \qquad \text{as} \quad s \to 0,$$

then $\mathbf{f}$ is *differentiable* at t, and $\mathbf{g}$ is the *derivative* of $\mathbf{f}$ at t. The standard notation for the derivative is

$$\dot{\mathbf{f}}(t) \equiv \mathbf{g}.$$

Thus the derivative $\dot{\mathbf{f}}(t)$ defines a linear function that approximates the function $\mathbf{f}(t + \cdot) - \mathbf{f}(t)$ near $s = 0$. For a tensor-valued function, a similar definition and notation may be used.

There are simple rules for interchanging the order of differentiation and other operations. A few of these, in a notation which may confuse functions with their values, are listed below.

$$(\mathbf{u} \otimes \mathbf{v})^{\cdot} = \dot{\mathbf{u}} \otimes \mathbf{v} + \mathbf{u} \otimes \dot{\mathbf{v}}.$$

$$(\mathbf{LM})^{\cdot} = \dot{\mathbf{L}}\mathbf{M} + \mathbf{L}\dot{\mathbf{M}}.$$

$$(\mathbf{L}^{\mathrm{T}})^{\cdot} = (\dot{\mathbf{L}})^{\mathrm{T}}.$$

$$(\mathbf{L}^{m})^{\cdot} = \sum_{k=1}^{m} \mathbf{L}^{k-1}\dot{\mathbf{L}}\mathbf{L}^{m-k}.$$

$$(\mathbf{L}^{-1})^{\cdot} = -\mathbf{L}^{-1}\dot{\mathbf{L}}\mathbf{L}^{-1}.$$

Of course, for the last rule to hold it is necessary that $\mathbf{L}$ be invertible.

If $\mathbf{Q}$ is a function whose values are orthogonal tensors, then the values of $\dot{\mathbf{Q}}\mathbf{Q}^{\mathrm{T}}$ are skew.

3. Gradients

A function $\mathbf{f}$ that maps points in a Euclidean space $\mathscr{E}_n$ into a vector space $\mathscr{W}_m$ is called a *vector field*. A vector field is said to be *differentiable* at $\mathbf{x}$ if there is a linear mapping $\nabla\mathbf{f}(\mathbf{x})$ of n-dimensional vectors into m-dimensional vectors such that

$$\mathbf{f}(\mathbf{x} + \mathbf{u}) = \mathbf{f}(\mathbf{x}) + \nabla\mathbf{f}(\mathbf{x})(\mathbf{u}) + \mathbf{o}(\mathbf{u}) \qquad \text{as} \quad \mathbf{u} \to \mathbf{0}.$$

The mapping $\nabla\mathbf{f}(\mathbf{x})$ is called the *gradient* of $\mathbf{f}$ at $\mathbf{x}$. Equivalently, the gradient $\nabla\mathbf{f}(\mathbf{x})$, if it exists, is a linear map such that

$$\lim_{a\to 0} \frac{\mathbf{f}(\mathbf{x} + a\mathbf{u}) - \mathbf{f}(\mathbf{x})}{a} = \nabla\mathbf{f}(\mathbf{x})(\mathbf{u}).$$

Two special cases deserve notice. First, if $\mathscr{W} = \mathscr{V}$, the translation space of $\mathscr{E}$, then $\nabla\mathbf{f}(\mathbf{x})$ is a tensor, and $\nabla\mathbf{f}(\mathbf{x})\mathbf{u}$ is written for $\nabla\mathbf{f}(\mathbf{x})(\mathbf{u})$. Second, if $\mathscr{W}$ is the set of real numbers, the field f is called a *scalar field*. By the representation theorem for scalar-valued functions of vectors we know that $\nabla f(\mathbf{x})(\mathbf{u})$ equals the inner product of some vector and $\mathbf{u}$. In this sense we say that the gradient of a scalar field at a point is a vector. Writing $\nabla f(\mathbf{x})$ for that vector, we have

$$f(\mathbf{x} + \mathbf{u}) = f(\mathbf{x}) + \nabla f(\mathbf{x}) \cdot \mathbf{u} + o(\mathbf{u}).$$

Similar definitions can be framed for point-valued, vector-valued, and tensor-valued functions of points.

Among the rules for taking the gradients of products of various kinds are

$$\nabla(fg) = f\,\nabla g + g\,\nabla f,$$

$$\nabla(\mathbf{f}\cdot\mathbf{g}) = (\nabla\mathbf{f})^{\mathrm{T}}\mathbf{g} + (\nabla\mathbf{g})^{\mathrm{T}}\mathbf{f},$$

$$\nabla(f\mathbf{g}) = \mathbf{g}\otimes\nabla f + f\,\nabla\mathbf{g}.$$

There is also the *chain rule* for taking the gradient of a composite function. If $\mathbf{f}\circ\mathbf{g}$ denotes the composition of $\mathbf{g}$ with $\mathbf{f}$, the rule can be written symbolically as

$$\nabla(\mathbf{f}\circ\mathbf{g}) = ((\nabla\mathbf{f})\circ\mathbf{g})\,\nabla\mathbf{g}.$$

If χ maps points into points, and if f is a scalar-valued function of points, then

$$\nabla(f\circ\chi) = (\nabla\chi)^{\mathrm{T}}((\nabla f)\circ\chi).$$

4. Other Differential Operators

The repeated or *second* gradient is the mapping that results from taking the gradient twice. It is denoted by ∇^2. If f is scalar-valued, the value of $\nabla^2 f$ is a symmetric tensor.

The operators *divergence* div and *Laplacian* Δ upon vector fields and scalar fields, respectively, are defined as follows:

$$\operatorname{div}\mathbf{f} \equiv \operatorname{tr}\nabla\mathbf{f},$$

$$\Delta f \equiv \operatorname{div}\nabla f = \operatorname{tr}\nabla^2 f.$$

If $\mathbf{L}$ is a tensor field and $\mathbf{a}$ is a fixed vector, then $\mathbf{L}^{\mathrm{T}}\mathbf{a}$ is a vector field, and it is easy to see that the values of $\operatorname{div}(\mathbf{L}^{\mathrm{T}}\mathbf{a})$ are scalar-valued linear functions of the vector $\mathbf{a}$. Thus the divergence $\operatorname{div}\mathbf{L}$ of a tensor field $\mathbf{L}$ can be defined by the requirement that

$$\mathbf{a}\cdot\operatorname{div}\mathbf{L} = \operatorname{div}(\mathbf{L}^{\mathrm{T}}\mathbf{a}).$$

Among the rules for calculating divergences and gradients of products are the following:

$$\operatorname{div}(f\mathbf{g}) = \mathbf{g}\cdot\nabla f + f\operatorname{div}\mathbf{g},$$

$$\operatorname{div}(\mathbf{L}\mathbf{g}) = (\operatorname{div}\mathbf{L}^{\mathrm{T}})\cdot\mathbf{g} + \operatorname{tr}(\mathbf{L}\,\nabla\mathbf{g}),$$

$$\operatorname{div}(\nabla\mathbf{g})^{\mathrm{T}} = \nabla\operatorname{div}\mathbf{g},$$

$$\operatorname{div}[\nabla\mathbf{g} \pm (\nabla\mathbf{g})^{\mathrm{T}}] = \Delta\mathbf{g} \pm \nabla\operatorname{div}\mathbf{g}.$$

A vector field $\mathbf{f}$ is *lamellar* if there is a scalar field P such that for every sufficiently short curve $\mathscr{C}$ that connects two sufficiently near points $\mathbf{x}_1$ and $\mathbf{x}_2$

$$\int_{\mathscr{C}} \mathbf{f} \cdot d\mathbf{x} = P(\mathbf{x}_1) - P(\mathbf{x}_2),$$

the sense of $\mathscr{C}$ being from $\mathbf{x}_1$ to $\mathbf{x}_2$. The function P is a *potential* of $\mathbf{f}$. Conversely, if there is a scalar field P such that

$$\mathbf{f} = -\nabla P,$$

the field $\mathbf{f}$ is lamellar. If $\mathbf{f}$ is differentiable, it is lamellar if and only if $\nabla \mathbf{f}$ is symmetric:

$$(\nabla \mathbf{f})^{\mathrm{T}} = \nabla \mathbf{f}.$$

If the domain of $\mathbf{f}$ is a simply connected open set, the restriction to sufficiently near points and sufficiently short curves is unnecessary.

If the domain of the lamellar field $\mathbf{f}$ is multiply connected, of course a potential exists locally, but the line integral $\int_{\mathscr{C}} \mathbf{f} \cdot d\mathbf{x}$ is not generally independent of the path $\mathscr{C}$ connecting two given points. If the two curves $\mathscr{C}_1$ and $\mathscr{C}_2$ connect $\mathbf{x}_1$ to $\mathbf{x}_2$, then

$$\int_{\mathscr{C}_1} \mathbf{f} \cdot d\mathbf{x} - \int_{\mathscr{C}_2} \mathbf{f} \cdot d\mathbf{x} = \sum_{k=1}^{q} n_k K_k;$$

the "cyclic constants" K_k are determined by $\mathbf{f}$ and its domain alone, and the numbers n_k are integers. The concept of potential may be extended to lamellar fields on multiply connected domains by introducing "cyclic functions", which map each point onto a set[1] of the form $\{P_0 + \sum_{k=1}^{q} n_k K_k\}$.

5. *Co-ordinates*

The Cartesian n-space $\mathscr{R}_n$ is the n-dimensional vector space whose elements are ordered lists of n real numbers:

$$(x^1, x^2, \ldots, x^n),$$

provided addition and scalar multiplication be defined by the corresponding operations on the entries x^k.

A *co-ordinate system* on an open set of an n-dimensional Euclidean space

[1] The classical treatment of cyclic potentials, due to KELVIN, is most easily available in §§49–51 of H. LAMB's *Hydrodynamics*, Cambridge, Cambridge University Press, 2nd–6th editions, 1895/1932. It is not easy to find a simple treatment that satisfies modern standards of rigor.

An elegant, rigorous treatment of lamellar fields that need not be differentiable, and also of solenoidal fields, may be found in a paper by H. WEYL, "The method of orthogonal projection in potential theory," *Duke Mathematical Journal* 7, 411–444 (1940).

is a one-to-one mapping of that set into $\mathscr{R}_n$, a mapping which has an invertible gradient and a continuous second gradient. If $\bar{\mathbf{x}}$ is such a mapping,

$$\bar{\mathbf{x}}(\mathbf{x}) = (\bar{x}^1(\mathbf{x}), \bar{x}^2(\mathbf{x}), \ldots, \bar{x}^n(\mathbf{x})),$$

where $\bar{x}^k$ is a scalar field having the same degree of smoothness as that assumed for $\bar{\mathbf{x}}$. The number $\bar{x}^k(\mathbf{x})$ is the kth *co-ordinate* of the point $\mathbf{x}$ in the co-ordinate system $\bar{\mathbf{x}}$.

If $\hat{\mathbf{x}}$ denotes the inverse of $\bar{\mathbf{x}}$, then

$$\bar{x}^k(\hat{\mathbf{x}}(x^1, x^2, \ldots, x^n)) = x^k, \qquad k = 1, 2, \ldots, n,$$

for all lists $(x^1, x^2, \ldots, x^n)$ that lie in the range of $\bar{\mathbf{x}}$. We set

$$\mathbf{e}^k(\mathbf{x}) \equiv \nabla \bar{x}^k(\mathbf{x}), \qquad \mathbf{e}_m(\mathbf{x}) \equiv \partial_{\bar{x}^m} \hat{\mathbf{x}}(x^1, x^2, \ldots, x^n)\Big|_{x^j = \bar{x}^j(\mathbf{x})},$$

where ∂_{x^m} denotes the partial derivative with respect to x^m of a point-valued function of the n real variables $x^1, x^2, \ldots, x^n$. The vector $\mathbf{e}^k(\mathbf{x})$ is normal at $\mathbf{x}$ to the *co-ordinate surface* $x^k(\mathbf{y}) = \text{const.}$ that passes through $\mathbf{x}$. The vector $\mathbf{e}_m(\mathbf{x})$ is tangent to the mth *co-ordinate curve* at $\mathbf{x}$, that curve being the set of points near $\mathbf{x}$ for which every co-ordinate but x^m has the same value as it does at $\mathbf{x}$.

The sets of vectors $\mathbf{e}^1(\mathbf{x}), \mathbf{e}^2(\mathbf{x}), \ldots, \mathbf{e}^n(\mathbf{x})$ and $\mathbf{e}_1(\mathbf{x}), \mathbf{e}_2(\mathbf{x}), \ldots, \mathbf{e}_n(\mathbf{x})$ are reciprocal bases of the translation space of $\mathscr{E}$. The basis $\mathbf{e}_1(\mathbf{x}), \mathbf{e}_2(\mathbf{x}), \ldots, \mathbf{e}_n(\mathbf{x})$ is called the *natural basis* of the co-ordinate system $\bar{\mathbf{x}}$ at $\mathbf{x}$, and $\mathbf{e}^1(\mathbf{x}), \mathbf{e}^2(\mathbf{x}), \ldots, \mathbf{e}^n(\mathbf{x})$ is the *reciprocal natural basis* there. As the point $\mathbf{x}$ varies over the domain of $\bar{\mathbf{x}}$, fields of natural bases and their reciprocals are obtained. In general, these bases are not orthonormal. If co-ordinate surfaces are mutually orthogonal, the co-ordinate curves are normal to the co-ordinate surfaces, so $\mathbf{e}^k$ is parallel to $\mathbf{e}_k$, but generally the two are not the same. Indeed, the natural basis field is orthonormal only if it is a constant field, in which case the co-ordinates are called *rectangular Cartesian.* The values of the Cartesian co-ordinate fields may be interpreted as distances from a particular set of n mutually orthogonal hyperplanes, or as distances measured parallel to a particular set of n mutually orthogonal lines, as we please.

Two other systems are commonly used in 3-dimensional space. The *cylindrical co-ordinates* (r, θ, z) of $\mathbf{x}$, are, respectively, the distance of $\mathbf{x}$ from a certain line called the *axis*, the angle subtended upon a particular plane through that line by the plane through the axis at $\mathbf{x}$, and the distance of $\mathbf{x}$ from a particular plane perpendicular to the axis. Hence

$$\mathbf{e}_r = \frac{\partial \mathbf{x}}{\partial r} = \mathbf{e}^r, \qquad \mathbf{e}_\theta = \frac{\partial \mathbf{x}}{\partial \theta} = r^2 \mathbf{e}^\theta, \qquad \mathbf{e}_z = \frac{\partial \mathbf{x}}{\partial z} = \mathbf{e}^z.$$

The *spherical co-ordinates* (r, θ, φ) are respectively, the distance of $\mathbf{x}$ from a certain point, an angle between planes through an axis through that particular point, and an angle subtended upon the axis by a line from the particular point to $\mathbf{x}$. Hence

$$\mathbf{e}_r = \frac{\partial \mathbf{x}}{\partial r} = \mathbf{e}^r, \qquad \mathbf{e}_\theta = \frac{\partial \mathbf{x}}{\partial \theta} = r^2 \mathbf{e}^\theta, \qquad \mathbf{e}_\varphi = \frac{\partial \mathbf{x}}{\partial \varphi} = r^2 \sin^2 \theta \mathbf{e}^\varphi.$$

6. *Contravariant, Covariant, and Mixed Components Relative to a Co-ordinate System*

The value $\mathbf{v}(\mathbf{x})$ of a vector field at $\mathbf{x}$ is a vector and hence has unique components relative to any basis, and in particular relative to the natural and reciprocal bases of a co-ordinate system $\bar{\mathbf{x}}$. Thus

$$\mathbf{v} = v^k \mathbf{e}_k = v_k \mathbf{e}^k.$$

The scalar fields $v^1, v^2, \ldots, v^n$ are the *contravariant component fields of* $\mathbf{v}$ *relative to the co-ordinate system* $\bar{\mathbf{x}}$; likewise, the fields $v_1, v_2, \ldots, v_k$ are the *covariant component fields* relative to that system. When a particular co-ordinate system is set down for use, we usually speak simply of *contravariant and covariant components*, respectively.

The covariant and contravariant *metric components*, g_{km} and g^{km}, are the scalar fields defined as follows:

$$g_{km} \equiv \mathbf{e}_k \cdot \mathbf{e}_m, \qquad g^{km} \equiv \mathbf{e}^k \cdot \mathbf{e}^m,$$

so

$$\mathbf{e}_k = g_{km} \mathbf{e}^m, \qquad \mathbf{e}^k = g^{km} \mathbf{e}_m, \qquad g_{hk} g^{kq} = \delta_h^q.$$

For cylindrical co-ordinates

$$\|g_{km}\| = \begin{Vmatrix} 1 & 0 & 0 \\ 0 & r^2 & 0 \\ 0 & 0 & 1 \end{Vmatrix}, \qquad \|g^{km}\| = \begin{Vmatrix} 1 & 0 & 0 \\ 0 & r^{-2} & 0 \\ 0 & 0 & 1 \end{Vmatrix},$$

and for spherical ones

$$\|g_{km}\| = \begin{Vmatrix} 1 & 0 & 0 \\ 0 & r^2 & 0 \\ 0 & 0 & r^2 \sin^2 \theta \end{Vmatrix}, \qquad \|g^{km}\| = \begin{Vmatrix} 1 & 0 & 0 \\ 0 & r^{-2} & 0 \\ 0 & 0 & r^{-2} \sin^2 \theta \end{Vmatrix}.$$

In terms of the metric components, it is easy to relate covariant and contravariant components of one and the same vector field $\mathbf{v}$:

$$v^k = g^{km} v_m, \qquad v_m = g_{mk} v^k.$$

Similar definitions and rules hold for the components of tensor fields, *e.g.*

$$L^{km} = g^{kp}L_p{}^m = g^{pm}L^k{}_p = g^{kp}g^{mq}L_{pq}.$$

Let $\bar{\mathbf{x}}$ and $\tilde{\mathbf{x}}$ be co-ordinate systems. Then the co-ordinates of $\mathbf{x}$ with respect to these two systems are functionally related:

$$\tilde{x}^k(\mathbf{x}) = f^k(\bar{x}^1, \bar{x}^2, \ldots, \bar{x}^n),$$
$$\bar{x}^q(\mathbf{x}) = g^q(\tilde{x}^1, \tilde{x}^2, \ldots \tilde{x}^n).$$

Let $\tilde{\mathbf{e}}_1(\mathbf{x}), \tilde{\mathbf{e}}_2(\mathbf{x}), \ldots, \tilde{\mathbf{e}}_n(\mathbf{x})$ be the natural basis of the co-ordinate system $\tilde{\mathbf{x}}$ at $\mathbf{x}$. From the definition of natural basis it follows that

$$\begin{aligned}\tilde{\mathbf{e}}_m &= \partial_{\tilde{x}^m}\hat{\mathbf{x}}(g^1(\tilde{x}^1, \ldots \tilde{x}^n), \ldots, g^n(\tilde{x}^1, \ldots, \tilde{x}^n)),\\ &= [\partial_{\bar{x}^k}\hat{\mathbf{x}}(\bar{x}^1, \ldots, \bar{x}^n)]\partial_{\tilde{x}^m}g^k(\tilde{x}^1, \ldots, \tilde{x}^n),\\ &= (\partial_{\tilde{x}^m}g^k)\bar{\mathbf{e}}_k.\end{aligned}$$

Thus if we set

$$A^k_m = \partial_{\tilde{x}^m}g^k, \qquad \bar{A}^p_q = \partial_{\bar{x}^q}f^p.$$

(usually denoted by $\partial\bar{x}^k/\partial\tilde{x}^m$ and $\partial\tilde{x}^p/\partial\bar{x}^q$), from the transformation laws in §AII.1 we may read off the relations between components of various kinds relative to different co-ordinate systems. *E.g.*, if the components of a vector $\mathbf{u}$ with respect to the two systems are distinguished by superimposed bars and tildes, then

$$\tilde{u}_k = \frac{\partial\bar{x}^m}{\partial\tilde{x}^k}\bar{u}_m, \qquad \tilde{u}^k = \frac{\partial\tilde{x}^k}{\partial\bar{x}^m}\bar{u}^m,$$

and so on for tensors.

These transformation laws for components were used to define vector fields in some of the older literature. *E.g.*, the scalar functions $\bar{A}^p{}_r{}^q{}_s$ and $\tilde{A}^k{}_u{}^h{}_v$ are said to be components, in the co-ordinate systems $\bar{\mathbf{x}}$ and $\tilde{\mathbf{x}}$, of a tensor of order four (contravariant order two and covariant order two), if they are related as follows:

$$\tilde{A}^k{}_u{}^h{}_v = \frac{\partial\tilde{x}^k}{\partial\bar{x}^p}\frac{\partial\bar{x}^r}{\partial\tilde{x}^u}\frac{\partial\tilde{x}^h}{\partial\bar{x}^q}\frac{\partial\bar{x}^s}{\partial\tilde{x}^v}\bar{A}^p{}_r{}^q{}_s$$

the functions on the left-hand side being evaluated at the argument $\tilde{\mathbf{x}}(\mathbf{x})$, and those on the right-hand side at $\bar{\mathbf{x}}(\mathbf{x})$. The other approaches to tensors of order greater than 2 which were mentioned above in §§AII.4 and AII.5 may be extended to fields in a straightforward way.

Whatever be the definitions chosen, there is no doubt that specific calculations are performed most easily by means of the transformation laws. For example, it is obvious that for a rectangular Cartesian co-ordinate system $g_{km} = \delta_{km} = g^{km}$. The covariant metric components $\tilde{g}_{km}$ in the co-ordinate system $\tilde{\mathbf{x}}$, therefore, are obtained as follows:

$$\tilde{g}_{km} = \frac{\partial x^p}{\partial \tilde{x}^k}\frac{\partial x^q}{\partial \tilde{x}^m}\delta_{pq} = \sum_{p=1}^{n} \frac{\partial x^p}{\partial \tilde{x}^k}\frac{\partial x^p}{\partial \tilde{x}^m},$$

where the rectangular Cartesian co-ordinates x^p are presumed given as functions of the general co-ordinates $\tilde{x}^k$:

$$x^p = f^p(\tilde{x}^1, \tilde{x}^2, \ldots, \tilde{x}^n), \qquad p = 1, 2, \ldots, n.$$

For example, in cylindrical co-ordinates, if we write r, θ, and z, respectively, for $\tilde{x}^1, \tilde{x}^2, \tilde{x}^3$, then

$$x^1 = x = r \cos\theta,$$

$$x^2 = y = r \sin\theta,$$

$$x^3 = z.$$

It is a trivial matter to obtain in this way the matrices $\|\tilde{g}_{km}\|$ and $\|\tilde{g}^{pq}\|$ for cylindrical co-ordinates. Likewise, the components of vectors and tensors relative to any co-ordinate system may be calculated in a routine way from their components relative to a rectangular Cartesian system.

Tensors of order greater than 2 occur rarely in this book. A student who does not possess a technique of handling them should be able to follow all developments by simply referring them to components relative to rectangular Cartesian co-ordinates. Of course, this procedure, while often inelegant, is perfectly rigorous.

7. *Physical Components Relative to an Orthogonal Co-ordinate System*

The vectors and tensors that occur in physical problems usually are assigned physical dimensions. For example, a velocity has the dimensions of length divided by time. The components of a velocity field with respect to a co-ordinate system do not necessarily have these same dimensions, since the dimensions of the different members of natural basis are not usually all the same. For example, in a cylindrical system $\mathbf{e}^r$ is dimensionless, but $\mathbf{e}_\theta$ has the dimension of length, and $\mathbf{e}^\theta$ has the dimension of reciprocal length. In physical problems it is often desirable to be able to interpret each component of a vector in the same terms as the vector itself, and for this reason *physical components* are used. For an orthogonal co-ordinate system, these components are defined unambiguously as being the components with respect to the following orthonormal basis field:

$$\mathbf{i}_k \equiv \frac{\mathbf{e}_k}{|\mathbf{e}_k|} = \frac{\mathbf{e}^k}{|\mathbf{e}^k|}, \qquad k = 1, 2, \ldots, n;$$

here $\mathbf{e}_1, \mathbf{e}_2, \ldots, \mathbf{e}_n$ is the natural basis field of the co-ordinate system, and $\mathbf{e}^1, \mathbf{e}^2, \ldots, \mathbf{e}^n$ is its reciprocal natural basis field. The orthonormal basis field $\mathbf{i}_1, \mathbf{i}_2, \ldots, \mathbf{i}_n$ is everywhere tangent to the co-ordinate curves and normal to the co-ordinate surfaces. Physical components are denoted by indices at middle height, neither subscript nor superscript, thus:

$$v\,r, \qquad v\,\theta, \qquad v\,z.$$

A similar notation is used for tensors, *e.g.*

$$T\,rr, \qquad T\,r\theta, \qquad etc.$$

In some problems it is useful to take components with respect to a basis field, the elements of which are not proportional to the natural basis of any co-ordinate system. Such fields of bases are called *anholonomic*, and the same term is applied to components taken with respect to them. An example of this kind occurs in §V.4 of the text. In that particular case the basis is orthonormal.

8. Christoffel Components

The gradients $\boldsymbol{\Gamma}_{(k)}$ of the natural basis of a co-ordinate system exist and are continuous tensor fields:

$$\boldsymbol{\Gamma}_{(k)} \equiv \nabla \mathbf{e}_k, \qquad k = 1, 2, \ldots, n.$$

The mixed components $\Gamma_p{}^k{}_q$ of these fields, namely

$$\Gamma_p{}^k{}_q \equiv \Gamma_{(p)}{}^k{}_q = \mathbf{e}^k \cdot \boldsymbol{\Gamma}_{(p)} \mathbf{e}_q,$$

are the *Christoffel components* of the given co-ordinate system. It is possible to prove that

$$\Gamma_p{}^k{}_q = \Gamma_q{}^k{}_p$$

and that

$$\partial_{x^s} \mathbf{e}_r = \Gamma_r{}^k{}_s \mathbf{e}_k.$$

Furthermore, the Christoffel components can be calculated as follows from the metric components g^{kp} and g_{qr} of the co-ordinate system:

$$\Gamma_u{}^k{}_v = \tfrac{1}{2} g^{kp} (\partial_{x^v} g_{pu} + \partial_{x^u} g_{pv} - \partial_{x^p} g_{uv}).$$

It can be shown that the Christoffel components of a co-ordinate system vanish identically if and only if its natural basis field is constant. Such a co-ordinate system is called *Cartesian.*

9. *Covariant Derivatives, Differential Operators*

If $\mathbf{f}$ is a vector field, its gradient $\nabla\mathbf{f}$ is a tensor field. The four kinds of components of $\nabla\mathbf{f}$ are called *covariant derivatives* of $\mathbf{f}$. These are defined as components always are:

$$\begin{aligned} f^k{}_{,m} &\equiv \mathbf{e}^k \cdot (\nabla\mathbf{f})\mathbf{e}_m, \\ f_{k,m} &\equiv \mathbf{e}_k \cdot (\nabla\mathbf{f})\mathbf{e}_m, \\ f_k{}^{,m} &\equiv \mathbf{e}_k \cdot (\nabla\mathbf{f})\mathbf{e}^m, \\ f^{k,m} &\equiv \mathbf{e}^k \cdot (\nabla\mathbf{f})\mathbf{e}^m. \end{aligned}$$

Each covariant derivative is thus a scalar field.

To calculate the covariant derivatives of $\mathbf{f}$ in terms of the components of $\mathbf{f}$, we note first that

$$\nabla\mathbf{f} = \nabla(f^p\mathbf{e}_p) = \mathbf{e}_p \otimes \nabla f^p + f^p\,\nabla\mathbf{e}_p .$$

Hence

$$\begin{aligned} f^k{}_{,m} &= \mathbf{e}^k \cdot (\mathbf{e}_p \otimes \nabla f^p + f^p \boldsymbol{\Gamma}_{(p)})\mathbf{e}_m \\ &= \partial_{x^m} f^k + f^p \Gamma_p{}^k{}_m . \end{aligned}$$

Likewise

$$f_{k,m} = \partial_{x^m} f_k - f_p \Gamma_k{}^p{}_m .$$

The covariant derivatives equal the corresponding partial derivatives for all f if and only if the co-ordinate system is Cartesian.

Similar results hold for tensors of all orders. In particular, covariant derivatives of all tensors reduce to the corresponding partial derivatives if the co-ordinate system is Cartesian.

The values of all differential operators can be calculated in terms of covariant derivatives or Christoffel components. For example

$$\begin{aligned} (\operatorname{div} \mathbf{L})^k &= \mathrm{L}^{km}{}_{,m}, \\ &= \frac{1}{\sqrt{g}}\partial_{x^m}(\sqrt{g}\,L^{km}) + L^{pm}\Gamma_p{}^k{}_m, \end{aligned}$$

where $g \equiv \det g_{pq}$.

The easiest way to get expressions in terms of physical components is to derive them first in terms of contravariant or covariant components, which is a simple routine matter, and then convert the results. We record here the physical components of the divergence of a symmetric tensor **L** in cylindrical co-ordinates:

$$(\text{div } \mathbf{L})^{\langle r \rangle} = \partial_r L^{\langle rr \rangle} + \frac{1}{r} \partial_\theta L^{\langle r\theta \rangle} + \partial_z L^{\langle rz \rangle} + \frac{L^{\langle rr \rangle} - L^{\langle \theta\theta \rangle}}{r},$$

$$(\text{div } \mathbf{L})^{\langle \theta \rangle} = \partial_r L^{\langle r\theta \rangle} + \frac{1}{r} \partial_\theta L^{\langle \theta\theta \rangle} + \partial_z L^{\langle \theta z \rangle} + \frac{2}{r} L^{\langle r\theta \rangle},$$

$$(\text{div } \mathbf{L})^{\langle z \rangle} = \partial_r L^{\langle rz \rangle} + \frac{1}{r} \partial_\theta L^{\langle \theta z \rangle} + \partial_z L^{\langle zz \rangle} + \frac{1}{r} L^{\langle rz \rangle}.$$

In spherical co-ordinates they are

$$(\text{div } \mathbf{L})^{\langle r \rangle} = \partial_r L^{\langle rr \rangle} + \frac{1}{r} \partial_\theta L^{\langle r\theta \rangle} + \frac{1}{r \sin \theta} \partial_\varphi L^{\langle r\varphi \rangle} + \frac{1}{r} (2L^{\langle rr \rangle} - L^{\langle \theta\theta \rangle} - L^{\langle \varphi\varphi \rangle} + L^{\langle r\theta \rangle} \cot \theta),$$

$$(\text{div } \mathbf{L})^{\langle \theta \rangle} = \partial_r L^{\langle \theta r \rangle} + \frac{1}{r} \partial_\theta L^{\langle \theta\theta \rangle} + \frac{1}{r \sin \theta} \partial_\varphi L^{\langle \theta\varphi \rangle} + \frac{1}{r} [(L^{\langle \theta\theta \rangle} - L^{\langle \varphi\varphi \rangle}) \cot \theta + 3L^{\langle r\theta \rangle}],$$

$$(\text{div } \mathbf{L})^{\langle \varphi \rangle} = \partial_r L^{\langle \varphi r \rangle} + \frac{1}{r} \partial_\theta L^{\langle \varphi\theta \rangle} + \frac{1}{r \sin \theta} \partial_\varphi L^{\langle \varphi\varphi \rangle} + \frac{1}{r} (3L^{\langle r\varphi \rangle} + 2L^{\langle \theta\varphi \rangle} \cot \theta).$$

Appendix III

Solutions of the Exercises

Note: Exercises that can be solved immediately by merely following the directions given in their statements are omitted from this list.

I.2.1 Use $(\text{I.2-5})_3$, Axiom B3, and the definition of meet to prove the first implication. Use $(\text{I.2-5})_4$ and the definition of join to prove the second.

I.2.2 Adopting the first of the two possible hypotheses, we set

$$\mathscr{P}_1 \equiv \mathscr{B} \curlywedge \mathscr{C}, \qquad \mathscr{P}_2 \equiv \mathscr{C} \curlywedge \mathscr{D}, \qquad \mathscr{P}_3 \equiv (\mathscr{B} \curlywedge \mathscr{C}) \curlywedge \mathscr{D}.$$

By the definition of meet,

$$\mathscr{P}_3 \prec \mathscr{B}, \qquad \mathscr{P}_3 \prec \mathscr{C}, \qquad \mathscr{P}_3 \prec \mathscr{D}.$$

Thus $\mathscr{P}_3$ is a part of $\mathscr{B}$, $\mathscr{C}$, and $\mathscr{D}$. Now suppose that

$$\mathscr{X} \prec \mathscr{B}, \qquad \mathscr{X} \prec \mathscr{C}, \qquad \mathscr{X} \prec \mathscr{D}.$$

Then, again by the definition of meet,

$$\mathscr{X} \prec \mathscr{P}_1, \qquad \mathscr{X} \prec \mathscr{P}_2, \qquad \mathscr{X} \prec \mathscr{P}_3.$$

Thus any part $\mathscr{X}$ of $\mathscr{B}$, $\mathscr{C}$, and $\mathscr{D}$ is a part of $\mathscr{P}_3$. By the definition of the meet of three bodies, then,

$$\mathscr{B} \curlywedge \mathscr{C} \curlywedge \mathscr{D} = \mathscr{P}_3.$$

From this last, we see that

$$\mathscr{P}_3 \prec \mathscr{B}, \qquad \mathscr{P}_3 \prec \mathscr{C} \curlywedge \mathscr{D}.$$

Suppose now that

$$\mathscr{Y} \prec \mathscr{B}, \qquad \mathscr{Y} \prec \mathscr{C} \curlywedge \mathscr{D}.$$

Since $\mathscr{B} \curlywedge \mathscr{C} \curlywedge \mathscr{D}$ exists,

$$\mathscr{Y} \prec \mathscr{B} \curlywedge \mathscr{C} \curlywedge \mathscr{D} = \mathscr{P}_3.$$

The definition of meet shows that

$$\mathscr{P}_3 = \mathscr{B} \curlywedge (\mathscr{C} \curlywedge \mathscr{D}).$$

A similar proof holds if we assume that $\mathscr{B} \curlywedge (\mathscr{C} \curlywedge \mathscr{D})$ exists.

I.2.3 Use $(\text{I.2-8})_1$ and $(\text{I.2-11})_2$.

I.2.4 Use $(\text{I.2-8})_1$, $(\text{I.2-15})_1$, and $(\text{I.2-11})_2$.

I.3.1 Let $\mathscr{B}$, $\mathscr{C}$, and $\mathscr{D}$ be bodies, and let $\mathscr{D} = \mathscr{B} \curlywedge \mathscr{C}$. Then $\mathscr{B} = \overline{\mathring{\mathscr{B}}}$, $\mathscr{C} = \overline{\mathring{\mathscr{C}}}$, $\mathscr{D} = \overline{\mathring{\mathscr{D}}}$. Since the interior of $\mathscr{B} \cap \mathscr{C}$ is $\mathring{\mathscr{B}} \cap \mathring{\mathscr{C}}$, we know that $\overline{\mathring{\mathscr{B}} \cap \mathring{\mathscr{C}}}$ is a body. $\mathscr{D} \prec \mathscr{B}$ means that $\overline{\mathring{\mathscr{D}}} \subset \overline{\mathring{\mathscr{B}}}$, and hence $\mathring{\mathscr{D}} \subset \mathring{\mathscr{B}}$. Likewise, $\mathscr{D} \prec \mathscr{C}$ implies that $\mathring{\mathscr{D}} \subset \mathring{\mathscr{C}}$. Hence $\mathring{\mathscr{D}} \subset \mathring{\mathscr{B}} \cap \mathring{\mathscr{C}}$. Therefore, $\mathscr{D} = \overline{\mathring{\mathscr{D}}} \subset \overline{\mathring{\mathscr{B}} \cap \mathring{\mathscr{C}}}$.

I.4.1 Use (I.2-37) to write $\mathscr{B} \curlyvee \mathscr{C}$ as the join of separate bodies. Then use Axiom M3, (I.4-3), and Axiom M1 (if necessary).

I.5.1 Substitute (I.5-16) into (I.5-26), then use (I.5-20).

I.5.3 Expand $\mathbf{f}(\mathscr{B} \curlyvee \mathscr{C}, \mathscr{B} \curlyvee \mathscr{C})$ with the aid of Axioms FE2 and FE3, then similarly expand the results.

I.8.1 $(\text{I.8-6})_1$ follows simply by differentiating (I.8-4). To derive $(\text{I.8-6})_2$, use the definition of $\mathbf{M}_{\mathbf{x}_1}$ so as to get $\dot{\mathbf{M}}_{\mathbf{x}_1}$, and adjust the terms.

I.8.2 Write $\mathbf{F}(\mathscr{B}, \mathscr{B}^e) + \mathbf{F}(\mathscr{B}, \mathscr{B})$ explicitly, and use (I.5-26).

I.8.3 In (I.8-6) take $\mathbf{x}_c$ for $\mathbf{x}_1$, and use (I.8-28). The second term in (I.8-29) is the rate of change of rotational momentum with respect to the fixed place $\mathbf{x}_0$ of a mass-point located at the center of mass of $\chi(\mathscr{B}, t)$ and endowed with the linear momentum of $\mathscr{B}$.

I.9.1

$$\begin{aligned} \mathbf{x}^* &= \mathbf{x}_0^*(t) + \mathbf{Q}(t)(\mathbf{x} - \mathbf{x}_0), \\ &= \mathbf{x}_0^*(t) + \mathbf{Q}(t)(\mathbf{x} - \bar{\mathbf{x}}_0(t) + \bar{\mathbf{x}}_0(t) - \mathbf{x}_0), \\ &= \mathbf{x}_0^*(t) + \mathbf{Q}(t)(\bar{\mathbf{x}}_0(t) - \mathbf{x}_0) + \mathbf{Q}(t)(\mathbf{x} - \bar{\mathbf{x}}_0(t)), \\ &= \bar{\mathbf{x}}_0^*(t) + \mathbf{Q}(t)(\mathbf{x} - \bar{\mathbf{x}}_0(t)), \end{aligned}$$

say. Since a transformation of this kind, along with $t^* = t + a$, preserves the metrics in $\mathscr{E}$ and $\mathscr{R}$, it defines a change of frame. To prove the group property, use the fact that the orthogonal tensors form a group.

I.9.2 Differentiate $\mathbf{Z} = \mathbf{Y}\mathbf{Y}^{\mathrm{T}}$ and use (I.9-16) and (I.9-15). A solution of (I.9-17) is furnished by $\mathbf{Z}(t) = \mathbf{1}$; by uniqueness it is the only solution satisfying $\mathbf{Z}(t_0) = \mathbf{1}$.

I.9.3 $\mathbf{A}^*$ is formed from the tensor $\mathbf{Q}^*$ that enters the inverse of (I.9-4). Since $\mathbf{Q}^* = \mathbf{Q}^{\mathrm{T}}$, (I.9-18) follows. To derive (I.9-19), write out the equations for the three changes of frame. $\mathbf{A}_3 = \dot{\mathbf{Q}}_3 \mathbf{Q}_3^{\mathrm{T}}$; simplification by $(\text{I.9-19})_1$ yields $(\text{I.9-19})_2$. When ϕ_3 and ϕ_2 coincide at some instant, $\mathbf{Q}_2 = \mathbf{Q}_2^{\mathrm{T}} = \mathbf{1}$ at that instant.

I.9.4 If $\mathbf{e}$ is a unit vector on the axis of rotation, $\mathbf{Qe} = \pm\mathbf{e}$. Hence $\dot{\mathbf{Q}}\mathbf{e} + \mathbf{Q}\dot{\mathbf{e}} = \pm\dot{\mathbf{e}}$; that is,

$$\mathbf{Ae} + (\mathbf{Q} \mp \mathbf{1})\dot{\mathbf{e}} = \mathbf{0}.$$

Thus if $\dot{\mathbf{e}} = \mathbf{0}$, it follows that $\mathbf{Ae} = \mathbf{0}$. The relation between ω and $\dot{\theta}$ can be proved by use of the explicit representation of the orthonormal components of rotations over a 3-dimensional space:

$$[\mathbf{Q}] = \begin{Vmatrix} \cos\theta & -\sin\theta & 0 \\ \sin\theta & \cos\theta & 0 \\ 0 & 0 & 1 \end{Vmatrix},$$

where θ is the angle of rotation and the basis vector $\mathbf{e}_3$ lies in the axis of rotation.

I.9.5

$$\begin{aligned} \operatorname{tr}[\dot{\mathbf{R}}\mathbf{R}^{\mathrm{T}}(\mathbf{R} - \mathbf{R}^{\mathrm{T}})] &= \operatorname{tr}[\dot{\mathbf{R}} + (\dot{\mathbf{R}}\mathbf{R}^{\mathrm{T}})^{\mathrm{T}}\mathbf{R}^{\mathrm{T}}], \\ &= 2 \operatorname{tr} \dot{\mathbf{R}}. \end{aligned}$$

In a space of 3 dimensions

$$\operatorname{tr} \dot{\mathbf{R}} = -2 \sin\theta\dot{\theta}.$$

Also, by appeal to the explicit representation given in §AII.14,

$$\begin{aligned} \operatorname{tr}[\dot{\mathbf{R}}\mathbf{R}^{\mathrm{T}}(\mathbf{R} - \mathbf{R}^{\mathrm{T}})] &= -2 \sin\theta W^{KL}\varepsilon_{LKP}e^P, \\ &= -4 \sin\theta\boldsymbol{\omega}\cdot\mathbf{e}. \end{aligned}$$

If $\theta = 0$, no relation between $\boldsymbol{\omega}\cdot\mathbf{e}$ and $\dot{\theta}$ can hold, because $\mathbf{e}$ can be any unit vector. If $\theta = \pi$, the foregoing argument delivers no result, but $\boldsymbol{\omega}\cdot\mathbf{e}$ is a continuous function of $\mathbf{e}$, so we may infer (I.9-20) by a passage to the limit, using the result established for values of θ near π.

I.10.1 The remark just preceding the exercise solves half of it, for in any framing that gives rise to a spin $\overline{\mathbf{W}}$ having the same axis as $\mathbf{W}$ at each t the points on that axis will maintain their mutual distances. Conversely, suppose that

$$\bar{\mathbf{c}} + \overline{\mathbf{W}}(\mathbf{x} - \bar{\mathbf{x}}_0) = \mathbf{c} + \mathbf{W}(\mathbf{x} - \mathbf{x}_0) \qquad \forall\mathbf{x} \in \chi(\mathscr{B}, t).$$

Then

$$(\overline{\mathbf{W}} - \mathbf{W})(\mathbf{x} - \mathbf{x}_0) = \overline{\mathbf{W}}(\mathbf{x}_0 - \overline{\mathbf{x}}_0) + \mathbf{c} - \overline{\mathbf{c}}.$$

Thus $\overline{\mathbf{W}} - \mathbf{W}$ is a skew tensor such that

$$(\overline{\mathbf{W}} - \mathbf{W})(\mathbf{x} - \mathbf{y}) = \mathbf{0} \quad \text{if} \quad \mathbf{x} \in \chi(\mathscr{B}, t) \quad \text{and} \quad \mathbf{y} \in \chi(\mathscr{B}, t).$$

For any given $\mathbf{x}$ and $\mathbf{y}$ there are infinitely many skew tensors $\mathbf{S}$ such that $\mathbf{S}(\mathbf{x} - \mathbf{y}) = \mathbf{0}$. Such $\mathbf{S}$ belong in common to all $\mathbf{x}$ and $\mathbf{y}$ that lie upon the same straight line. If $\chi(\mathscr{B}, t)$ contains three non-collinear places, $\overline{\mathbf{W}} - \mathbf{W}$ is a skew tensor whose nullspace contains two distinct straight lines; since the nullspace of a non-null skew tensor is 1-dimensional, $\overline{\mathbf{W}} - \mathbf{W} = \mathbf{0}$.

I.10.2 $(\text{I.10-1})_2$ shows that $\dot{\mathbf{x}}_0 = \mathbf{c}$. Therefore $\dot{\mathbf{p}}_i = \mathbf{W}\mathbf{p}_i$, $i = 1, 2$. Compute $\frac{d}{dt}(\mathbf{p}_1 \cdot \mathbf{p}_2)$, and use $\mathbf{W} = -\mathbf{W}^{\mathrm{T}}$.

I.10.4 Use $(\text{I.8-2})_1$, $(\text{I.10-1})_2$, (I.10-5), and the fact that $\mathbf{c} = \text{const}$. To obtain $(\text{I.10-8})_2$, note that $\mathbf{Q}(\mathbf{a} \wedge \mathbf{b})\mathbf{Q}^{\mathrm{T}} = \mathbf{Q}\mathbf{a} \wedge \mathbf{Q}\mathbf{b}$, and use (I.10-3) and (I.10-7).

I.10.5 Use $\mathbf{W} = -\mathbf{Q}^{\mathrm{T}}\mathbf{A}\mathbf{Q}$ and (I.9-15).

I.10.7 If $\dot{\mathbf{e}} = \mathbf{0}$, then $\mathbf{W}\mathbf{e} = \mathbf{0}$ if and only if $(\dot{\mathbf{W}} + \mathbf{W}^2)\mathbf{e} = \mathbf{0}$.

I.10.8 Use (I.8-3), (I.10-1), and (I.10-4) to get (I.10-14). If $\mathbf{x}_0(t)$ is chosen as directed, then $\dot{\mathbf{x}}_0(t) = \mathbf{c}$, and $\overline{\mathbf{p}} = \mathbf{0}$. (I.10-15) follows with the aid of a little manipulation in the third term on the right-hand side of (I.10-14).

I.11.1 Since under Galilean transformations $\ddot{\mathbf{x}}^* = \mathbf{Q}\ddot{\mathbf{x}}$ and $M^* = M$, $(\text{I.8-5})_1$ shows that $\dot{\mathbf{m}}^* = \mathbf{Q}\mathbf{m}$.

I.11.2 The first statement follows at once from the chain rule. The fourth statement is proved as follows from (I.11-2):

$$\begin{aligned}\det \mathbf{T}^* &= \det(\mathbf{Q}\mathbf{T}\mathbf{Q}^{\mathrm{T}}) = (\det \mathbf{Q})(\det \mathbf{T})(\det \mathbf{Q}^{\mathrm{T}}),\\ &= \det \mathbf{T}.\end{aligned}$$

This being so, $\det(\mathbf{T} - t\mathbf{1})$ is frame-indifferent, no matter what be the number t. Therefore, $\mathbf{T}^*$ and $\mathbf{T}$ have the same latent roots. Consequently they have the same trace and the same proper numbers. If $\mathbf{e}$ is a proper vector corresponding to the proper number t, then

$$\mathbf{T}\mathbf{e} = t\mathbf{e},$$

so

$$(\mathbf{Q}^{\mathrm{T}}\mathbf{T}^*\mathbf{Q})\mathbf{e} = t\mathbf{e},$$

and hence

$$\mathbf{T}^*(\mathbf{Q}\mathbf{e}) = t(\mathbf{Q}\mathbf{e}).$$

Thus $\mathbf{Q}\mathbf{e}$ is the corresponding proper vector of $\mathbf{T}^*$.

I.11.3 A given smooth surface can be imbedded in a smooth family of surfaces $f = \text{const.}$, where f is a frame-indifferent scalar. Then by results of Exercise I.11.2, both ∇f and $|\nabla f|$ are frame-indifferent, and of course $\mathbf{n} = \nabla f / |\nabla f|$. A better proof can be constructed by writing an equation for a single surface as

$$\mathbf{x} - \mathbf{x}_0 = \mathbf{f}(a, b),$$

where a and b are parameters. For each a and b, the left-hand side is a frame-indifferent vector, so the vector-valued function $\mathbf{f}$ is frame-indifferent. Accordingly, $\partial_a \mathbf{f}$ and $\partial_b \mathbf{f}$ are frame-indifferent. They span the tangent plane at (a, b). The line normal to the tangent plane contains exactly two unit vectors. Their construction as above shows that both are frame-indifferent.

I.12.1 Use Axiom A2 and results from Exercise I.11.2.

I.13.1 We note that (I.5-22) holds as long as the system of forces is balanced. Since the axioms of inertia as applied to analytical dynamics respect the requirement that the forces be balanced, they do not alter the requirement (I.5-22).

I.13.2 Note that $(\mathbf{f} \wedge \mathbf{g})^{\cdot} = \mathbf{0}$. The spectral decomposition of $\mathbf{E}^*_{\mathbf{x}^*}$ is

$$\mathbf{E}^*_{\mathbf{x}_0^*} = E_1 \mathbf{e} \otimes \mathbf{e} + E_2 \mathbf{f} \otimes \mathbf{f} + E_3 \mathbf{g} \otimes \mathbf{g}.$$

It is easy to show that

$$\mathbf{E}^*_{\mathbf{x}_0^*} \mathbf{A}^2 - \mathbf{A}^2 \mathbf{E}^*_{\mathbf{x}_0^*} = \mathbf{0},$$

and

$$\mathbf{E}^*_{\mathbf{x}_0^*} \dot{\mathbf{A}} + \dot{\mathbf{A}} \mathbf{E}^*_{\mathbf{x}_0^*} = \dot{\omega}(E_2 + E_3) \mathbf{f} \wedge \mathbf{g}.$$

Thus (I.13-21) reduces to (I.13-23).

I.14.1 The heating $\mathbf{Q}$ obeys an identity of the type (I.5-2).

I.14.2 That $V_{\mathrm{m}}(\mathbf{v}^*) = V_{\mathrm{m}}(\mathbf{v}) \Leftrightarrow V_{\mathrm{m}}(\mathbf{v}) = V_{\mathrm{m}}(|\mathbf{v}|)$ follows from a theorem of CAUCHY [NFTM, p. 29]. Thus by (I.14-9)

$$\mathbf{f}_{ij} = \left(\frac{\partial V_{\mathrm{m}}(|\mathbf{v}|)}{\partial |\mathbf{v}|} \bigg|_{\mathbf{v} = \mathbf{x}_j - \mathbf{x}_i} \right) (\mathbf{x}_i - \mathbf{x}_j),$$

which proves incidentally that the mutual forces are central. That is, the requirement of frame-indifference makes conservative mutual forces necessarily central:

$$V_0(\mathbf{x}^*, \mathbf{x}_0^*) = V_0(\mathbf{x}, \mathbf{x}_0) \quad \Leftrightarrow \quad V_0(\mathbf{x}, \mathbf{x}_0) = V_0(|\mathbf{x} - \mathbf{x}_0|).$$

Therefore $\mathbf{f}_i^0$ as defined by $(\text{I.14-9})_1$ is parallel to $\mathbf{x}_i - \mathbf{x}_0$. The result follows immediately if one notes that

$$V_0(\mathbf{x}^* - \mathbf{x}_0^*, \mathbf{n}^*) = V_0(\mathbf{x} - \mathbf{x}_0, \mathbf{n})$$
$$V_0(\mathbf{x} - \mathbf{x}_0, \mathbf{n}) = V_0(|\mathbf{x} - \mathbf{x}_0|, (\mathbf{x} - \mathbf{x}_0) \cdot \mathbf{n}).$$

I.14.3 In a rest frame for the given rigid motion, $K = 0$ and $V = \text{const.}$ Since $K + V$ is frame-indifferent, (I.14-15) follows trivially.

II.5.1 The adjugate adj $\mathbf{L}$ of a tensor $\mathbf{L}$ is that tensor whose components are the cofactors[1] of the elements of $[\mathbf{L}]$. Then

$$\mathbf{L}\,\text{adj}\,\mathbf{L} = (\det \mathbf{L})\mathbf{1},$$

so if $\mathbf{L}$ is invertible, adj $\mathbf{L} = (\det \mathbf{L})\mathbf{L}^{-1}$. If $\mathbf{L}$ is a differentiable function of a parameter t, then it is easy to show that

$$(\det \mathbf{L}) = \text{tr}(\dot{\mathbf{L}}\,\text{adj}\,\mathbf{L}) = \det \mathbf{L}\,\text{tr}(\dot{\mathbf{L}}\mathbf{L}^{-1}),$$

for the latter form it being presumed that $\mathbf{L}$ is invertible. Take $\mathbf{L} = \mathbf{F}$, and use the chain rule to show that $\dot{\mathbf{F}} = \mathbf{G}\mathbf{F}$. (In fact, this formula is equivalent to (II.11-5) below, which is a simple special case of results to be obtained in §II.8.)

II.5.2 (An easier problem of this kind is given below as Exercise II.6.2). Consider the linear partial differential equation

$$\sum_{i=0}^{n} P_i\,\partial_{x_i} Z = R, \qquad \text{(L)}$$

where $P_0, P_1, \ldots, P_n, R$ are given functions of $x_0, x_1, \ldots, x_n$. The *characteristics* of (L) are the integral curves of the system

$$\frac{dx_0}{P_0} = \frac{dx_1}{P_1} = \cdots = \frac{dZ}{R}. \qquad \text{(C)}$$

A *characteristic integral* is a function $f_i(x_0, x_1, \ldots, x_n, Z)$ such that $f_i = \text{const.}$ on every characteristic curve. The formal statement of LAGRANGE's theorem is that if $f_1, \ldots, f_n$ are any n functionally independent characteristic integrals of (C), then the general solution of (L) is

$$F(f_1, f_2, \ldots, f_n, Z) = 0.$$

To treat (II.5.7) in n dimensions, let $x_0 = t$, write $\mathbf{x}$ for $(x_1, x_2, \ldots, x_n)$, and set $Z \equiv \log \rho$. Then $P_0 = 1$, $P_i = \dot{x}_i$, and, by (II.5-6), $R = -\dot{J}/J$. Hence n members of (C) can be written in the form $d\mathbf{x} = \dot{\mathbf{x}}\,dt$, so n families of characteristic curves are provided by the path-lines of the body-points. Thus χ_κ^{-1} denotes n characteristic integrals. An $(n+1)^{\text{th}}$ integral can be obtained

[1] Formally, if the components of $\mathbf{L}$ are $\mathbf{L}^k_m$,

$$\det \mathbf{L} = \varepsilon^{k_1 k_2 \ldots k_n} L^1_{k_1} L^2_{k_2} \cdots L^n_{k_n} = \varepsilon_{k_1 k_2 \ldots k_n} L^{k_1}_1 L^{k_2}_2 \cdots L^{k_n}_n.$$
$$(\text{adj}\,\mathbf{L})^p_q = \varepsilon_{k_1 k_2 \ldots k_n} L^{k_1}_1 \cdots L^{k_{p-1}}_{p-1}\,\delta^{k_p}_q L^{k_{p+1}}_{p+1} \cdots L^{k_n}_n.$$

by integrating $dt = dZ/R = -d \log \rho / d \log J$, the resulting integral being ρJ. Thus the general solution of (II.5-7) is

$$F(\chi_\kappa^{-1}, \rho J) = 0,$$

and this is (II.5-4).

Note: The method of characteristics for linear partial differential equations of first order was invented by LAGRANGE on the basis of this example and the one in Exercise II.6.2, both of these having arisen in hydrodynamics. The trivial generalization of the particular case (II.5-7) from 3 dimensions to n was obtained by LIOUVILLE; it is the only one of the several statements physicists call "Liouville's theorem in statistical mechanics" that has any connection with LIOUVILLE.

A rigorous treatment of LAGRANGE's theory in the large is intricate. Most modern books on partial differential equations omit it. A clear and precise treatment of the local theory may be found in Chapter IX of Volume 2 of C. DE LA VALLÉE POUSSIN's *Cours d'Analyse Infinitésimale*, 7th ed., Louvain, Librairie Universitaire, and Paris, Gauthier-Villars, 1937.

II.5.3 By the theorem of integral calculus used to derive (II.2-6), the volume of $\chi_1(\mathscr{P}, t)$ is given by

$$\int_{\chi_1(\mathscr{P},t)} dV = \int_{\chi_2(\mathscr{P},t)} J\, dV.$$

The condition of isochoric motion is therefore locally equivalent to $J = 1$. To complete the exercise, use (II.5-6) and (II.5-7).

II.5.4 For a plane motion $(\text{II.5-8})_1$ becomes

$$\partial_x \dot{x} + \partial_y \dot{y} = 0,$$

where x, y are rectangular Cartesian co-ordinates and $\dot{x}$, $\dot{y}$ are the corresponding components of the velocity field. This is a necessary and sufficient condition that in each simply connected region there be a single-valued function q such that

$$\dot{x} = -\partial_y q, \qquad \dot{y} = \partial_x q,$$

and this is (II.5-9). (A proof that q is single-valued for a broad class of multiply connected regions may be found in CFT, §161.) Clearly $\dot{\mathbf{x}} \cdot \nabla q = 0$, so the stream lines are normal to the normals of the curves $q(\cdot, t) = \text{const.}$

II.5.5

$$n_k\, dA(\mathbf{x}) = \varepsilon_{kpq}\, dx^p\, dx^q = \varepsilon_{kpq} F^p{}_\alpha F^q{}_\beta\, dX^\alpha\, dX^\beta.$$

(An interpretation for this transformation law is given in §II.13.) The result follows by comparing both sides of (II.5-10).

II.6.1 As (II.6-11) suggests, take $\mathfrak{A}/\rho$ for Ψ in (II.6-9), then use (II.5-7), (II.6-3), and the divergence theorem. The value of left-hand side of (II.6-10) is the time derivative of $\int \mathfrak{A}\, dV$ for a given part $\mathscr{P}$ of $\mathscr{B}$; the operation denoted by a prime is the time derivative of $\int \mathfrak{A}\, dV$ obtained if, neglecting the motion χ, we confuse $\mathscr{P}$ with its present shape $\chi(\mathscr{P}, t)$. The difference between these is explained and evaluated by the third term, which gives the rate of increase of $\int \mathfrak{A}\, dV$ for $\mathscr{P}$ effected by the motion of body-points out of or into the present shape of $\mathscr{P}$. To complete the exercise, refer to the definitions (I.8-1) and (I.8-2), and take for $\mathfrak{A}$ first $\rho\dot{\mathbf{x}}$ and then $(\mathbf{x} - \mathbf{x}_0) \wedge \rho\dot{\mathbf{x}}$.

II.6.2 For a moving surface $\mathscr{S}$, choose a particular parametric representation:

$$\mathbf{x} = \mathbf{g}(\mathbf{A}, t),$$

and think of $\mathbf{A}$ as being attached permanently to a point on $\mathscr{S}$ as it progresses. Then the velocity $\mathbf{u}$ at $\mathbf{A}$ is given by

$$\mathbf{u} \equiv \partial_t \mathbf{g}.$$

Of course the field $\mathbf{u}$ so defined on $\mathscr{S}$ depends upon the particular parametrization used to describe $\mathscr{S}$. Now suppose the parameter $\mathbf{A}$ to have been eliminated, so that an equation for $\mathscr{S}$ is

$$f(\mathbf{x}, t) = 0.$$

All the infinitely many different parametrizations of $\mathscr{S}$ will lead to one and the same set of points satisfying a relation of this kind, and this relation characterizes $\mathscr{S}$ over an interval of time:

$$f(\mathbf{g}(\mathbf{A}, t), t) = 0 \tag{1}$$

for each fixed $\mathbf{A}$ in any parametrization. If $h(\mathbf{x}, t) = 0$ is another equation for $\mathscr{S}$, then h is an invertible function of f. Differentiation of (1) yields

$$f' + (\operatorname{grad} f) \cdot \mathbf{u} = 0. \tag{2}$$

Now the unit normal to $\mathscr{S}$ in the direction of increasing f is given by

$$\mathbf{n} = \frac{\operatorname{grad} f}{|\operatorname{grad} f|}.$$

Therefore (2) asserts that

$$\mathbf{u} \cdot \mathbf{n} = -\frac{f'}{|\operatorname{grad} f|} = -\frac{h'}{|\operatorname{grad} h|},$$

h being any differentiable function of f. Because the right-hand side is independent of the parametrization, so is the left-hand side. Thus what we have defined as the speed of displacement S_n is in fact the common normal speed of advance of all possible assignments of velocity to points on $\mathscr{S}$.

II.6.3 For the method of characteristics, see Exercise II.5.2. In the present instance $R = 0$, and f is the unknown function. Thus the general solution is

$$F(\chi_\kappa^{-1}, f) = 0,$$

so the body-points $\mathbf{X}$ that lie upon $f(\mathbf{x}, t) = \text{const.}$ at any one time lie upon it always.

II.6.4 Note that $g(\mathbf{x}, t) = g(\chi_\kappa(\mathbf{X}, t), t) \equiv G(\mathbf{X}, t) = 0$, and

$$\mathbf{n}_\kappa = \frac{\text{Grad } G(\mathbf{X}, t)}{|\text{Grad } G(\mathbf{X}, t)|} = \frac{\mathbf{F}^{\mathrm{T}} \text{ grad } g}{|\text{Grad } G(\mathbf{X}, t)|}.$$

A little simplification gives (II.6-20). If $\mathscr{S}_\kappa$ is not a material surface, then at different times different body points will lie on $\mathscr{S}_\kappa$. Of course (II.6-21) is merely an application of (II.6-16). To get (II.6-22), use $(\text{II.6-3})_1$ and (II.6-16).

II.9.1 The common proof starts from the assumption $\mathbf{F}\mathbf{F}^{\mathrm{T}} = \mathbf{1}$ and by differentiating it and using the fact that $F^k_{\alpha,\beta} = F^k_{\beta,\alpha}$ concludes that $\mathbf{F} = \text{const.}$ GURTIN & WILLIAMS have found an elegant proof that does not require $\mathbf{F}$ to be differentiable. Let $\mathbf{f}$ be a differentiable function of place $\mathbf{z}$ in some open connected set $\mathscr{T}$ on which $(\nabla \mathbf{f})(\nabla \mathbf{f})^{\mathrm{T}} = \mathbf{1}$. Then $\det \nabla \mathbf{f} = \pm 1$. If $\mathbf{x}_0 \in \mathscr{T}$, there is an open ball $\mathscr{S}$ such that $\mathbf{x}_0 \in \mathscr{S} \subset \mathscr{T}$ and that $\mathbf{f}$ is invertible in $\mathscr{S}$. If $\mathbf{x} \in \mathscr{S}$ and $\mathbf{y} \in \mathscr{S}$, let $\mathscr{C}$ be the line segment from $\mathbf{y}$ to $\mathbf{x}$. Then

$$\mathbf{f}(\mathbf{x}) - \mathbf{f}(\mathbf{y}) = \int_{\mathscr{C}} \nabla \mathbf{f}(\mathbf{z}) \, d\mathbf{z}.$$

Because $|\mathbf{Q}\mathbf{u}| = |\mathbf{u}|$ for any orthogonal tensor $\mathbf{Q}$ and any vector $\mathbf{u}$,

$$|\mathbf{f}(\mathbf{x}) - \mathbf{f}(\mathbf{y})| \leqq \int_{\mathscr{C}} |\nabla \mathbf{f}(\mathbf{z}) \, d\mathbf{z}| = \int_{\mathscr{C}} ds = |\mathbf{x} - \mathbf{y}|.$$

Just the same argument applies to $\mathbf{f}^{-1}$:

$$|\mathbf{x} - \mathbf{y}| = |\mathbf{f}^{-1}(\mathbf{f}(\mathbf{x})) - \mathbf{f}^{-1}(\mathbf{f}(\mathbf{y}))| \leqq |\mathbf{f}(\mathbf{x}) - \mathbf{f}(\mathbf{y})|.$$

Comparison of these two results yields

$$|\mathbf{f}(\mathbf{x}) - \mathbf{f}(\mathbf{y})| = |\mathbf{x} - \mathbf{y}|.$$

Therefore **f** preserves distances in $\mathscr{S}$. Since $\mathscr{T}$ is connected, the assertion follows. If $\bar{\mathbf{U}} = \mathbf{U}$, then grad $(\bar{\chi}_\kappa \circ \chi_\kappa^{-1}) = \bar{\mathbf{F}}\mathbf{F}^{-1} = \bar{\mathbf{R}}\mathbf{R}^{\mathrm{T}}$, which must be constant in virtue of the preceding.

II.9.2 $\mathbf{C}(\tau) = \mathbf{F}^{\mathrm{T}}(\tau)\mathbf{F}(\tau)$. Use (II.8-7), and simplify.

II.9.3 The principal stretches v are the roots of $\det(\mathbf{B} - v^2\mathbf{1}) = 0$. $\mathbf{B} = \mathbf{RCR}^{\mathrm{T}}$, and

$$[\mathbf{R}] = \begin{Vmatrix} \cos\theta & \sin\theta & 0 \\ -\sin\theta & \cos\theta & 0 \\ 0 & 0 & 1 \end{Vmatrix}$$

since the principal axes are rotated about the x_3-axis.

II.9.4

$$\|g_{\alpha\beta}\| = \begin{Vmatrix} 1 & 0 & 0 \\ 0 & R^2 & 0 \\ 0 & 0 & 0 \end{Vmatrix}, \qquad \|g^{\alpha\beta}\| = \begin{Vmatrix} 1 & 0 & 0 \\ 0 & R^{-2} & 0 \\ 0 & 0 & 1 \end{Vmatrix}, \qquad \text{etc.}$$

II.9.5 Calculate the physical components of **B** by employing

$$B_{\langle km \rangle} = B^{km}\sqrt{g_{kk}\,g_{mm}} \qquad \text{(no summation)},$$

and compare with $(\text{II.9-13})_1$.

II.9.6 Use (II.6-5) and $(\text{II.9-5})_4$ to get (II.9-19).

II.11.1 EULER proved the result by first differentiating the component equations $\dot{x}_{k,m} + \dot{x}_{m,k} = 0$. The elegant proof of GURTIN & WILLIAMS does not require that **D** be differentiable. Let the notations be as in Exercise II.9.1. Then

$$[\mathbf{f}(\mathbf{x}) - \mathbf{f}(\mathbf{y})] \cdot [\mathbf{x} - \mathbf{y}] = \int_{\mathscr{C}} [\mathbf{x} - \mathbf{y}] \cdot \nabla\mathbf{f}(\mathbf{z})\, d\mathbf{z}.$$

Since $\mathscr{C}$ is a straight line, $d\mathbf{z}$ is parallel to $\mathbf{x} - \mathbf{y}$, so the integrand is 0 if $\nabla\mathbf{f}$ is skew. Therefore

$$[\mathbf{f}(\mathbf{x}) - \mathbf{f}(\mathbf{y})] \cdot [\mathbf{x} - \mathbf{y}] = 0. \tag{A}$$

This condition is equivalent to (I.10.1) in the present notation:

$$\mathbf{f} = \mathbf{c} + \mathbf{W}(\mathbf{x} - \mathbf{x}_0), \qquad \mathbf{W}^{\mathrm{T}} = -\mathbf{W} = \text{const.} \tag{B}$$

Indeed, that (B) ⇒ (A) is immediate. Conversely, by differentiating (A) with respect to **x** we obtain

$$\nabla\mathbf{f}(\mathbf{x})^{\mathrm{T}}(\mathbf{x} - \mathbf{y}) + \mathbf{f}(\mathbf{x}) - \mathbf{f}(\mathbf{y}) = \mathbf{0}.$$

Differentiation of this result with respect to **y** yields

$$-\nabla\mathbf{f}(\mathbf{x})^{\mathrm{T}} - \nabla\mathbf{f}(\mathbf{y}) = \mathbf{0}.$$

Thus $\nabla\mathbf{f}$ is both constant and skew.

II.11.2 $\mathbf{G} = \partial_u \mathbf{F}(u)\mathbf{F}^{-1}(t)|_{u=t}$. Use the polar decomposition of $\mathbf{F}(u)$ and $\mathbf{F}(t)$, and then carry out the indicated differentiation to obtain

$$\mathbf{G} = \dot{\mathbf{R}}\mathbf{R}^{-1} + \mathbf{R}\dot{\mathbf{U}}\mathbf{U}^{-1}\mathbf{R}^{-1}.$$

This can be written as

$$\begin{aligned}\mathbf{D} + \mathbf{W} = \dot{\mathbf{R}}\mathbf{R}^{-1} &+ \tfrac{1}{2}\mathbf{R}(\dot{\mathbf{U}}\mathbf{U}^{-1} - \mathbf{U}^{-1}\dot{\mathbf{U}})\mathbf{R}^{\mathrm{T}} \\ &+ \tfrac{1}{2}\mathbf{R}(\dot{\mathbf{U}}\mathbf{U}^{-1} + \mathbf{U}^{-1}\dot{\mathbf{U}})\mathbf{R}^{\mathrm{T}}.\end{aligned}$$

Use uniqueness of the additive decomposition of a tensor into symmetric and skew parts to get $(\text{II.11-13})_{2,3}$. To get $(\text{II.11-13})_1$, start from $(\text{II.11-13})_3$, and use $\mathbf{C} = \mathbf{U}^2$ and the polar decomposition theorem. (In fact, $(\text{II.11-13})_1$ is easy to derive directly, but from it as a starting point there seems to be no obvious way to reach $(\text{II.11-13})_2$.) The last relation follows from $\mathbf{B} = \mathbf{F}\mathbf{F}^{\mathrm{T}}$ and $\dot{\mathbf{F}}|_{\mathbf{F}=\mathbf{1}} = \mathbf{G}$.

II.11.4 To get (II.11-18), use Leibniz's rule to differentiate $\mathbf{F}_t(\tau)^{\mathrm{T}}\mathbf{F}_t(\tau)$. To get (II.11-19), first prove that

$$\overset{(n)}{\mathbf{C}} = \mathbf{F}^{\mathrm{T}}\mathbf{A}_n\mathbf{F}.$$

(A prescription for proving this formula is given below in §II.14, where it is listed as (II.14-15).) Hence

$$\begin{aligned}\overset{(n+1)}{\mathbf{C}} &= \mathbf{F}^{\mathrm{T}}\mathbf{A}_{n+1}\mathbf{F}, \\ &= \mathbf{F}^{\mathrm{T}}\dot{\mathbf{A}}_n\mathbf{F} + \dot{\mathbf{F}}^{\mathrm{T}}\mathbf{A}_n\mathbf{F} + \mathbf{F}^{\mathrm{T}}\mathbf{A}_n\dot{\mathbf{F}}.\end{aligned}$$

Now use (II.11-5).

II.11.5 A formula for the derivative of the determinant of an invertible tensor was given in Exercise II.5.1. Differentiating it yields

$$\begin{aligned}(\det \mathbf{L})^{\cdot\cdot} &= (\det \mathbf{L})\,\mathrm{tr}[\ddot{\mathbf{L}}\mathbf{L}^{-1} - (\dot{\mathbf{L}}\mathbf{L}^{-1})^2] + (\det \mathbf{L})\,\mathrm{tr}(\dot{\mathbf{L}}\mathbf{L}^{-1}), \\ (\det \mathbf{L})^{\cdot\cdot\cdot} &= (\det \mathbf{L})\,\mathrm{tr}[\dddot{\mathbf{L}}\mathbf{L}^{-1} - 3\ddot{\mathbf{L}}\mathbf{L}^{-1}\dot{\mathbf{L}}\mathbf{L}^{-1} + 2(\dot{\mathbf{L}}\mathbf{L}^{-1})^3] \\ &\quad + (\det \mathbf{L})^{\cdot}(\cdots) + (\det \mathbf{L})^{\cdot\cdot}(\cdots),\end{aligned}$$

etc. If $\det \mathbf{L} = 1$ always, these relations reduce to

$$\begin{aligned}\mathrm{tr}[\ddot{\mathbf{L}}\mathbf{L}^{-1} - (\dot{\mathbf{L}}\mathbf{L}^{-1})^2] &= 0, \\ \mathrm{tr}(\dddot{\mathbf{L}}\mathbf{L}^{-1} - 3\ddot{\mathbf{L}}\mathbf{L}^{-1}\dot{\mathbf{L}}\mathbf{L}^{-1} + 2(\dot{\mathbf{L}}\mathbf{L}^{-1})^3] &= 0,\end{aligned}$$

etc. In an isochoric motion, we may substitute $\mathbf{C}_t(u)$ for $\mathbf{L}$. Putting $u = t$, followed by use of the definition (II.11-17), yields $(\text{II.11-20})_{2,3}$. The term involving the time derivative of highest order in the formula for $(\det \mathbf{L})^{(n)}$ is $(\det \mathbf{L})\,\mathrm{tr}(\overset{(n)}{\mathbf{L}}\mathbf{L}^{-1})$, so the general assertion of the exercise follows.

II.11.6 Let $\mathbf{x}$ be a point of $\mathscr{S}$, and let $\mathbf{k}$ be a vector in the tangent plane of $\mathscr{S}$ at $\mathbf{x}$. Then there are points $\mathbf{y}(h)$ on $\mathscr{S}$ such that

$$\mathbf{y}(h) = \mathbf{x} + h\mathbf{k} + \mathbf{o}(h) \qquad \text{as} \quad h \to 0.$$

Therefore

$$\mathbf{Gk} \equiv \lim_{h\to 0} \frac{\dot{\mathbf{x}}(\mathbf{x} + h\mathbf{k}) - \dot{\mathbf{x}}(\mathbf{x})}{h},$$
$$= \lim_{h\to 0} \frac{\dot{\mathbf{x}}(\mathbf{y}(h)) - \dot{\mathbf{x}}(\mathbf{x})}{h}.$$

If $\dot{\mathbf{x}}$ vanishes on $\mathscr{S}$, the difference quotient on the right-hand side vanishes, so (II.11-26) follows. Thus

$$\mathbf{Ge} = \mathbf{0}, \qquad \mathbf{Gf} = \mathbf{0},$$

$\mathbf{e}$ and $\mathbf{f}$ being defined as in the preceding text. Replacing $\mathbf{G}$ by $\mathbf{D} + \mathbf{W}$ and then using (II.11.24) shows that

$$\mathbf{De} = -\mathbf{We} = \mathbf{0},$$
$$\mathbf{Df} = -\mathbf{Wf} = -\tfrac{1}{2}W\mathbf{n}.$$

Now since

$$\mathbf{Dn} = (\mathbf{n}\cdot\mathbf{Dn})\mathbf{n} + (\mathbf{e}\cdot\mathbf{Dn})\mathbf{e} + (\mathbf{f}\cdot\mathbf{Dn})\mathbf{f},$$
$$E = \mathbf{n}\cdot\mathbf{Dn} + \mathbf{e}\cdot\mathbf{De} + \mathbf{f}\cdot\mathbf{Df},$$

the two preceding equations show that

$$\mathbf{Dn} = (\mathbf{n}\cdot\mathbf{Dn})\mathbf{n} - \tfrac{1}{2}W\mathbf{f},$$
$$E = \mathbf{n}\cdot\mathbf{Dn}.$$

Hence (II.11-27) follows. Because $\mathbf{De} = \mathbf{0}$, $\mathbf{e}$ is a principal axis of stretching, the corresponding principal stretching is 0, and $\det \mathbf{D} = 0$. The second principal invariant of $\mathbf{D}$ is $-\frac{1}{4}W^2$. Thus the characteristic equation of $\mathbf{D}$ is

$$D(D^2 - ED - \tfrac{1}{4}W^2) = 0,$$

the solutions of which are (II.11-28).

II.11.7 $(\mathbf{F}^{\mathrm{T}}\dot{\mathbf{F}})^{\cdot} = \mathbf{F}^{\mathrm{T}}\ddot{\mathbf{F}}$ + a symmetric tensor. By choosing n successively as 1 and 2 in (II.11-15) show that $(\mathbf{F}^{\mathrm{T}}(\nabla\dot{\mathbf{x}})\mathbf{F})^{\cdot} = \mathbf{F}^{\mathrm{T}}(\nabla\ddot{\mathbf{x}})\mathbf{F}$ + a symmetric tensor. Take the skew part of this relation to get (II.11-29).

II.11.8 By (II.11-32)

$$\mathbf{W} = \mathbf{0} \quad \Leftrightarrow \quad \mathbf{W}_{\kappa} = \mathbf{0}.$$

II.11.9 Use (II.11-15) to show that

$$\mathbf{G}_2 = \ddot{\mathbf{F}}\mathbf{F}^{-1} = (\mathbf{GF})^{\cdot}\mathbf{F}^{-1} = \dot{\mathbf{G}} + \mathbf{G}^2.$$

The skew part is (II.11-34).

II.11.10 By (II.11-34)

$$(\tfrac{1}{2}|\mathbf{W}|^2)^{\cdot} = \mathbf{W} \cdot [\mathbf{W}_2 - \mathbf{DW} - \mathbf{WD}].$$

If dim $\mathscr{V} = 3$, verify the following identity for any skew tensor $\mathbf{W}$ and any symmetric tensor $\mathbf{D}$:

$$\mathbf{W} \cdot (\mathbf{DW} + \mathbf{WD}) = |\mathbf{W}|^2(\text{tr}\,\mathbf{D} - \mathbf{n} \cdot \mathbf{Dn}),$$

$\mathbf{n}$ being any unit vector in the nullspace of $\mathbf{W}$. Use of (II.11-10) yields (II.11-35), from which the conclusion of the exercise is obvious.

II.12.1 The ellipsoid in $\boldsymbol{\kappa}(\mathscr{B})$ is swept out by the termini of vectors $\mathbf{m}_\kappa$ such that

$$\text{const.} = |\mathbf{m}|^2 = |\mathbf{Fm}_\kappa \cdot \mathbf{Fm}_\kappa| = \mathbf{m}_\kappa \cdot \mathbf{Cm}_\kappa.$$

Let $\mathbf{e}_1$, $\mathbf{e}_2$, $\mathbf{e}_3$ be an orthonormal set of unit proper vectors of $\mathbf{C}$, so that $\mathbf{Ce}_j = v_j^2\mathbf{e}_j$, where v_j is the principal stretch corresponding to $\mathbf{e}_j$. Let the co-ordinates of $\mathbf{m}_\kappa$ with respect to this basis be m^i. Then the above equation for the ellipsoid assumes the co-ordinate form

$$\sum_{i=1}^{3} (m^i)^2 v_i^2 = \text{const.}$$

Therefore, the principal axes of the ellipsoid are the principal axes of strain at $\mathbf{X}$, and the lengths of the semi-axes are inversely proportional to the corresponding squared principal stretches. The extremal properties of the principal stretches correspond inversely to the extremal properties of the lengths of vectors to points on the ellipsoid.

That the principal axes are not sheared, is the same as the statement $\cos\theta_{(\mathbf{e}_i,\, \mathbf{e}_j)} = \delta_{ij}$, which is an immediate consequence of (II.12-6).

Since (II.12-1) can be written in the form

$$\begin{aligned}\chi_\kappa(\mathbf{X}, t) &= \mathbf{X}_0 + (\mathbf{x}_0(t) - \mathbf{X}_0) + \mathbf{R}(t)\mathbf{U}(t)(\mathbf{X} - \mathbf{X}_0),\\ &= \mathbf{X}_0 + (\mathbf{x}_0(t) - \mathbf{X}_0) + \mathbf{V}(t)\mathbf{R}(t)(\mathbf{X} - \mathbf{X}_0),\end{aligned}$$

the last statement follows immediately by aid of (II.9-4).

II.12.2 Differentiate (II.12-6) after writing it as

$$v_{\mathbf{n}_\kappa} v_{\mathbf{m}_\kappa} \cos \theta_{(\mathbf{n}_\kappa, \mathbf{m}_\kappa)} = \mathbf{n}_\kappa \cdot \mathbf{U}^2 \mathbf{m}_\kappa .$$

II.13.1 $\int_{\chi(\mathscr{C}, t)} \mathbf{f} \cdot d\mathbf{x} = \int_{\kappa(\mathscr{C})} \mathbf{f}(\mathbf{X}, t) \cdot \mathbf{F}(\mathbf{X}, t)\, d\mathbf{X}$. Now on the right-hand side differentiation can be performed under the integral sign.

II.13.2 The volume V of a tetrahedron whose vertices are $\mathbf{x}_0(t)$ and the termini of $\mathbf{p}_1$, $\mathbf{p}_2$, and $\mathbf{p}_3$, is given in terms of the components p_a^k as follows:

$$V = \varepsilon_{mnq} p_1^m p_2^n p_3^q .$$

Hence

$$\begin{aligned} \dot{V} &= \varepsilon_{mnq}[G_r^m p_1^r p_2^n p_3^q + G_r^n p_1^m p_2^r p_3^q + G_r^q p_1^m p_2^n p_3^r], \\ &= (\operatorname{tr} \mathbf{G})V. \end{aligned}$$

II.13.3 Put $d\mathbf{x} = \mathbf{t}\, ds$, $\mathbf{f} = f\mathbf{t}$ in (II.13-1).

II.13.4 By (II.13-6) and (II.11-8)

$$\mathbf{e}^c = \dot{\mathbf{e}} - (\mathbf{D} + \mathbf{W})\mathbf{e}. \tag{A}$$

Since $\mathbf{e} \cdot \dot{\mathbf{e}} = 0$ and $\mathbf{W}$ is skew,

$$\mathbf{e} \cdot \mathbf{e}^c = -\mathbf{e} \cdot \mathbf{D}\mathbf{e}.$$

Substituting this into (II.13-14) and then putting the result into (A) yields (II.13-15). If $\mathbf{D}\mathbf{e} = d\mathbf{e}$, (II.13-15) reduces to

$$\dot{\mathbf{e}} = \mathbf{W}\mathbf{e}.$$

Cf. the discussion of rigid motion in §I.10. For any vector $\mathbf{m}$ we obtain from (II.13-15)

$$\mathbf{m} \cdot \dot{\mathbf{e}} = \mathbf{m} \cdot \mathbf{D}\mathbf{e} + \mathbf{m} \cdot \mathbf{W}\mathbf{e} - (\mathbf{e} \cdot \mathbf{D}\mathbf{e})(\mathbf{m} \cdot \mathbf{e}).$$

Hence (II.13-16) follows. All these results apply to the position vectors $\mathbf{p}$ in a homogeneous motion because $\mathbf{p}^c = \mathbf{0}$. Since $\mathbf{m} \cdot \mathbf{n} = \cos \theta_{(\mathbf{m}, \mathbf{n})}$, $(\text{II.13-16})_1$ reduces to (II.12-10) if $\mathbf{m} \cdot \mathbf{n} = 0$ at the instant in question. Likewise $(\text{II.13-16})_2$ reduces to (II.12-15).

II.13.5 Referring to the notations and formulae in §AII.15, show that if $\mathbf{f} = \mathbf{S}_\times$, then

$$\mathbf{f}^c = (\dot{\mathbf{S}} + \mathbf{S}\mathbf{G} + \mathbf{G}^T\mathbf{S} - E\mathbf{S})_\times .$$

Then apply the condition (II.13-12) and recall that if $\mathbf{S}$ and $\mathbf{U}$ are skew and non-null, then $\mathbf{S}_\times \| \mathbf{U}_\times$ if and only if $\mathbf{S}$ and $\mathbf{U}$ commute.

II.13.6 The argument is phrased in terms of *vortex tubes*. These are surfaces swept out by the vortex lines through the points of some circuit nowhere tangent to the axes of spin. The flux of spin has the same value, at a given instant, for all like-oriented surfaces bounded by circuits embracing the tube just once (HELMHOLTZ's *First Vorticity Theorem*). KELVIN's argument, a classic example of conceptual mathematics, may be found in §128 of CFT.

II.13.7 POINCARÉ's Theorem makes the hypothesis equivalent to $\mathbf{W}\mathbf{W}_2 = \mathbf{W}_2\mathbf{W}$. Exercise II.11.10 makes the differential relation equivalent to $\mathbf{W} \cdot \mathbf{W}_2 = 0$. If neither $\mathbf{W}$ nor $\mathbf{W}_2$ vanishes, the two requirements are incompatible, for one requires the axes of the two tensors to coincide and the other requires that they be perpendicular to each other. If $\mathbf{W} = \mathbf{0}$, then $\mathbf{W}_2 = \mathbf{0}$; *cf.* (VI.6-10), below. Thus $\mathbf{W}_2 = \mathbf{0}$ is the only possibility, and clearly it is sufficient that the two conditions be compatible.

II.14.1 By (II.8-7) and (II.14-6),

$$\begin{aligned} \mathbf{F}_t^*(\tau) &= \mathbf{F}^*(\tau)(\mathbf{F}^*(t))^{-1}, \\ &= \mathbf{Q}(\tau)\mathbf{F}(\tau)\mathbf{F}^{-1}(t)\mathbf{Q}(t)^{\mathrm{T}}, \\ &= \mathbf{Q}(\tau)\mathbf{F}_t(\tau)\mathbf{Q}(t)^{\mathrm{T}}. \end{aligned}$$

Hence

$$\mathbf{R}_t^*(\tau)\mathbf{U}_t^*(\tau) = \mathbf{Q}(\tau)\mathbf{R}_t(\tau)\mathbf{Q}(t)^{\mathrm{T}}[\mathbf{Q}(t)\mathbf{U}_t(\tau)\mathbf{Q}(t)^{\mathrm{T}}].$$

The result (II.14-20) follows by the uniqueness of a polar decomposition.

III.1.1

$$\mathbf{f}_{\mathrm{B}}^{\mathrm{a}} = \int_{\chi(\mathscr{B},\,t)} \mathbf{b}\, dM = M\mathbf{b},$$

$$\begin{aligned} \mathbf{F}_{\mathrm{B}}^{\mathrm{a}} &= \int_{\chi(\mathscr{B},\,t)} (\mathbf{x} - \mathbf{x}_0) \wedge \mathbf{b}\, dM, \\ &= \int_{\chi(\mathscr{B},\,t)} \mathbf{p}\, dM \wedge \mathbf{b} = \bar{\mathbf{p}} \wedge M\mathbf{b}, \end{aligned}$$

by (I.8-27).

III.1.2 By (III.1-37) and $(\mathrm{III}.1.38)_2$,

$$\int_{\mathscr{S}} \mathbf{t}_{\mathscr{S}}\, dA = -\int_{-\mathscr{S}} \mathbf{t}_{-\mathscr{S}}\, dA.$$

The Lebesgue differentiation theorem gives $\mathbf{t}_{\mathscr{S}} = -\mathbf{t}_{-\mathscr{S}}$ a.e.

III.2.1 Expand $\rho(\dot{\mathbf{x}} - \mathbf{v}) \otimes (\dot{\mathbf{x}} - \mathbf{v})\mathbf{n}$ and $\mathbf{p} \wedge [\rho(\dot{\mathbf{x}} - \mathbf{v}) \otimes (\dot{\mathbf{x}} - \mathbf{v})\mathbf{n}]$; integrate over the shape of $\mathscr{B}$; use the divergence theorem; note that

$$\operatorname{div}(\rho\mathbf{p} \otimes \dot{\mathbf{x}}) = \rho\dot{\mathbf{x}} + \mathbf{p}\operatorname{div}(\rho\dot{\mathbf{x}});$$

use (II.6-6) for a motion with steady density, and note that $\dot{\mathbf{x}} \cdot \mathbf{n} = 0$ on the obstacle.

III.3.1 Choose Δr such that $A(\Delta\mathscr{A}) \leqq A(\Delta\mathscr{A}') + A(\Delta\mathscr{A}^*)$. Then (III.3-11) and $(\text{III.3-9})_2$ yield (III.3-12). Hence

$$A(\partial\Delta\mathscr{D}) = A(\Delta\mathscr{A}) + A(\Delta\mathscr{A}^*) + A(\Delta\mathscr{A}'),$$
$$V(\Delta\mathscr{D}) = o(\Delta r^3) \qquad \text{as} \quad \Delta r \to 0.$$

III.3.2 Let $\mathscr{T}'$ be the tangent plane to $\mathscr{S}$ and $\mathscr{T}$ at $\mathbf{x}$. With respect to rectangular Cartesian co-ordinates (x, y, z) with x-axis and y-axis in $\mathscr{T}'$, let $z = f(x, y)$ and $z = g(x, y)$ be the local representations of $\mathscr{S}$ and $\mathscr{T}$ near $\mathbf{x}$. Choose Δr such that when $x^2 + y^2 \leqq \Delta r^2$, $\mathscr{S}$ and $\mathscr{T}$ lie entirely between two paraboloids $z = \pm K(x^2 + y^2)$, where

$$K \equiv \max(|\partial_x^2 f|, |\partial_x\partial_y f|, |\partial_y^2 f|, |\partial_x^2 g|, |\partial_x\partial_y g|, |\partial_y^2 g|).$$

Follow the same procedure as before.

III.4.1 Let $\mathscr{C}$ be a cube the length of whose edges is ε, and let two of its faces be normal to $\mathbf{k}$. Then

$$\int_{\mathscr{C}} \mathbf{t}\, dA = 2\varepsilon^2\mathbf{k},$$

and

$$\frac{\int_{\mathscr{C}} \mathbf{t}\, dA}{A(\mathscr{C})} = \frac{1}{3}\mathbf{k}, \qquad \frac{\int_{\mathscr{C}} \mathbf{t}\, dA}{V(\mathscr{C})} = \frac{2}{\varepsilon}\mathbf{k},$$

so (III.1-14) is violated.

III.4.2 Immediate from (III.4-1) and the definition of the transpose.

III.6.1 Prove the identity

$$\operatorname{div}(\mathbf{v} \otimes \mathbf{S}) = \mathbf{v} \otimes \operatorname{div}\mathbf{S} + (\nabla\mathbf{v})\mathbf{S}^{\mathrm{T}},$$

take the skew part, set $\mathbf{v} = \mathbf{x} - \mathbf{x}_0$, and apply the divergence theorem.

III.6.2 Hold $\mathbf{x}$ fixed, and drop it from the notation; do not assume that $\mathbf{T}^{\mathrm{T}} = \mathbf{T}$. Trivially (III.6-9)$\Leftrightarrow$(III.6-10), and

$$(\text{III.6-10}) \quad \Rightarrow \quad (\text{III.6-7}) \;\&\; (\text{III.6-8}).$$

Write (III.6-8) in the form $\mathbf{n} \cdot \mathbf{Tn} = -p$ for all unit vectors $\mathbf{n}$; let $\mathbf{n} = \cos\theta \mathbf{n}_1 + \sin\theta \mathbf{n}_2$, $\mathbf{n}_1$ and $\mathbf{n}_2$ being unit vectors, and show that $\mathbf{n}_1 \cdot \mathbf{Tn}_2 = -\mathbf{n}_2 \cdot \mathbf{Tn}_1$. Hence conclude that

$$\text{(III.6-8)} \quad \Leftrightarrow \quad \mathbf{T} = -p\mathbf{1} + \mathbf{S},$$

$\mathbf{S}$ being a skew tensor. If $\mathbf{T}$ is symmetric, (III.6-10) follows. Otherwise it does not.

(III.6-7) requires that

$$|\mathbf{t}(\mathbf{n})|^2 = \mathbf{n} \cdot \mathbf{T}^{\mathrm{T}}\mathbf{Tn} = p^2.$$

The result just obtained in regard to (III.6-8) shows that $\mathbf{T}^{\mathrm{T}}\mathbf{T} = p^2\mathbf{1}$ + a skew tensor, but this latter is $\mathbf{0}$ because $\mathbf{T}^{\mathrm{T}}\mathbf{T}$ is symmetric. If $p = 0$, (III.6-10) holds trivially; if $p \neq 0$, we have shown that $p^{-1}\mathbf{T}$ is an orthogonal tensor, say $-\mathbf{Q}$. Then

$$\mathbf{n} \cdot \mathbf{t}(\mathbf{n}) = -p\mathbf{n} \cdot \mathbf{Qn}.$$

If $\mathbf{R}$ is the rotation such that $\mathbf{Q} = \pm\mathbf{R}$, show that

$$\mathbf{n} \cdot \mathbf{Rn} = 1 - 2n_\perp^2 \sin^2 \tfrac{1}{2}\theta,$$

$n_\perp$ being the magnitude of the component of $\mathbf{n}$ normal to the axis of $\mathbf{R}$, and θ being the angle of $\mathbf{R}$. In order that $\mathbf{n} \cdot \mathbf{Rn} > 0$ $\forall\mathbf{n}$ it is necessary and sufficient that $1 - 2\sin^2 \tfrac{1}{2}\theta > 0$. Thus (III.6-7) is equivalent to

$$\mathbf{T} = -p\mathbf{R}, \qquad 0 \leqq \theta < \tfrac{1}{2}\pi \quad \text{or} \quad \tfrac{3}{2}\pi < \theta \leqq 2\pi.$$

If $\mathbf{T}$ is symmetric, $\mathbf{R} = \mathbf{R}^{\mathrm{T}}$, and hence $\theta = 0$ or π. Since the latter alternative is excluded by the result just obtained, $\mathbf{R} = \mathbf{1}$. Thus (III.6-10) follows from (III.6-7) if $\mathbf{T}$ is symmetric. Otherwise it does not.

III.6.3 The result is really a special case of NOLL's theorem in §I.12 but is more than a century older. For an independent proof, hold t fixed and consider the rigid motion defined as follows by a constant vector $\mathbf{v}_0$, the position vector $\mathbf{p}$, and a constant skew tensor $\mathbf{W}_0$: $\mathbf{v} = \mathbf{v}_0 + \mathbf{W}_0\mathbf{p}$. Since $\mathbf{a} \cdot \mathbf{Sb} = \mathbf{S} \cdot (\mathbf{a} \otimes \mathbf{b})$ for any skew tensor $\mathbf{S}$,

$$P = \mathbf{v}_0 \cdot \left[\int_{\partial\chi(\mathscr{P},t)} \mathbf{t}\, dA + \int_{\chi(\mathscr{P},t)} \rho\mathbf{b}\, dV\right]$$
$$+ \mathbf{W}_0 \cdot \left[\int_{\partial\chi(\mathscr{P},t)} \mathbf{t} \otimes \mathbf{p}\, dA + \int_{\chi(\mathscr{P},t)} \rho\mathbf{b} \otimes \mathbf{p}\, dV\right].$$

In order that $P = 0$ for all choices of $\mathbf{v}_0$ and $\mathbf{W}_0$, it is necessary and sufficient that the first bracket vanish and the second bracket be symmetric.

III.6.4 Use (I.14-1), (III.6-11), the divergence theorem, (III.6-1), and (II.11-8). In a rigid motion $\mathbf{D} = \mathbf{0}$. In an isochoric motion $\operatorname{tr} \mathbf{D} = 0$.

III.6.5 Substitute from (III.1-7) and (III.6-14) into (III.6-11) and use (II.6-8) to obtain (III.6-16). Then (III.6-17) and (III.6-18) are easy to obtain.

III.7.1

$$\left(\int_{\mathscr{A}} \mathbf{p} \otimes \mathbf{Te} \, dA\right)\mathbf{e} = \int_{\mathscr{A}} (\mathbf{e} \cdot \mathbf{Te})\mathbf{p} \, dA,$$

$$= \frac{F}{A(\mathscr{A})} \int_{\mathscr{A}} \mathbf{p} \, dA,$$

$$= F\mathbf{p}_0(\mathscr{A}).$$

Using subscripts 1 and 2 to refer to quantities associated with the two plane, parallel faces, if $\mathbf{n}_1 = \mathbf{e}$ we must take $\mathbf{n}_2$ as $-\mathbf{e}$, so (III.7-4) yields

$$\mathbf{e} \cdot \mathbf{Te} = \frac{1}{V(\mathscr{S})}[F_1\mathbf{p}_0(\mathscr{A}_1) - F_2\mathbf{p}_0(\mathscr{A}_2)] \cdot \mathbf{e}.$$

Equilibrium of forces requires that $F_1 = F_2$; equilibrium of moments, that $\mathbf{p}_0(\mathscr{A}_1) - \mathbf{p}_0(\mathscr{A}_2)$ be parallel to $\mathbf{e}$.

III.7.2 Taking Ψ as $\mathbf{p} \otimes \mathbf{p}$ in (III.7.3) yields at once

$$\overline{T_{rk} p_q} + \overline{T_{rq} p_k} = M_{kqr}.$$

Forming the combination indicated by (III.7-10)$_1$ and then using Cauchy's Second Law yields

$$\overline{p_k T_{qr}} = L_{kqr}.$$

III.8.1 Let the constant g denote the gravitational acceleration, let ρ denote the density of the heavy liquid, and let z denote the distance downward from the surface of the liquid. Then $p = \rho g z$ on the surface of the submerged part of the body, say $\mathscr{S}$. The resultant surface force and surface torque upon $\chi(\mathscr{B}, t)$ are, respectively,

$$\mathbf{f}_{\mathrm{C}}^{\mathrm{a}} = -\rho g \int_{\mathscr{S}} z\mathbf{n} \, dA,$$

$$\mathbf{F}_{\mathrm{C}}^{\mathrm{a}} = -\rho g \int_{\mathscr{S}} (\mathbf{x} - \mathbf{x}_0) \wedge z\mathbf{n} \, dA.$$

The formulae used to denote the two integrands serve to extend them smoothly to the interior of the part of $\chi(\mathscr{B}, t)$ below the plane $z \equiv 0$. (This fact expresses STEVIN's *Principle of Solidification*: The load exerted by one part of a heavy fluid body upon another is unchanged if either is replaced by a rigid solid.) Thus we can apply Green's transformation to express the two surface integrals as volume integrals over the submerged part $\mathscr{V}$. Since grad $z = \mathbf{k}$, a unit vector pointing downwards,

$$\int_{\mathscr{S}} z\mathbf{n}\, dA = \left(\int_{\mathscr{V}} dV\right)\mathbf{k} = V(\mathscr{V})\mathbf{k}.$$

Likewise

$$\int_{\mathscr{S}} (\mathbf{x} - \mathbf{x}_0) \wedge z\mathbf{n}\, dA = \int_{\mathscr{V}} (\mathbf{x} - \mathbf{x}_0)\, dV \wedge \mathbf{k},$$

$$= \mathbf{p}_{\mathrm{m}} \wedge V(\mathscr{V})\mathbf{k},$$

$\mathbf{p}_{\mathrm{m}}$ being the position vector of the center of buoyancy.[1]

To consider a heavy body, invoke the result of Exercise III.1.1 to conclude that the load on that body is equipollent to two parallel forces: the weight of the body, acting downward at its center of mass, and the weight of the displaced fluid, acting upward at the center of buoyancy. Consideration of a simple vector diagram suffices to conclude the exercise.

III.8.2 DAY's proof is as follows. Since $\partial\chi(\mathscr{B}, t)$ has a differentiable unit normal field $\mathbf{n}$, that field can be extended smoothly into a small region containing $\partial\chi(\mathscr{B}, t)$ in its interior. A standard theorem of differential geometry asserts then that

$$k = \tfrac{1}{2} \operatorname{div} \mathbf{n}. \tag{A}$$

If $\mathbf{n}$ is any differentiable field of unit vectors, and if $\mathbf{c}$ is a constant vector field, then in a 3-dimensional space

$$\mathbf{n} \cdot \operatorname{curl}(\mathbf{n} \times \mathbf{c}) = -\mathbf{c} \cdot (\operatorname{div} \mathbf{n})\mathbf{n}, \tag{B}$$

$$\mathbf{n} \cdot \operatorname{curl}\{\mathbf{n} \times [\mathbf{c} \times (\mathbf{x} - \mathbf{x}_0)]\} = -\mathbf{c} \cdot [(\mathbf{x} - \mathbf{x}_0) \times (\operatorname{div} \mathbf{n})\mathbf{n}].$$

By Kelvin's transformation the integral of $\mathbf{n} \cdot \operatorname{div}(\mathbf{p} \wedge \mathbf{q})$ over a surface $\mathscr{S}$ is equal to the value of a line integral around $\partial\mathscr{S}$. If $\mathscr{S}$ is a surface without boundary, that value is 0. Thus the integrals of the right-hand sides of (B) over $\partial\chi(\mathscr{B}, t)$ both equal 0. Use of (A) completes the proof.

[1] To define the centroid of a region, in (I.8-27) replace M by V and $\mathscr{B}$ by $\mathscr{V}$.

IV.4.1 By hypothesis, $\mathbf{QK} = \mathbf{KQ}\ \forall \mathbf{Q}$. For $\mathbf{Q}$ take the reflection $\mathbf{R_e}$ in the plane normal to $\mathbf{e}$. Then $\mathbf{R_e v} = -\mathbf{v}$ if and only if $\mathbf{v}$ is proportional to $\mathbf{e}$. But $\mathbf{R_e K e} = \mathbf{K R_e e} = -\mathbf{Ke}$, so $\mathbf{Ke}$ is proportional to $\mathbf{e}$, no matter what be $\mathbf{e}$. Suppose now that $\mathbf{Ke} = \alpha\mathbf{e}$, $\mathbf{Kf} = \beta\mathbf{f}$, $\mathbf{K}(\mathbf{e}+\mathbf{f}) = \gamma(\mathbf{e}+\mathbf{f})$. Then $\alpha\mathbf{e} + \beta\mathbf{f} = \gamma(\mathbf{e}+\mathbf{f})$. Choosing $\mathbf{e}$ and $\mathbf{f}$ as linearly independent shows that $\alpha = \beta = \gamma$.

Note. We may ask if the condition $\mathbf{RK} = \mathbf{KR}$ for all rotations $\mathbf{R}$ implies (IV.4-23). The answer is no if the dimension of the vector space is 2, for then all rotations commute. If the dimension of the vector space is odd, the answer is obviously yes, since the tensors $\pm\mathbf{R}$ exhaust the orthogonal tensors. The answer is yes also for vector spaces of even dimension greater than 2 but is not so obvious. It is easy to prove that a symmetric tensor which commutes with every rotation is proportional to $\mathbf{1}$.

IV.4.2 Follow the procedure given in the proof of (IV.4-2).

IV.4.3 If $\mathfrak{g}$ in (IV.4-1) is an affine function of $\mathbf{F}$,

$$\mathbf{T} = \mathbf{A} + \mathsf{B}[\mathbf{F}],$$

$\mathbf{A}$ and B being constants. In order for this constitutive equation to satisfy the Principle of Material Frame-Indifference, it is necessary and sufficient that

$$\mathbf{Q}(\mathbf{A} + \mathsf{B}[\mathbf{F}])\mathbf{Q}^{\mathsf{T}} = \mathbf{A} + \mathsf{B}[\mathbf{QF}]$$

for all invertible $\mathbf{F}$ and $\mathbf{Q}$. Put $\mathbf{F} = C\mathbf{1}$, $C \neq 0$, to obtain

$$\mathbf{QAQ}^{\mathsf{T}} - \mathbf{A} = C\mathbf{f}(\mathsf{B}, \mathbf{Q})$$

for all non-zero C. If $\mathbf{f}$ vanishes, then $\mathbf{QAQ}^{\mathsf{T}} = \mathbf{A}$; if $\mathbf{f}$ does not vanish, the same conclusion follows by giving C two distinct values such as 1 and 2. Because $\mathbf{A}$ commutes with all orthogonal tensors, $\mathbf{A} = A\mathbf{1}$. The functional equation becomes

$$\mathbf{Q}(\mathsf{B}[\mathbf{F}])\mathbf{Q}^{\mathsf{T}} = \mathsf{B}[\mathbf{QF}].$$

Taking $\mathbf{Q} = -\mathbf{1}$ shows that $\mathsf{B}[\mathbf{F}] = \mathbf{0}$. (Note. $\mathbf{T} = \alpha\mathbf{1} + \beta\mathbf{V}$ does satisfy the principle, but the right-hand side is not an affine function of $\mathbf{F}$.)

IV.6.1 (IV.6-2) and (IV.6-3) are straightforward. A frame-indifferent constraint equivalent to (IV.6-1) is of the form $\mu(\mathbf{C}) = 0$, where μ vanishes if and only if λ vanishes. This statement is logically equivalent to the last sentence of the exercise. (Note that μ and λ are not claimed to be functionally dependent, though of course they may be.)

IV.7.1 Only (IV.7-13) requires care, because CAUCHY's Theorem in §IV.4 refers to a function whose domain is the space of all symmetric tensors. If the domain of $\mathfrak{g}$ is the subspace of traceless symmetric tensors, we define as follows a function $\mathfrak{f}$ on all symmetric tensors:

$$\mathfrak{f}(\mathbf{D}) \equiv \mathfrak{g}(\mathbf{D} - \tfrac{1}{3}(\operatorname{tr} \mathbf{D})\mathbf{1}).$$

If $\mathfrak{g}$ is affine and isotropic, so is $\mathfrak{f}$. Thus CAUCHY's Theorem applies to $\mathfrak{f}$. Specializing the result to traceless tensors $\mathbf{D}$ yields (IV.7-13).

IV.7.2 Since $\mathbf{t} = -p\mathbf{n}$, the last statement follows at once from the result proved in Exercise III.6.5. More generally, (III.6-17) becomes

$$\dot{K} + \dot{V} = P_c = -\int_{\partial\chi(\mathscr{P},\, t)} p\dot{\mathbf{x}} \cdot \mathbf{n}\, dV.$$

If $p =$ const. on $\partial\chi(\mathscr{P}, t)$, the right-hand side becomes

$$-p\int_{\chi(\mathscr{P},\, t)} \operatorname{div} \dot{\mathbf{x}}\, dV,$$

which vanishes since the flow is isochoric.

IV.9.1 A glance at (II.11-5) and $(\text{IV.9-7})_1$ gives the result.

IV.9.2

$$\det \mathbf{F}(t) = \det[\mathbf{F}_0(1 + t\mathbf{F}_1)] = \det \mathbf{F}_0 \det(1 + t\mathbf{F}_1),$$
$$\det(\mathbf{1} + t\mathbf{F}_1) = 1 + t(\operatorname{tr} \mathbf{F}_1) + \tfrac{1}{2}t^2[(\operatorname{tr} \mathbf{F}_1)^2 - \operatorname{tr} \mathbf{F}_1^2] + t^3 \det \mathbf{F}_1,$$

since the second principal invariant of a tensor $\mathbf{S}$ equals $\frac{1}{2}[(\operatorname{tr} \mathbf{S})^2 - \operatorname{tr} \mathbf{S}^2]$.

IV.9.3 Let $v_i(t)$ and $\mathbf{e}_i(t)$ be the proper numbers and corresponding proper vectors of $\mathbf{U}$. Note that $\dot{\mathbf{U}}\mathbf{U} = \mathbf{U}\dot{\mathbf{U}} \Leftrightarrow \dot{\mathbf{e}}_i = \mathbf{0}$. Thus (IV.9-16) follows. Write (IV.9-2) as

$$\mathbf{x} = \mathbf{x}_0(t) + \sum_{i=1}^{3} v_i(t)\mathbf{e}_i \otimes \mathbf{e}_i(\mathbf{X} - \mathbf{X}_0),$$

and conclude the required result. Taking the $\mathbf{e}_i$ as the axes, let the block be the region included by the planes $X_i = \pm a_i$. Show that it is deformed into a similar block. Since it is already proved that $v_i = a_i + b_i t$, where $a_i > 0$ and $b_i > 0$, the motion will be isochoric if and only if

$$\prod_{i=1}^{3} (a_i + b_i t) = 1 \qquad \forall t.$$

This condition holds if and only if $a_1 a_2 a_3 = 1$, $b_i = 0$.

IV.10.2 For a pure stretch $\dot{\mathbf{R}} = \mathbf{0}$, $\mathbf{R} = \mathbf{1}$. Use (II.11-13)$_2$ and (II.11-32).

IV.10.3 For an unconstrained simple body $\mathbf{T} = \mathfrak{G}(\mathbf{F}^t)$, and for the corresponding incompressible simple body $T = \rho(\varpi - h)\mathbf{1} + \mathfrak{G}(\mathbf{F}^t)$. $\mathbf{F}(t)$ is given by (IV.9-13). In the unconstrained body, every component of $\mathbf{T}$ is determined uniquely. In the incompressible body, the function h is arbitrary. For example, if $\varpi = 0$, then by choice of h we may give any one of the tractions T_{xx}, T_{yy}, and T_{zz} any value we please, *e.g.* 0.

IV.12.1 If $\mathbf{H}_1, \mathbf{H}_2 \in \mathscr{g}_\kappa$, then

$$\mathfrak{G}_\kappa(\mathbf{F}^t\mathbf{H}_1\mathbf{H}_2) = \mathfrak{G}_\kappa(\mathbf{F}^t\mathbf{H}_1) = \mathfrak{G}_\kappa(\mathbf{F}^t),$$

where the first step follows because $\mathbf{H}_2 \in \mathscr{g}_\kappa$, and the second because $\mathbf{H}_1 \in \mathscr{g}_\kappa$. Thus $\mathbf{H}_1\mathbf{H}_2 \in \mathscr{g}_\kappa$. Similar arguments verify the other axioms of a group.

IV.12.2 Since $\mathfrak{G}_\kappa$ satisfies Axiom N3,

$$\mathfrak{G}_\kappa(\mathbf{Q}\mathbf{F}^t) = \mathbf{Q}\mathfrak{G}_\kappa(\mathbf{F}^t)\mathbf{Q}^T.$$

This when combined with (IV.12-6) implies that $\mathbf{Q}^T \in \mathscr{g}_\kappa$.

IV.15.1 By NOLL's rule $\mathbf{Q}^* = \mathbf{P}\mathbf{Q}\mathbf{P}^{-1}$, which can be written as

$$\mathbf{Q}^*\mathbf{R}_0\mathbf{U}_0 = \mathbf{R}_0\mathbf{Q}\mathbf{Q}^T\mathbf{U}_0\mathbf{Q}.$$

Use the uniqueness of the polar decomposition.

IV.15.2 Since in general $\mathscr{g}_\kappa \neq \mathscr{o}$, $\mathscr{g}_\kappa$ will not be an invariant subgroup of $\mathscr{o}$. Thus $\mathscr{g}_\kappa^*(= \mathbf{R}\mathscr{g}_\kappa\mathbf{R}^{-1})$ will not be equal to $\mathscr{g}_\kappa$.

IV.15.3 Put $\mathbf{U}_0 = K\mathbf{1}$ in Theorem 2.

IV.16.2 Note that (IV.16-3) depends upon $\boldsymbol{\kappa}$ only through ρ_κ/J.

IV.18.1 Substitute (IV.18-3) in (II.8-8) to get (IV.18-13). The other relations follow easily from the definitions (II.11-2) and (II.11-17). The condition tr $\mathbf{G} = 0$ is necessary and sufficient that the motion be isochoric, so (IV.18-16) follows.

IV.18.2 If $\mathbf{A}$ and $\mathbf{B}$ commute, then $e^{\mathbf{A}}e^{\mathbf{B}} = e^{\mathbf{A}+\mathbf{B}}$.

IV.18.3 The most general form of $\mathbf{A}_2$ is

$$[\mathbf{A}_2] = \begin{Vmatrix} u & a & b \\ a & v & c \\ b & c & w \end{Vmatrix}.$$

Remembering that (22) does not hold, show that this $\mathbf{A}_2$ commutes with $\mathbf{M} - \overline{\mathbf{M}}$ as given by (IV.18-27) if and only if $x = 0$. When $\mathbf{A}_1 = \alpha\mathbf{1}$, by the lemma $\mathbf{A}_1$ commutes with every skew tensor. Therefore $(\mathbf{M} - \mathbf{M}^T)\mathbf{A}_1 = \mathbf{A}_1(\mathbf{M} - \mathbf{M}^T)$. Using (IV.18-15)$_3$, conclude that $\mathbf{M}\mathbf{M}^T = \mathbf{M}^T\mathbf{M}$, and then arrive at (IV.18-26).

IV.18.4 Use the power series for the exponential function.

IV.18.5 Use $(\text{IV}.18\text{-}15)_{4,7}$ to get (IV.18-30). Then (IV.18-31) follows by use of $(\text{IV}.18\text{-}14)_2$ and $(\text{IV}.18\text{-}15)_7$.

IV.18.6 Use (II.8-3) and (II.8-4) to obtain the relative description of the motion whose spatial velocity field is (IV.18-33):

$$\xi_1 = x_1, \qquad \xi_2 = (\tau - t)\kappa x_1 + x_2,$$
$$\xi_3 = (\tau - t)(\lambda x_1 + \nu x_2) + \tfrac{1}{2}(t - \tau)^2 \kappa \nu x_1 + x_3.$$

Hence show that $\mathbf{F}_t(\tau)$ assumes the form (IV.18-13) with the special values $\mathbf{Q} = \mathbf{1}$ and

$$[\kappa \mathbf{N}_0] = \begin{Vmatrix} 0 & 0 & 0 \\ \kappa & 0 & 0 \\ \lambda & \nu & 0 \end{Vmatrix},$$

components being taken with respect to the Cartesian co-ordinate basis. Since $(\kappa \mathbf{N}_0)^3 = \mathbf{0}$ and $(\kappa \mathbf{N}_0)^2 = \mathbf{0} \Leftrightarrow \kappa\nu = 0$, the first two assertions of the exercise follow. Show that the relative description of the motion whose spatial velocity field is (IV.18-33) is

$$\xi_k = x_k\, e^{a_k(\tau - t)}, \qquad k = 1, 2, 3.$$

Hence show that $\mathbf{F}_t(\tau)$ assumes the form (IV.18-13) with the special values $\mathbf{Q} = \mathbf{1}$ and

$$[\kappa \mathbf{N}_0] = \begin{Vmatrix} a_1 & 0 & 0 \\ 0 & a_2 & 0 \\ 0 & 0 & a_3 \end{Vmatrix}.$$

If we select the placement at $t = 0$ as the reference placement, the referential description of (IV.18-33) is

$$x_k = X_k\, e^{a_k t}, \qquad k = 1, 2, 3.$$

Thus

$$[\mathbf{F}] = \begin{Vmatrix} e^{a_1 t} & 0 & 0 \\ 0 & e^{a_2 t} & 0 \\ 0 & 0 & e^{a_3 t} \end{Vmatrix} = [\mathbf{U}];$$

(IV.9-15) is satisfied; and the last sentence of the exercise follows by the result of Exercise IV.10.2.

Index

Names of persons are cited only when they serve to identify theorems or concepts, not for attributions. The contents of the appendices are not indexed.

P

R

S